Stabilisation/Solidification
of Contaminated Soil and Waste:
Practice

The practice behind this versatile remedial technology

Edward Bates & Colin Hills

Published by
Hygge Media

www.hyggemedia.com

Stabilisation and Solidification of Contaminated Soil and Waste:
Practice

Copyright © 2015 Edward Bates and Colin Hills.
All rights reserved. First paperback edition printed 2015 in Great Britain.

A catalogue record for this book is available from the British Library.

ISBN 978-0-9932729-3-6
ISBN 978-0-9932729-4-3

Published by Hygge Media
For more copies of this book, please email: info@hyggemedia.com

Edward Bates and Colin Hills retain copyright to this document, but grant exclusive rights to Hygge Media to publish and distribute copies of this document. Permission is granted to refer to or quote from this publication with the customary acknowledgment of the source.

Although every precaution has been taken in the preparation of this document, the authors assume no responsibility for errors or omissions. Neither is any liability assumed for damages resulting from the use of this information contained herein.

Disclaimer: Whilst every effort has been taken to use the highest quality Photographs in this book, some of the images are low resolution. Many of the photographs were taken in the field mostly for personal records and some predate digital photography and were not originally intended for publishing. Where low resolution photographs have been used, it was felt by the publisher that these were the only images available to illustrate specific processes or projects.

www.hyggemedia.com

PREFACE

Stabilisation/Solidification (S/S) is a remedial technology that has attracted widespread use around the world. S/S employs readily available cement-based binders that can be formulated into a targeted 'system' that cements soil or waste into a hardened engineering material. The application of S/S results in the physical and chemical stabilisation of contaminated material over the long term, enabling disposal or re-use on or off-site.

In this second volume on S/S, the authors present the state of the art of the application of S/S in the field. As such, this 'Manual of Practice' is designed for use by regulators, professionals and informed individuals. Considering the multinational contributions to this work, the reader may note that sometimes words may use the United Kingdom English spelling, for example "stabilisation" rather than the conventional American English spelling "stabilization". However, in all cases, the meaning should be clear.

This 'Manual' has been prepared in close consultation with experienced practitioners and vendors of S/S who apply this remedial technology on a day-to-day basis. As such, this 'Manual' will provide an authoritative reference source for all who are involved in managing the risks associated with contaminated soil and waste.

The authors are indebted to the hard work and professionalism of the contributors to this work, and it is their invaluable contributions that have made this Manual of Practice possible.

The companion volume to this work entitled: *Stabilisation and Solidification of Contaminated Soil and Waste: Science* draws extensively from the scientific literature and personal experience of the authors, and is intended to deliver a readily accessible account of the science underpinning S/S. Both volumes should be seen as complimentary, providing invaluable insight into why S/S is a versatile risk management strategy, that remains fit for the 21st Century.

Colin Hills and Edward Bates*
University of Greenwich, UK
*Cincinnati, USA
August 2015

ACKNOWLEDGMENTS

The contributing editors wish to express their most sincere gratitude to all the expert authors who have given their valuable time to the preparation of material for this document.

Each author is gratefully acknowledged and is listed in the table below with the subject matter of their contribution. Information on each contributing author can also be found in Section 11 "Meet the Authors", where their particular expertise, experience and present professional affiliation are given.

The Contributing Editors particularly wish to acknowledge the substantial contribution made by Dr Peter Gunning to the completion of this Manual of Practice. Peter's unselfish support and encouragement throughout was exemplary, and was a major factor in the successful and timely completion of this work.

The authors would also like to offer special thanks to Miss Esme Hills for her 'steady eye', and to Mr Stany Pensaert for his constructive comments and invaluable insights gained as a practitioner of S/S of Contaminated Soil and Waste.

SUMMARY

This manual of practice brings together the collective experience of leading practitioners in the design and execution of remedial actions employing S/S.

This manual presents the state of the art in the application of S/S for the remediation of contaminated land and treatment of waste.

The information presented is, to the contributing editors' knowledge, not available in any single document anywhere at the current time.

The simple structure of this manual is designed to inform the reader on the background of S/S, and then to the site assessment of risk, the design and selection of binders and the choice of equipment, to field-execution and product quality control/assurance, and finally to capping and post-remedial management of sites that have been treated using S/S.

It is intended that this manual will provide an accessible reference source for the planning and execution of S/S under most circumstances, and includes an inventory of over 200 completed S/S operations in the USA and elsewhere, and more than 40 detailed 'case studies' where S/S has been successfully used to treat both organic and inorganic contaminated materials.

Each section of this manual has encapsulated the world experience of recognised experts in his/her particular expertise. However, it should be noted that the subject material presented here represents the considered views of the authors and should not be taken to necessarily represent the views of the author's employer, or any government or private organisation, or of the other contributors to this work.

The mention of trade names or commercial products are for illustrative purposes only, and does not constitute an endorsement or recommendation for use in any application, and no public or private funding was provided for preparing this document.

CONTRIBUTING AUTHORS

Section	Title	Contributors
1.1-1.4	Introduction and overview	Colin Hills & Ed Bates
1.5	Long-term performance	Peter Gunning & Colin Hills
1.6	History of use of cementitious binders	Peter Gunning & Colin Hills
1.7	External Information Sources	Peter Gunning
2.1	Use of S/S in the USA (Superfund and RCRA)	Ed Bates
2.2	Use of S/S in the UK and Europe	Colin Hills & Paula Carey
2.3	Use of S/S in Canada	Colin Dickson & Cameron Ells
2.4-2.5	Risk management by S/S, S/S and the 'Risk Framework'	Kate Canning
2.6	Future trends	Ed Bates & Colin Hills
3.0-3.3	Applicability of S/S	Paul Lear
3.4	Geotechnical considerations	Craig Lake
3.5	Ex-situ and in-situ application of S/S	Paul Lear
4.0-4.6	Ex-situ S/S equipment and application	Paul Lear
5.0-5.1	In-situ auger mixing	Dan Ruffing & Ken Andromalos
5.2-5.3	Injection tillers and rotary drum mixers, In-situ bucket mixing	Robert Schindler & Dan Ruffing
6.0	Quality assurance and quality control	Roy Wittenberg
6.1	Quality assurance and quality control during S/S	Roy Wittenberg
6.2-6.4	Sample collection - the performance sampling plan, Sample preparation, curing and testing, variation and failures	Ed Bates
7.0	Performance specifications/tests	Tom Plante & Ed Bates
7.1	Setting overall goals for treatment	Tom Plante
7.2	Importance of site characterisation and conceptual site model	Tom Plante
7.3	Performance parameters, test and methods	Tom Plante & Stany Pensaert
7.4	Impact of other protective measures	Ed Bates & Craig Lake
8.1-8.2	Common reagents, Reagent sourcing	Paul Lear

8.3	Bench-scale treatability study, design and execution	Ed Bates & Paul Lear
8.4	Pilot-scale tests	Tom Plante
8.5	Selecting samples for treatability testing	Paul Lear
9.1	Overview of factors to consider for post-treatment management of S/S material	Roy Wittenberg
9.2	Post-treatment management of S/S	Roy Wittenberg
9.3	Capping technologies	Roy Wittenberg
9.4	Capping strategies based on slope, drainage, and climate	Roy Wittenberg
9.5	Post-construction monitoring	Ed Bates
Appendix A	S/S Completed Projects List	Stany Pensaert, Aiman Naguib, Paul Lear, Ken Andromalos, Steve Birdwell, Tom Plante, Bob Garrett and others
Appendix B	Selected S/S Case Studies	Stany Pensaert, Paul Lear, Ken Andromalos, Steve Birdwell, Dan Ruffing and others

CONTENTS

	List of Tables	*xviii*
	List of Figures	*xx*
	List of Abbreviations / Acronyms	*xxvi*
	Table of Unit Conversion	*xxix*

Part One	**Introduction and overview**	**1**
1.1	Purpose	1
1.2	Target audiences	1
1.3	Definition of terms	2
1.4	Appropriate uses of S/S technology	3
1.5	Long term performance	5
1.6	History of use of cementitious binders	8
1.7	Information sources on S/S systems	10
1.7.1	General information	10
1.7.2	Technical information	10
1.7.3	Government departments and agency information resources	10
1.7.4	Research, committees and associations	10
1.7.5	Forums, networks, and discussion groups	10
Part Two	**History of S/S as a risk management tool**	**17**
2.1	Use of S/S in the USA (Superfund and RCRA)	17
2.1.1	The regulatory impetus for the remediation of sites in the USA	17
2.1.2	Use of S/S in the EPA Superfund Program	18
2.1.3	Conclusions	21
2.2	Use of S/S in the UK and Europe	22
2.2.1	Treatment of waste and soil by S/S	22
2.2.2	Legislative background	23
2.2.3	Public and private initiatives	24
2.3	Use of S/S in Canada	25
2.3.1	Performance (risk)-based versus prescriptive regulations	25
2.3.2	Relative use of S/S in Canadian provinces	26
2.3.3	Sydney Tar Ponds and coke ovens project – Nova Scotia	26

2.4	Risk management by S/S	26
2.4.1	Definition of risk	26
2.4.2	Risk management frameworks	27
2.4.3	Summary	28
2.5	S/S and the 'Risk Framework'	28
2.5.1	Water environment	28
2.5.2	Human health	29
2.5.3	Other receptors	30
2.5.4	Durability and integrity	30
2.6	Future trends	31
2.6.1	USA	31
2.6.2	UK and Europe	32
2.6.3	Conclusions	32

Part Three — Applicability of S/S — 35

3.1	Metals	35
3.1.1	Chemistry	35
3.2	Other inorganic compounds	38
3.2.1	Chemistry	38
3.3	Organics	39
3.3.1	Chemistry and physics of organic stabilisation	39
3.3.2	Physical aspects of organic stabilisation	43
3.4	Geotechnical considerations	43
3.4.1	Soil classification	43
3.4.2	Site Characterization of Soil and Groundwater	46
3.4.3	Approaches to site variability	48
3.5	Ex-situ and in-situ application of S/S	49
3.5.1	Technical considerations	49
3.5.2	Physical material handling	54
3.5.3	Economics	56

Part Four — Ex-situ S/S equipment and application — 59

4.1	Mixing Chambers	59
4.1.1	Pug-mill mixers	61
4.1.2	Screw Mixers	64
4.1.3	Ribbon Blenders	65
4.1.4	Quality control using mixing chambers	67
4.2	Excavator bucket mixing	69
4.2.1	Quality Control using excavator bucket mixing	72
4.3	Tillers and Other Ex-situ Mixers	72
4.3.1	Rototillers	73
4.3.2	Asphalt millers	74
4.3.3	Paddle aerators/compost turners	76

4.3.4	Quality control using tillers and paddle mixers	76
4.4	Mixing pits	79
4.4.1	Earthen Pits	80
4.4.2	Open-top tanks	81
4.4.3	Enclosures	82
4.5	Ancillary equipment for ex-situ mixing	82
4.5.1	Screening equipment	82
4.5.2	Crushing equipment	84
4.5.3	Shredding equipment	84
4.5.4	Magnetic separation equipment	85
4.6	On-site placement of ex-situ treated material	85
4.6.1	Timing	86
4.6.2	Placement	86
4.6.3	Compaction	86
Part Five	**In-situ S/S equipment and its application**	**93**
5.1	In-situ auger mixing	93
5.1.1	Equipment	93
5.1.2	Staffing requirements	97
5.1.3	Treatment metrics and considerations	101
5.1.4	Treatment plan	102
5.1.5	Quality control	102
5.1.6	Operational Issues	102
5.1.7	Summary of limitations, advantages, and disadvantages	106
5.1.8	Costs	108
5.2	Injection tillers and rotary drum mixers	108
5.2.1	Equipment	108
5.2.2	Staffing requirements	111
5.2.3	Treatment metrics and considerations	111
5.2.4	Quality control	113
5.2.5	Operational issues	115
5.2.6	Advantages & disadvantages of rotary tillers & injection drum mixers	115
5.2.7	Costs	115
5.3	In-situ bucket mixing	115
5.3.1	Equipment	116
5.3.2	Staffing requirements	117
5.3.3	Treatment metrics and considerations	117
5.3.4	Quality control	118
5.3.5	Operational Issues	119

Part Six — Quality assurance and quality control — 121

- 6.1 Quality assurance and quality control during S/S — 121
  - 6.1.1 Quality assurance and quality control during S/S — 121
  - 6.1.2 The importance of CQA in the S/S design and construction process — 121
  - 6.1.3 Development of an effective CQA program — 122
  - 6.1.4 CQA objectives — 123
  - 6.1.5 Roles and responsibilities in the CQA process — 123
  - 6.1.6 General categories for the CQA process — 124
  - 6.1.7 Standards and decision processes for S/S performance — 127
- 6.2 Sample collection - the performance sampling plan — 132
  - 6.2.1 Frequency of sampling — 132
  - 6.2.2 Number of samples per sampling event — 133
  - 6.2.3 Sample collection methods — 134
- 6.3 Sample preparation and curing — 137
  - 6.3.1 Sample preparation — 137
- 6.4 Variation and failures — 143
  - 6.4.1 Normal variation — 143
  - 6.4.2 Recognising failure — 144
  - 6.4.3 Handling failure — 145
  - 6.4.4 Coring of in-place S/S material — 149

Part Seven — Performance specifications for S/S — 151

- 7.1 Setting overall goals for treatment — 151
- 7.2 The importance of site characterisation & the conceptual model — 156
- 7.3 Performance parameters, tests/methods — 160
  - 7.3.1 Performance parameters — 160
  - 7.3.2 Performance tests and methods — 161
  - 7.3.3 Phases of performance testing — 164
  - 7.3.4 Performance criteria in specifications — 165
- 7.4 Impact of other protective measures — 165

Part Eight — Developing effective S/S formulations — 171

- 8.1 Common reagents — 171
  - 8.1.1 Portland cement — 171
  - 8.1.2 Quicklime and hydrated lime — 172
  - 8.1.3 Cement kiln dust — 173
  - 8.1.4 Lime kiln dust — 174
  - 8.1.5 Fly ash — 174
  - 8.1.6 Bottom ash — 176
  - 8.1.7 Magnesium oxide — 176
  - 8.1.8 Phosphates — 177
  - 8.1.9 Sulfides — 178
  - 8.1.10 Carbon-based reagents — 178

8.2	Reagent sourcing	180
8.2.1	Proprietary reagents	180
8.2.2	Mineral and waste reagents	180
8.2.3	Importance of location	181
8.2.4	Reagent samples for testing	181
8.3	Bench-scale treatability design and testing	182
8.3.1	Bench-scale treatability study objectives	183
8.3.2	Importance of the conceptual implementation plan	183
8.3.3	Reagent selection considerations	184
8.3.4	Bench-scale treatability testing approaches	184
8.3.5	Initial sample characterisation	191
8.3.6	Laboratory procedures	191
8.4	Pilot-scale tests	193
8.4.1	Pilot test objectives	195
8.4.2	Confirming the conceptual implementation plan	195
8.4.3	Implementing the pilot testing	196
8.4.4	S/S safety considerations	203
8.5	Selecting samples for treatability testing	204
8.5.1	Option 1: Maximum Contamination Sample	204
8.5.2	Option 2: high contamination sample	205
8.5.3	Option 3: average contamination sample	205
8.5.4	Rationale for using the high contamination sample	206
8.5.5	Sites with multiple materials requiring treatment	206

Part Nine — Post-treatment capping and monitoring — 207

9.1	Overview of post-treatment management of S/S material	207
9.1.1	Climatologic and geographic considerations	208
9.1.2	Site geometry	208
9.1.3	Integration with existing site development	208
9.1.4	Future site use	209
9.1.5	Regulatory requirements	209
9.2	Post treatment management of S/S	211
9.2.1	Key engineering parameters S/S post treatment management	211
9.2.2	Engineering approach for preparing S/S material prior to capping	212
9.3	Capping technologies	215
9.3.1	Overview of design considerations & general types of caps applicable for S/S monoliths	215
9.3.2	Clay and/or other low permeability earthen materials	217
9.3.3	Geo-membranes	219
9.3.4	Geo-synthetic clay liner (GCLs)	221
9.3.5	Clay with geo-membrane	223
9.3.6	Geo-composite clay liner with geo-membrane	225
9.3.7	Evapotranspiration (ET) caps	225

9.3.8	Asphalt and concrete pavement applications	226
9.3.9	Comparison of capping technologies	227
9.4	Capping strategies based on slope, drainage, and climate	229
9.4.1	Slope and drainage considerations	229
9.4.2	Capping strategies based on climatologic conditions	229
9.5	Post-construction monitoring	231
9.5.1	Regulatory impetus and guidance for post-construction monitoring	231
9.5.2	Specific features common in monitoring programs	235

Consolidated Reference List   237

Contributing author biographies   241

Appendix A: Remediated sites employing S/S 249

Appendix B: Case studies employing S/S 295

Case Study 1: Abex Superfund Site 298
Case Study 2: American Creosote Superfund Site 300
Case Study 3: Bayou Trepagnier 304
Case Study 4: BHAD Chromium Pond 307
Case Study 5: Cambridge 310
Case Study 6: Former Camden Gas Works Site 312
Case Study 7: Camp Pendleton Scrap Yard 315
Case Study 8: Chevron Pollard Landfill Site 318
Case Study 9: Carnegie 321
Case Study 10: Columbus 322
Case Study 11: CSX Benton Harbor Scrapyard 323
Case Study 12: Double Eagle Refinery Site 325
Case Study 13: Dundalk 327
Case Study 14: East Rutherford 328
Case Study 15: Foote Minerals Superfund Site 331
Case Study 16: Ghent 334
Case Study 17: Guernsey 338
Case Study 18: Hercules 009 Landfill Site 341
Case Study 19: Hoedhaar Lokeren 344
Case Study 20: Irving 346
Case Study 21: Johnston Atoll Solid Waste Burn Pit 348
Case Study 22: Kingston 350
Case Study 23: London Olympic Site 353
Case Study 24: Martinsville 357
Case Study 25: Milwaukee, Wisconsin 359
Case Study 26: Nederland 362

Case Study 27: New Bedford 365
Case Study 28: Nyack 366
Case Study 29: NYSEG Norwich NY MGP Site 368
Case Study 30: Obourg 371
Case Study 31: Perth Amboy 373
Case Study 32: Portsmouth 375
Case Study 33: Rieme 377
Case Study 34: Roma Street Station 382
Case Study 35: Sag Harbor 385
Case Study 36: Sanford MGP Superfund Site 387
Case Study 37: Söderhamn 391
Case Study 38: Southeast Wisconsin 393
Case Study 39: St Louis 396
Case Study 40: Sunflower Army Ammunition Depot 397
Case Study 41: Sydney Tar Ponds 399
Case Study 42: Umatilla Ammunition Depot 409
Case Study 43: Valero Paulsboro Refinery 411
Case Study 44: Waukegan 413
Case Study 45: West Doane Lake Site 415
Case Study 46: X-231B Pilot Study 418
Case Study 47: Zwevegem 420

Index   ***424***

LIST OF TABLES

Table	Title	Source
1.1	S/S applications examined in 2010	PASSiFy (2010)
1.2	General information sources on S/S	Gunning
1.3	Technical information sources on S/S	Gunning
1.4	Information on S/S from government departments and agencies	Gunning
1.5	Information on S/S from research, committees and associations	Gunning
1.6	Information on S/S from forums, networks, and discussion groups	Gunning
3.1	Considerations for choosing test pits versus boreholes	Lake
4.1	Advantages and disadvantages of pug-mill mixers	Lear & Gunning
4.2	Advantages and disadvantages of screw mixers	Lear & Gunning
4.3	Advantages and disadvantages of ribbon blenders	Lear & Gunning
4.4	Advantages and disadvantages of excavator bucket mixing	Lear & Gunning
4.5	Advantages and disadvantages of rototillers	Lear & Gunning
4.6	Advantages and disadvantages of asphalt millers	Lear & Gunning
4.7	Advantages and disadvantages of paddle aerators	Lear & Gunning
4.8	Ex-situ mixer manufacturers	Lear & Gunning
4.9	Advantages and disadvantages of earthen pits	Lear & Gunning
4.10	Advantages and disadvantages of open-top tanks	Lear & Gunning
4.11	Advantages and disadvantages of enclosures	Lear & Gunning
5.1	Examples of in-situ auger treatments including auger diameter and depth treated	Geo-Solutions & Gunning
5.2	Quality control planning for auger-based S/S mixing	Geo-Solutions & Gunning
5.3	Advantages and disadvantages of auger mixing	
5.4	Quality control planning for BOSS-based S/S mixing	Geo-Solutions & Gunning
5.5	Advantages and disadvantages of rotary tillers and drum mixers	Geo-Solutions & Gunning

5.6	Quality control planning for bucket-based S/S mixing	Geo-Solutions & Gunning
6.1	Example of a simple decision making framework agreed between the site engineer and contractor	Wittenberg
6.2	Pb in leachates from construction performance samples at 14, 21, and 28 days of curing, United Metals Site (µg/L)	Ed Bates
6.3	Results from duplicate performance samples, American Creosote site	Ed Bates
7.1	Typical S/S Site assessment considerations	T Plante
9.1	Minimum index properties for the use of clay	Wittenberg & Gunning
9.2	A comparison of capping technologies	NRT
9.3	Capping technologies vs slope and drainage	NRT
9.4	Capping technologies and climate	NRT

LIST OF FIGURES

Figure	Title	Source
Cover	Photo of two crane mounted ISS rigs in Winter	Ed Bates
1.1	Treating creosote, PCP, & dioxins in pug-mill	Ed Bates
1.2	Installing a grout slurry blanket to reduce emissions during in-situ S/S treatment	Ed Bates
1.3	Conceptual model of a site with environmental loading	After EA, 2004b
1.4	Longevity of S/S materials proposed by selected authors	PASSiFy (2010)
2.1	Cumulative source control technology selection over 30 years	After USEPA, 2013
2.2	Selection of S/S vs all other technologies	After USEPA, 2013
2.3	Waste types selected for S/S	After USEPA, 2000
2.4	S/S selection of in-situ vs ex-situ	After USEPA, 2013
2.5	Relative increase in selection of in-situ vs ex-situ S/S	After USEPA, 2013
3.1	Lead-contaminated soil and battery casings from a former battery-recycling site, subsequently chemically stabilised	Paul Lear
3.2	Chromium paint pigment-contaminated soil being chemically stabilised in-situ	Paul Lear
3.3	A contaminated wastewater sludge being excavated	Paul Lear
3.4	Treatment facility to chemically stabilise and solidify acid sludge tar	Paul Lear
3.5	General description and identification of soils	Alter BS5930
3.6	Description and identification of soils	Alter BS5930
3.7	Description and identification of soils	Alter BS5930
3.8	Ex-situ treatment in a roll-off box	Paul Lear
3.9	Typical ex-situ S/S equipment, including pug-mill, silos, reagent tank, water tank, stacker and haulage	Ed Bates
3.10	Ex-situ S/S treatment enclosure	Paul Lear
3.11	In-situ S/S treatment of oily sludge	Paul Lear
3.12	In-situ mixing equipment operating on top of S/S treated material	Paul Lear
3.13	Typical in-situ S/S equipment, batch plant with reagent silos and mix tanks: crane, Kelly bar, and auger	Ed Bates

3.14	A 3 ft limestone boulder excavated to facilitate in-situ treatment	Ed Bates
4.1	Ex-situ processes	Peter Gunning
4.2	A pug-mill with its ancillary equipment	Paul Lear
4.3	Ancillary equipment of a pug-mill	Paul Lear
4.4	Paddles inside a typical pug-mill	Ed Bates
4.5	Screw mixer	US Air Filtration, Temecula
4.6	Ribbon mixer	Phoenix Equipment Corporated
4.7	Control panel for waste feed conveyor	Paul Lear
4.8	Calibrating reagent feed from a screw auger	Ed Bates
4.9	Display panel for an automated pug-mill	Ed Bates
4.10	Modified excavator bucket for improved mixing	Paul Lear
4.11	Allu PMX Power Mixer™ attachment	Allu, Pennala, Finland
4.12	Rotating mixing head attachment	Paul Lear
4.13	Rotating rake mixing attachment	Paul Lear
4.14	Rototiller attachment	Alex Stacey
4.15	Asphalt miller	Bomag
4.16	Paddle aerator mixing windrowed soil	Paul Lear
4.17	Excavator mixing of Portland cement into Pb-contaminated soil and debris	Paul Lear
4.18	Bucket mixing hydrated lime and oily waste in earthen pit	Ed Bates
4.19	Stabilisation of hexavalent chromium-contaminated soil in a roll-off box	Paul Lear
4.20	Grizzly screen for separation of large debris, with oversized material, to the left and processed soil for ex-situ mixing, right	Paul Lear
4.21	Grizzly screen (foreground) combined with a trommel to remove battery casings and debris from excavated material	Paul Lear
4.22	A rotary shear shredder processing scrap metal	Paul Lear
4.23	Transport and placement of ex-situ S/S material	Ed Bates
4.24	Placed ex-situ S/S-treated material after multiple passes of the dozer	Ed Bates

4.25	Proctor compaction-testing results for ex-situ S/S treated material	Paul Lear
4.26	Sheeps-foot roller compacting S/S treated material (note the oval pads on the roller)	Ed Bates
4.27	Smooth roller compacting ex-situ S/S treated material	Ed Bates
5.1	In-situ processes	Peter Gunning
5.2	A crane mounted soil mixing rig	R Schindler
5.3	A crane mounted soil mixing rig	R Schindler
5.4	An excavator-mounted rig	D Ruffing
5.5	An excavator-mounted soil mixing rig	D Ruffing
5.6	Auger used at the USX Site, Duluth, MN	Ed Bates
5.7	Auger used at a coal gas plant site in FL	Ed Bates
5.8	An automated batch plant	D Ruffing
5.9	Self-propelled dry storage silo	Allu
5.10	Typical Column Layout Showing Overlapping Columns to Achieve 100 % Coverage	Geo-Solutions
5.11	Showing typical excavated overlapping columns	Geo-Solutions
5.12	In-situ mixing of soil under a live fibre optic cable	Ed Bates
5.13	In-situ mixing under a fibre optic cable	Ed Bates
5.14	Backhoe (Excavator) operated soil stabiliser (BOSS)	Geo-Solutions
5.15	Backhoe (Excavator) operated soil stabiliser	Geo-Solutions
5.16	Lang mixer-excavator, arm, and mixing head as one unit	Ed Bates
5.17	Allu mixer head attached to a standard excavator	Ed Bates
5.18	A batching plant for a large S/S project	R Schindler
5.19	Example of bucket mixing	Geo-Solutions
5.20	Bucket mixing of coal tar soils to depth of 15 ft (5 m)	Ed Bates
6.1	Flowchart for sampling and analysis-related decision making	NRT
6.2	Flowchart for design-related decision making	NRT
6.3	Flowchart for operational issues-related decision making	NRT
6.4	Collection of sample using piston tube inserted by excavator	Ed Bates

6.5	A sampler employing a hydraulic gate	Ed Bates
6.6	Preparing 3 in x 6 in (75 x 150 mm) molds	Ed Bates
6.7	A 0.5 in (12.5 mm) screen to remove oversize debris	Ed Bates
6.8	Labelling of specimen molds and cap	Ed Bates
6.9	Sample molds curing on-site in water bath	Ed Bates
6.10	Checking set/strength by a pocket penetrometer	Ed Bates
6.11	Detailed record keeping of properties of cured sample	Ed Bates
6.12	A slump test on an S/S bulk sample	Ed Bates
6.13	Poorly prepared field-specimens	T Plante
6.14	A well-prepared specimen with minimal air voids	T Plante
6.15	Increase in strength by S/S soil over time	T Plante
6.16	Using a field penetrometer to assess strength of field placed S/S material	Ed Bates
6.17	Excavating Top of In-situ Column	Ed Bates
6.18	Cores obtained by sonic drilling of S/S material	Ed Bates
7.1	Performance goals and performance specifications in the S/S process	Modified from ITRC
7.2	Strength-gain of bench vs field demonstration (FD) data	T Plante
7.3	A simplified conceptual site model	
7.4	Installing a slurry wall, Whitehouse, Florida	Ed Bates
7.5	Constructing a S/S sub-cap, Brunswick	Ed Bates
7.6	Constructing a vertical wall using a Bauer Panel Cutter, Brunswick	Ed Bates
7.7	Oily water emulsion, Stauffer	Ed Bates
7.8	Solidifying oily water emulsion, Stauffer	Ed Bates
7.9	Installing FML at Stauffer	Ed Bates
8.1	Hydrated lime added to waste in a mix pit	Ed Bates
8.2	Fly ash stockpile on a treatment site with its cover removed	Ed Bates
8.3	Granular activated carbon in a super sack	Ed Bates
8.4	Design of S/S formulations	Peter Gunning
8.5	Unconfined compressive strength testing	Paul Lear
8.6	Permeability	Paul Lear
8.7	A limited bench-scale field treatability study	Ed Bates

8.8	Reagents are measured on a weight/weight basis to the untreated soil	Paul Lear
8.9	Pocket penetrometer testing for approximate strength	Paul Lear
8.10	Eight-foot (2.5 m) In-situ auger and carrier assembly	Ed Bates
8.11	Pilot test using a 10 ft. (3 m) diameter in-situ auger	Ed Bates
8.12	Pilot test for in-situ bucket mixing	Ed Bates
8.13	Ex-situ pug-mill with support equipment	Ed Bates
8.14	Field quality control test specimen preparation	T Plante
8.15	Field slump test performance	T Plante
8.16	Grout density test by mud balance	T Plante
8.17	Field check to assure homogenous mixing	Ed Bates
8.18	Reagent silo electronic scale and calibration weight	T Plante
8.19	Trial pit in an S/S soil monolith	T Plante
8.20	In-situ auger test producing steam, sulfur dioxide emissions	Ed Bates
8.21	Collecting a bulk treatability sample from auger flights	Ed Bates
9.1	Typical profile for a Subtitle 'D' cap	NRT
9.2	A typical profile for a Subtitle 'C' cap	NRT
9.3	A smooth roller being used to compact ex-situ treated material	Ed Bates
9.4	Direct placement of S/S swell by an excavator for final grading	NRT
9.5	Final shaping/contouring of in-situ S/S using a dozer	Ed Bates
9.6	Reconditioning and placement of in-situ S/S swell	NRT
9.7	Example single component cap with a compacted clay layer	NRT
9.8	Cover layer comprising 2 ft of compacted soil, Schuylkill Metals, Florida	Ed Bates
9.9	Construction of a low permeability clay cap	NRT
9.10	Final grading to direct surface drainage	NRT
9.11	Installation of a geo-membrane showing the welded seams between sheets	Ed Bates

9.12	A single component cap with a geo-membrane	NRT
9.13	A typical single component cap with a geo-synthetic clay layer	NRT
9.14	GCL and soil cover with drainage channel, Peak Oil Site	Ed Bates
9.15	Cover with a GCL, 2 ft of soil, and a gravel surface	Ed Bates
9.16	Profile of a composite cap with a geo-membrane over a compacted clay layer	NRT
9.17	A composite cap with a geo-membrane over a GCL	NRT
9.18	A monolithic evapotranspiration cap	NRT
9.19	A capillary break evapotranspiration cap	NRT
9.20	A coarse-grained soil-based ET cap over an S/S monolith	NRT
9.21	Groundwater monitoring wells around an S/S Monolith at the Sanford Gasification Plant Site, FL	NRT
9.22	Groundwater monitoring wells around an S/S monolith at the Sanford gasification plant site	NRT

LIST OF ABBREVIATIONS / ACRONYMS

ADEME	Environment and Energy Management Agency
ANS/ANSI	American Nuclear Society/American National Standards Institute
ARAR	Applicable, or relevant, and appropriate requirement
ASTM	American Society for Testing and Materials
BC	British Columbia
BCA	British Cement Association
BDAT	Best Demonstrated Available Technology
BGS	Below Ground Surface
BS	British Standards
BTEX	Benzene, Toluene, Ethylbenzene and Xylene
BOSS	Backhoe operated soil stabilizer
CASSST	Codes and Standards for Stabilisation/Solidification Technology
CBR	California bearing ratio
CD	Chart Datum
CERCLA	Comprehensive Environmental Response, Compensation, and Liability Act
CEMBUREAU	The European Cement Association
CH	Clay, high-plasticity
CKD	Cement kiln dust
CL:AIRE	Contaminated Land: Applications in Real Environments
COC	Contaminant of Concern
COPA	Control of Pollution Act
CPT	Cone Penetration Testing
CQA	Construction quality assurance
CQC	Construction quality control
CL	Clay, low plasticity
C-S-H	Calcium silicate hydrate
CSM	Conceptual site model
cy	Cubic yard
DQO	Data quality objective
DNAPL	Dense non aqueous phase liquid
EA	Environment Agency
EPA	Environmental Protection Agency
ET	Evapotranspiration

EU	European Union
FL	Florida
FML	Flexible membrane liner
ft	Foot
FY	Fiscal year
GCL	Geo-synthetic clay liner
GPS	Global Positioning System
HDPE	High density polyethylene
HSWA	Hazardous and Solid Waste Amendments
IC	Institutional controls
IEA	International Energy Agency
ITRC	Interstate Technology & Regulatory Council
kPa	Kilopascal
L	Litre
lb	Pound
LDPE	Low density polyethylene
LEAF	Leaching Environmental Assessment Framework
LKD	Lime kiln dust
LTRA	Long-Term Response Action
m	Metre
MGP	Manufactured Gas Plant
MNA	Monitored natural attenuation
MPa	Mega Pascal
MPH	Miles per hour
NAPL	Non-aqueous phase liquids
NEN	The Netherlands Standardization Institute
O&M	Operation and maintenance
PAH	Polycyclic aromatic hydrocarbon
PCP	Pentachlorophenol
PCB	Polychlorobiphenyl
PVC	Polyvinyl chloride
PSI	Pounds per square inch
QA	Quality assurance
QC	Quality control

RCRA	Resource Conservation and Recovery Act
REACH	Registration, Evaluation, Authorisation and restriction of Chemicals
RPM	Revolutions per minute
SARA	Superfund Amendments and Reauthorization Act
SC	Sandy clay
SM	Sandy silt
SOP	Standard Operating Procedure
SP	Sand, poorly-graded
SPLP	Synthetic Precipitation Leaching Procedure
SPT	Standard Penetration Testing
S/S	Stabilisation/solidification
STARNET	Stabilisation/solidification Network
SW	Sand, well-graded
TCLP	Toxicity Characteristic Leaching Procedure
TOC	Total Organic Carbon
TPH	Total Petroleum Hydrocarbons
UCS	Unconfined Compressive Strength
USCS	Unified Soil Classification System
USEPA	United States Environmental Protection Agency
VOC	Volatile Organic Compound

TABLE OF UNIT CONVERSION

Speed

	mph to kph						kph to mph			
1	mph	↔	1.61	kph		1	kph	↔	0.62	mph

Length

	centimeter to inch						inch to centimeter			
1	cm	↔	0.39	in		1	in	↔	2.54	cm

	metres to feet						feet to metres			
1	m	↔	3.28	ft		1	ft	↔	0.30	m

Area

	Hectares to Square Metres						Square Metres to Hectares			
1	ha	↔	10,000	m²		1	m²	↔	0.0001	ha

Weight

	Short Tons to Tonnes						Tonnes to Short Tons			
1	Ton	↔	0.91	Tonne		1	Tonne	↔	1.1	Ton

	Long Tons to Tonnes						Tonnes to Long Tons			
1	Ton	↔	1.02	Tonne		1	Tonne	↔	0.98	Ton

	Pounds to Kilograms						Kilograms to Pounds			
1	lb	↔	0.45	kg		1	kg	↔	2.2	lb

	Ounce to Grams									
1	oz	↔	28.3	g		1	g	↔	0.03	oz

Volume

	Cubic yards to Cubic metres						Cubic metres to Cubic yards			
1	yd³	↔	0.76	m³		1	m³	↔	1.31	yd³

	US Gallon to litre						Litre to US Gallon			
1	gal	↔	3.78	l		1	l	↔	gal	0.26

	Imperial Gallon to litre						Litre to US Gallon			
1	gal	↔	4.54	l		1	l	↔	gal	0.22

Pressure

	PSI to MPa						MPa to PSI			
1	psi	↔	0.007	MPa		1	MPa	↔	145	psi

Density

	Pounds per cubic foot to Tonnes per m³						Tonnes per m³ to Pounds per cubic foot			
1	pcf	↔	0.02	t/m³		1	t/m³	↔	62.4	pcf

Torque

	Kilograms per force metre to Pounds per force foot						Pounds per force foot to Kilograms per force metre			
1	kgf-m	↔	7.23	lb.ft		1	lb.ft	↔	0.14	kgf.m

Currency (at time of writing, August 2015)

	US Dollars to Euros						Euros to US Dollars			
1	USD	↔	0.91	€		1	€	↔	1.10	USD

Stabilisation/Solidification of Contaminated Soil and Waste: Practice

The practice behind this versatile remedial technology

www.hyggemedia.com

PART ONE

Introduction & Overview

1.1 Purpose

This book presents a synthesis of the practical experience in the remediation of hazardous waste sites by stabilisation and solidification technology (S/S) professionals. The sites that typically require treatment are contaminated by industrial residues and present a risk to human health and the environment.

The practical experience contained herein is intended to act as a reference source on: the suitability of wastes for treatment, the design, application and quality aspects of S/S, and other key issues that (together) capture how S/S is applied in the field and validated as a risk-management strategy.

The issues facing the practitioner on how to implement S/S are discussed, and the 'know how' presented should provide valuable insight into the remediation of contaminated land and treatment of waste by S/S. No decision regarding treatment should be based solely on this document. Rather this document should be used in conjunction with other references and the knowledge of skilled professionals to reach reasoned decisions to fit the needs of any specific site.

Each contributing author is a recognised authority on the application of S/S and is responsible for the material presented in their respective sections. Contributing authors have experience gained from the application of S/S in the USA, Canada, the United Kingdom and Continental Europe – experience that the editors believe is not shared in any other published work.

1.2 Target audience

This 'Manual of Practice' has been designed as a practical reference for regulators, site owners, engineering firms, and others involved in selecting, designing, bidding, and providing oversight for the remediation of hazardous waste sites using S/S. This book provides guidance on applicable contaminants, site characteristics, project planning, equipment capabilities, production rates, performance specifications and the quality assurance of S/S treated materials.

This manual should provide stakeholder reassurance on the appropriateness of S/S as a viable and cost effective technology for managing the risks associated with contaminated soil and waste. Included are references to numerous case studies and an extensive reference list of completed projects that successfully employed S/S.

1.3 Definition of terms

The term Solidification/Stabilisation has been defined somewhat differently in various publications and promulgated regulations. This manual of practice follows the definitions for solidification and stabilization as presented in the Interstate Technology and Regulatory Council document 'Development of Performance Specifications for Solidification/Stabilization' (ITRC 2011):

"Although solidification and stabilization are defined separately, they are often implemented simultaneously through a single treatment process. The EPA defines each as follows (EPA 2000):

'Solidification' involves the processes that encapsulate contaminated material to form a solid material and restricts contaminant migration by decreasing the surface area exposed to leaching and/or by coating the contaminated material with low-permeability materials. Solidification is accomplished by mechanical processes, which mix the contaminated material with one or more reagents. Solidification entraps the contaminated material within a granular or monolithic matrix.

'Stabilisation' involves the processes where chemical reactions occur between the reagents and contaminated material to reduce the leachability of contaminated material into a stable insoluble form. Stabilization chemically binds free liquids and immobilises contaminated materials or reduces their solubility through a chemical reaction. The physical nature of the contaminated material may or may not be changed significantly by this process."

The following definitions for in-situ and ex-situ are used in this manual of practice. In-situ and ex-situ are defined based solely on the manner in which the treatment mixing is accomplished. Various regulatory agencies and authors have defined these terms differently and their definitions may have significance regarding specific requirements that must be met. For this Manual of Practice the following definitions are used.

'In-situ' is defined to mean that the contaminated soils/sludges are mixed with treatment reagents without removing the contaminated soils from the ground. After treatment the material may be left in its original location, placed in a different location on-site, or be sent off-site.

'Ex-situ' is defined to mean that the contaminated soil/sludge was excavated from its original location in the ground and then mixed with treatment reagents. After treatment the material may be returned to its original location, placed in a different location on-site, or be sent off-site.

As an example, reagents could be applied either as solids or as a grout directly to contaminated soils and then be mixed with an excavator. This would be in-situ treatment. If however the contaminated soils were removed from the ground and placed in a nearby mix pit, then reagents were added and mixed with an excavator, this would be ex-situ treatment.

These definitions are based on the mixing equipment and the approaches and are used in this document. Regulatory definitions may, however, differ from these definitions, and from agency to agency.

Other terms such as site remedial goals, material performance goals, specifications, etc., are defined below and discussed in Section 7 and essentially follow the definitions used by ITRC (ITRC, 2011).

- Remedial Goals – overall objectives of the remedy to address the identified risk pathways
- Material Performance Goals – expected behaviours of the treated S/S material to support meeting the remedial goals
- Material Performance Specifications – the collection of parameters, tests and criteria to be utilised in developing a mix design and in evaluating the ability of the mix design to meet the material performance goals
- Construction Performance Specifications – the collection of parameters, tests and criteria to be utilised to verify that the treated material created during implementation is consistent with the materials developed and characterised during the treatability testing and that key performance characteristics (e.g., strength, permeability, and possibly leaching reduction) are consistently met as the treatment progresses

1.4 Appropriate uses of S/S technology

With the development of an effective binder formulation and careful implementation, S/S can be used to treat a wide range of organic and inorganic contaminants in soils and sludges. Excavated sediments are also treatable. Section 3 discusses potential applications and considerations in detail, while the appendices provide numerous examples.

Generally, S/S can be used to treat metallic compounds, longer chain petroleum hydrocarbons and many of the larger chlorinated organics. Figure 1.1 is typical of a site 'set-up', where a binder system employing several reagents is being applied to excavated-material, in an ex-situ treatment step. The three reagent silos (cement, fly ash, activated carbon) and a pug-mill were used to treat soils contaminated with creosote, pentachlorophenol (PCP), dioxins, and metals, at the American Creosote Site in Tennessee. One of the considerable strengths of S/S is its versatility of application, using commonly available reagents in a formulation designed to manage the risks from contamination at a specific site. The successful treatment of soils at the American Creosote site is a good example of this (Bates et al., 2002).

Although dioxins can be treated quite easily with S/S, PCP can be difficult hence the use of activated carbon in the treatment formula employed at American Creosote. It should be noted that S/S often is not very effective for treating volatile organics or volatile inorganics, nor is very effective for treating liquid fuels or organic solvents such as benzene, toluene and xylene. Fortunately other technologies such as bioremediation and vapour extraction can be very effective for such contaminants.

However, if contaminants for which S/S is not well suited are present in a soil or sludge at very low concentrations, then often the material can be treated using a modified binder, incorporating activated carbon or organophilic clays into the formulation. Figure 1.2 depicts the installation of a grout slurry blanket over a bermed cell to minimise emissions during subsequent in-situ S/S treatment employing such a special binder.

It should be noted that S/S is generally easier to implement in sandy, silty or gravely soils, than in soils with high clay content as it is easier to achieve a uniformity of mixing in

Figure 1.1: Treating creosote, PCP, & dioxins in pug-mill

Figure 1.2: Installing a grout slurry blanket to reduce emissions during in-situ S/S treatment

the former, while the latter may tend to leave residual clay balls of unmixed and untreated material. Sections 3.1 and 3.2 of this document provide a more detailed discussion regarding appropriate contaminant types and soils for treatment by S/S.

The use of S/S for managing the risk associated with contamination arising from industrial processes has a long history, extending back more than half a century. The successful application of S/S is however dependent upon a detailed knowledge of the behaviour of contaminants in the environment, the soil-matrix to be treated, the appropriate choice of binder systems and its application to produce a rock-like monolithic or granular product in which contaminants are encapsulated.

The following sections of this document present the collective real-world experience of leading practitioners on how S/S is applied and validated as a versatile risk management strategy.

1.5 Long term performance

S/S stabilised soils and wastes are vulnerable to the same physical and chemical degradation processes that affect any other cement-bound material. Any impact on the binder system can affect its capacity to immobilise contaminants in the longer-term.

Stabilised/solidified material will differ from conventionally bound materials in significant ways. S/S systems incorporating cement as the primary binder are (to a certain extent) analogous to concrete, which is widely used as a construction material. Concrete is a carefully designed product with specific physical and mechanical properties. An S/S system however is inherently a more heterogeneous product, and therefore cannot be as precisely engineered as its final characteristics are not normally accurately predicted.

Figure 1.3 is a conceptual model of an S/S system, which illustrates the complexity of the processes and the impactors on performance with time. The key variables underlying the efficacy of S/S can be described as the nature of the soil/waste matrix, the contaminants, the interaction with the binder system and the transport properties of the waste form and surrounding 'containment system', whether that be engineered containment, or the inherent geology of the site of deposition/placement.

Acting upon the waste form after placement are the specific environmental loads, operating in the environment of service. These may include a hydraulic gradient, gaseous or saline infiltration, micro and macro-biological activity, freeze-thaw etc. Together, all of these variables will dictate the performance of the waste form with time – time that should ideally be measured in millennia. Despite the very different design specifications, and nearly 60 years of the use of S/S in the USA, it is encouraging to note that there are no reported major failures of S/S waste-forms in that country. Similarly, this observation has been supported by the findings of the PASSiFy project (PASSiFy, 2010), which examined samples of S/S materials taken directly from remedial operations in the USA, UK and France (see Table 1.1).

The PASSiFy study highlighted that a number of risk-indicators were present in the samples examined, but these did not indicate the onset of deleterious reactions. As waste forms appeared to behave much like cement-bound materials, the interactions between the soil-fraction, the waste and the binder system could be explained.

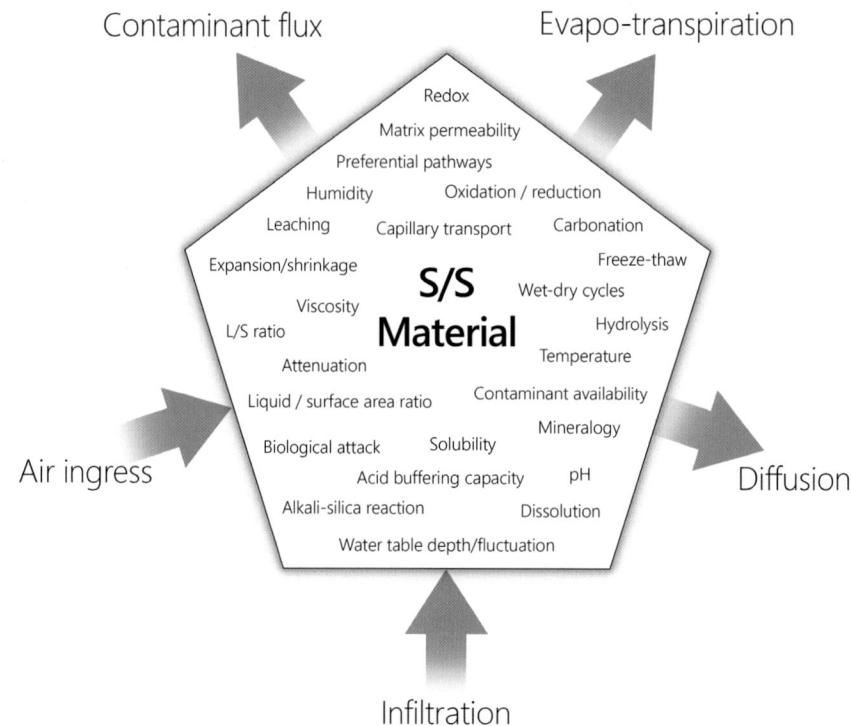

Figure 1.3: Conceptual model of a site with environmental loading

The life expectancy (the time in service where contaminants are not significantly released) of different S/S systems is predicted to extend from decades to thousands of years. Performance is dependent upon the binders being employed, the contaminants being treated and the environmental loads impacting upon the waste form. Figure 1.4 presents data taken from several studies to predict the long-term behaviour of S/S systems.

Site	Contaminants						Age (yrs)	S/S Method					Observations
	Metals	VOC	PAH	PCB	BTEX	TPH		In-Situ	Ex-Situ	Cement	Fly ash	Additive	Proprietary Mix	
American Creosote, Tennessee, USA	•	•	•				3		•	•	•	•		All chemical and physical targets met, apart from permeability. Evidence of natural weathering
Pepper Steel, Florida, USA	•			•			14		•	•	•			All original targets met. Evidence of natural weathering
South 8th Street, Arkansas, USA	•			•	•		4	•		•	•			All original targets met, apart from compressive strength. Evidence of natural weathering
Georgia Power Company, Georgia, USA			•		•		12	•		•				All original targets met, apart from compressive strength. Evidence of natural weathering
Selma, California, USA	•						5	•					•	All original targets met. No weathering data
Halton, UK	•						5	•		•		•		All original targets met. Evidence of natural weathering
Caerphilly, UK			•			•	1	•					•	All original targets met. Evidence of natural weathering
Quarry Dump, USA						•	10	•		•				All original targets met, apart from permeability

Table 1.1: S/S applications examined in 2010

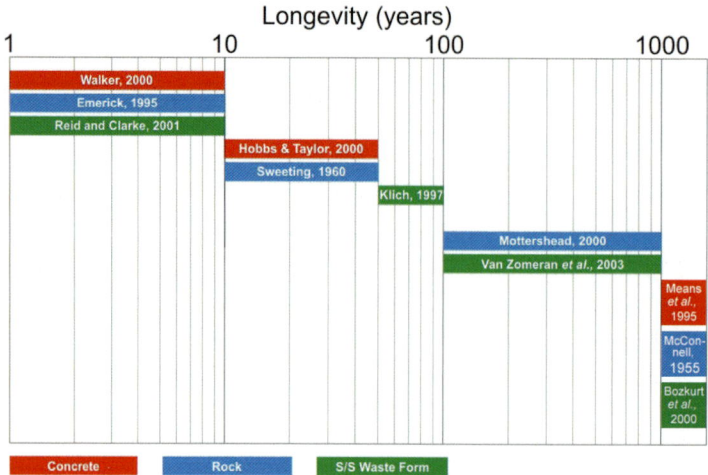

Figure 1.4: Longevity of S/S materials proposed by selected authors

1.6 History of use of cementitious binders

The history of modern cement begins with the endeavours of the British Engineer John Smeaton. Charged with rebuilding the Eddystone Lighthouse in Cornwall, England, during the mid-eighteenth century, his quest to find a durable water-resistant material led to the discovery that the best mortars were made by calcining limestone containing clay impurities (Miller, 2009; Hall, 1976). The resulting strong binder according to Smeaton "would equal the best merchantable Portland stone in solidity and durability" (MacLaren and White, 2003).

However, the story of cementitious binders actually spans 9 millennia, as described below. The secrets of making a perfect binder were lost for well over a thousand years after the fall of the Roman Empire. In the interim period leading up to the Industrial Revolution, simple lime mortars were used. These materials had little or no hydraulic properties (the ability to set underwater), and only fully set upon prolonged contact with carbon dioxide from the atmosphere, and consequently, from the surface inwards (Hall, 1976).

The Romans were accomplished at producing high quality hydraulic cements by blending quicklime, sand, volcanic ash and aggregate, and even used ox blood as a plasticising agent! Unlike modern concretes, the secret of these mixtures was the use of the pozzolanic volcanic ash, named after the Pozzuoli, near Naples in Italy where it is found.

Pozzolanic cements derive their strength from the rich silica and aluminate phases in the ash, which react with lime to form hydrated calcium silicates and aluminates. These compounds are similar to the cementing minerals in modern Portland cement (MacLaren and White, 2003; Hall, 1976).

In his 'Ten Books on Architecture' published around 25BC, Vitruvius advised on the use of different types of aggregate and binder ratios for specific applications. By the middle of the first century AD, Roman civil engineers had mastered the art of underwater concreting and constructed the harbour at the city of Caesarea. The importance of the raw materials quality, the detrimental effect of sand contaminated with earth and the problems associated with the excessive use of marine sand were then clearly understood (Delatte, 2004). The attention to detail, and use of specified ingredient is key to the endurance of many Roman structures, including the Pantheon, the Pont du Gard, and the Basilica of Constantinople (Wayman, 2011; MacLaren and White, 2003).

It is likely, however, that the Romans inherited cement technology from their Greek neighbours (Wayman, 2011). One of the best examples is a 3000 year-old cast water tank in the ancient city of Kamiros, on the Greek island of Rhodes. Analysis of this tank has shown that the Greeks had an excellent knowledge of concretes, practiced careful blending of different grades and types of aggregates, and mixed binders composed of volcanic earth and lime. The concrete on Rhodes still exhibits excellent compressive strength, elastic properties and low porosity despite exposure to weathering over millennia (US Department of Energy, 2000).

Whilst the Romans and Greeks were perfecting pozzolanic-based formulations, the Chinese explored alternative additives. The pyramids of Shaanxi, built during the Qin Dynasty around 200BC, contained conventional mixtures of lime and volcanic ash or clay, whereas more unusual materials were used in other applications (Miller, 2009). Lime was mixed with sticky rice to create a remarkable composite construction medium employed in all manner of important buildings, including tombs, urban constructions, and water conservancy facilities (Yang *et al.*, 2010).

From the third millennium BC, the Egyptians developed their own skills in the use of cementitious binders. The earliest construction utilised simple clay mortars to bind stone blocks. As their technology advanced, gypsum or lime-based mortars were adopted for the construction of the later pyramids. Like the Romans and the Greeks, the Egyptians were able to effectively work with the raw materials available to them, and formulated numerous different mortars from burnt lime and gypsum, in combination with marly limestone, kaolinitic clay, natron salt, sand, and diatomaceous earth.

An alternative theory proposed in the mid-1980s suggests the casing stones of the pyramids of Giza were in fact cast in-situ, using a granular limestone aggregate and an alkali alumino-silicate-based binder (MacKenzie *et al.*, 2011; Jana, 2007). On the Indian subcontinent, an equally wide variety of cement types were in use from the third millennia BC. Gypsum and lime cements, and bitumen mortars were utilised for wells, drains and building exteriors in early Bronze Age settlements such as the Mohenjo-Daro (Mound of the Dead) (Mays, 2010).

Between the third and first millennia BC, the kingdoms of Mesopotamia were also creating binders from local raw materials. Clay, bitumen or lime-based cements were used to bond adobe bricks or stone blocks in structures from courtyard houses to the immense ziggurats (Barbisan & Guardini, 2007; Rogers, 1900). Bitumen was also employed for waterproofing. Blended materials (e.g. bitumen with clay, or lime with bitumen) were routinely used for specific applications (Moorey, 1994; Johnson, 1987).

One of the earliest examples of the use of cementitious materials is at the early Bronze Age village of Yiftahel in Northern Israel. Yiftahel is one of the oldest permanent villages ever found, and contains a number of concrete floors. Carbon-14 dating of seeds embedded in the floor suggests that it was laid approximately 8850 years ago, in two layers, consisting of a roughly compacted lower layer, and a careful troweled surface. Analysis of samples from the floor showed that it is composed of nearly pure calcium carbonate and a small amount of silica which was probably sand, and exhibits compressive strengths equivalent to modern structural concrete (Kanare *et al.*, 2009; US Department of Energy, 2000).

When it comes to the use of cementitious binders, it seems there is little we could teach our ancient ancestors!

1.7 Information sources on S/S systems

There is a considerable amount of information in the literature about S/S, including in excess of 2000 journal papers and other documents. However, to obtain a balanced and authoritative view on the merits and difficulties of using S/S to treat soil and waste, the authors have identified a number of key information sources, which together provide a wealth of information on S/S technology. These include regulatory and industry guidance, the scientific literature and from networks and discussion groups, and are given in Tables 1.2 to 1.6 below.

1.7.1 General information

A number of general information resources on S/S in Table 1.2 are available to provide insight into all aspects of S/S.

1.7.2 Technical information

The documents identified in Table 1.3 provide more technical information on specific aspects of stabilisation/solidification technology.

1.7.3 Government departments and agency information resources

A number of agencies are responsible for the regulation and enforcement of environmental policies relating to contaminated land, pollution, and treatment. Information pertaining to these is given in Table 1.4.

1.7.4 Research, committees and associations

Listed in Table 1.5 are organisations that represent major industries related to S/S and/or conduct research into key topics connected to S/S.

1.7.5 Forums, networks, and discussion groups

The list given in Table 1.6 includes platforms, which allow contractors, planners, developers and researchers to share information and knowledge.

Publisher/Authors	Date	Title/Publisher
Environment Agency	2004	Guidance on the use of Stabilisation/Solidification for the Treatment of Contaminated Soil. Science Report SC980003/SR2.
Al-Tabbaa, A. and Perera, A.S.R	2006	UK Stabilisation/Solidification Treatment and Remediation – Parts 1-7. Advances in S/S for Waste and Contaminated Land – Proc. Int. Conf. on Stabilisation/Solidification Treatment and Remediation. pp 367-485.
BCA	2001	Cement-based stabilisation and solidification for the remediation of contaminated land. British Cement Association Publication 46.050.
BCA	2004	The essential guide to stabilisation/solidification for the remediation of brownfield land using cement and lime.
BCA Concrete Centre	2005	Remediating brownfield land using cement and lime. British Cement Association.
Environment Agency	2004	Review of Scientific Literature on the use of Stabilisation/solidification for the Treatment of contaminated soil, solid waste and sludges. Science Report SC980003/SR2. http://publications.environment-agency.gov.uk/pdf/SCHO0904BIFP-e-e.pdf?lang=e/pdf
EPA	1999	Solidification/stabilization Resource Guide. Report EPA/542-B-99-002. April 1999. http://www.epa.gov/tio/download/remed/solidstab.pdf
Construction Information Service		http://www.ihs.com/products/solutions/construction-information-service.aspx
Contaminated Site Clean-Up Information		http://clu-in.org/remediation
Portal for Soil and Water Management in Europe		http://www.eugris.info
Soil Environmental Services		http://www.soilenvironmentservices.co.uk

Table 1.2: General information sources on S/S

Organisation/Authors	Date	Title/Publisher
Barnett, F., S. Lynn, and D. Reisman	2009	Technology Performance Review: Selecting and Using Solidification/Stabilization Treatment for Site Remediation. EPA 600-R-09-148.
Army Environmental Policy Institute (AEPI	1998	Solidification Technologies for Restoration of Sites Contaminated with Hazardous Wastes.
Conner, J.R	1997	Guide to Improving the Effectiveness of Cement-Based Stabilization/Solidification. Portland Cement Association. PCA: EB211.
Conner, J.R	1990	Chemical Fixation and Solidification of Hazardous Wastes. New York, New York: Van Nostrand Reinhold.
Conner, J.R, and Hoeffner, S.L.	1998	The History of Stabilization/Solidification Technology, Critical Reviews in Environmental Science and Technology, 28 (4), pp 325-396.
EPA	1997	Innovative Site Remediation Design and Application, Volume 4: Stabilisation/Solidification. EPA 542-B-97-007.
EPA	2013	Superfund Remedy (Technology) Selection Reports, First thru Fourteenth Editions. http://clu-in.org/asr
EPRI	2012	State-of-the-Practice Liners and Caps for Coal Combustion Product Management Facilities, Electric Power Research Institute, EPRI, Palo Alto, CA, October 2012, Report 1023741.
Interstate Technology & Regulatory Council (ITRC)	2011	Development of Performance Specifications for Solidification/Stabilization. http://www.itrcweb.org/GuidanceDocuments/solidification_stabilization/ss-1.pdf
Ramboll Norge AS	2009	Cement Stabilisation and Solidification (STSO): Review of Techniques and Methods, Ramboll Norge AS, Oslo, Norway. Rap001-ld01, 57.2009.
Paria, S. and P.K. Yuet	2006	Solidification/stabilisation of organic and inorganic contaminants using Portland cement: A literature review. Environmental Reviews 14(4):217-255.
PASSiFy Project	2010	Performance Assessment of Solidified/Stabilised Waste-forms, An Examination of the Long-term Stability of Cement-treated Soil and Waste (Final Report), CL:AIRE, RP16. http://www.claire.co.uk/index.php?t&view=record&cat_id=23:stabilisation-solidification&id=298:performance-assessment-of-stabilisedsolidified-waste-forms-passify&Itemid=61
Spence, R.D. (Editor)	1993	Chemistry and Microstructure of Solidified Waste Forms. Lewis Publisher.
Taylor, H.F.W.	1997	Cement Chemistry. Thomas Telford Publishing, London.

Table 1.3: Technical information sources on S/S

Organisation/Body	Address
Environment Agency	http://www.environment-agency.gov.uk/
Department for Environment, Food and Rural Affairs	https://www.gov.uk/government/organisations/department-for-environment-food-rural-affairs
United Nations Environment Programme	http://www.unep.org/
European Environment Agency	http://www.eea.europa.eu/
Agency for Toxic Substances and Disease Registry	http://www.atsdr.cdc.gov/
Army Environmental Policy Institute	http://www.aepi.army.mil/
UK Government	https://www.gov.uk/contaminated-land
United States Department of Justice: Environment and Natural Resources Division	http://www.justice.gov/enrd/
European Union Environment	http://ec.europa.eu/environment/index_en.htm

Table 1.4: Information on S/S from government departments and agencies

Organisation/Body	Address
Portland Cement Association	http://www.cement.org/
Environmental Protection UK	http://www.environmental-protection.org.uk/
Contaminated Land: Applications in Real Environments (CL:AIRE)	http://www.claire.co.uk/
The Concrete Centre	http://www.concretecentre.com/
Construction Industry Research and Information Association	http://www.ciria.org/
Interstate Technology & Regulatory Council	http://www.itrcweb.org/
The Chartered Institution of Water and Environmental Management	http://www.ciwem.org/
Chartered Institute of Environmental Health	http://www.cieh.org/
British Cementitious Paving Association	http://www.britpave.org.uk/
Cement Association of Canada	http://www.cement.ca/en

Table 1.5: Information on S/S from research, committees and associations

Organisation/Body	Address
Network for Industrially Contaminated Land in Europe	http://www.nicole.org/
Stabilisation/solidification treatment and remediation network	http://www-starnet.eng.cam.ac.uk/
Brownfield Briefing: News, Views, Analysis	http://www.brownfieldbriefing.com/
Common Forum on Contaminated Land	http://www.commonforum.eu/
Association for Environmental Health & Sciences (AEHS)	http://www.aehsfoundation.org/
Environmental Knowledge Transfer Network	https://connect.innovateuk.org/web/sustainabilityktn

Table 1.6: Information on S/S from forums, networks, and discussion groups

PART TWO

History of S/S as a risk management tool

2.1 Use of S/S in the USA (Superfund and RCRA)[1]

The United States has been a leader in the application of treatment technologies for remediation of hazardous waste sites including the application of Solidification/Stabilisation (S/S). Sites are remediated under a number of regulatory programs, including the well-known "US Environmental Agency (USEPA) Superfund" program. Others include the "USEPA RCRA Corrective Action" program, State led clean-up programs, and private party voluntary clean-up actions (without USEPA or State oversight).

A comprehensive compilation of all site clean-up actions, or a comprehensive list of all sites remediated by S/S is not readily available. Perhaps the most authoritative data base available on the selection of S/S for site remediation is that compiled by the USEPA Superfund program which tabulated 280 S/S source treatment technology selections spanning the 30 fiscal years (FY) 1982-2011 (USEPA 2013, Appendix B).

2.1.1 The regulatory impetus for the remediation of sites in the USA

A strong driving factor for development of treatment technology and hazardous site remediation in the United States has been Federal legislation mandating responsibility for management of hazardous wastes and remediation for sites contaminated by past operations. In 1976 the US Congress passed The Resource Conservation and Recovery Act (RCRA) which "gives EPA the authority to control hazardous waste from the 'cradle-to-grave'. This includes the generation, transportation, treatment, storage, and disposal of hazardous waste" (USEPA 2011). In 1984 HSWA - the Federal Hazardous and Solid Waste Amendments focused on waste minimization and phasing out land disposal of hazardous waste (USEPA 2011).

Of considerable importance under RCRA regulations, is that a generator of hazardous wastes is not absolved of liability for proper management and disposal of these wastes by contracting these services to another party. RCRA however was designed to apply to current and future generators of hazardous wastes.

In 1980, the US Congress enacted the Comprehensive Environmental Response, Compensation, and Liability Act (CERCLA, or Superfund) to address the dangers of abandoned or uncontrolled hazardous waste sites. CERCLA provides the USEPA and other federal agencies the authority to respond to a release or a substantial threat of a release of a hazardous substance into the environment, or a release or substantial threat of a release of "any pollutant or contaminant, which may present an immediate and substantial danger to public health or welfare" (USEPA 2007, page 1-1). This legislation has been of great importance since it provides for taking action to remediate abandoned sites and/or compelling responsible

---

[1] The information presented in this section is based upon the authors' experience, and is not intended to present the policy of the USEPA, nor has it been reviewed and approved by the USEPA

parties to take such action even though the creation of the problem may predate regulations that prohibited such disposal. This retroactive provision within CERCLA has been responsible for stimulating the clean-up of many sites.

"The Superfund Amendments and Reauthorisation Act of 1986 (SARA) expressed a preference for permanent remedies (that is, treatment) over containment or removal and disposal, in remediation of Superfund sites" (USEPA 2007, page 1-1). This preference for treatment over containment or disposal has been a powerful driver for the development and application of treatment technologies, including S/S. Oversight of contaminated site clean-up is generally done under one, or more, of the following authorities:

- Federal USEPA
- Superfund - CERCLA
- RCRA
- State
- Delegated Federal Authority
- State Hazardous Waste Programs
- State Voluntary Programs
- Private Sector
- Unregulated Sites
- Licensed (Delegated) Professionals

The combination of RCRA (especially RCRA corrective action), for active sites and sites previously active under RCRA permits and which still have financially viable responsible parties, and the Superfund program for orphan sites and sites contaminated prior to the RCRA, has been a powerful stimulant for the development of treatment technology. A compiled database indicating the frequency of S/S use at RCRA sites is not available, though RCRA does recognise S/S as a BDAT (Best Demonstrated Available Technology) for treatment of many waste streams containing metals.

Although not specific to S/S, some appreciation of the magnitude of the RCRA corrective action clean up activity can be obtained from a recent USEPA publication (USEPA 2006-2008). This publication indicated that in 2008, there were 1968 RCRA high priority sites, with 96.2 % controlling human exposure, 83.4 % involved controlling the migration of contaminated groundwater, 43 % involved choice of final remedy, and 34.6 % reported the final remedy was under construction or had been completed.

As of January 7, 2014, the USEPA Superfund Program reported 1320 current sites, 53 proposed new sites, and 374 sites that had been deleted (http://www.epa.gov/superfund/sites/npl/index.htm). The Superfund program continues to provide frequent status reports that contain information on the frequency of remedy selection, including S/S, for Superfund sites, which can be accessed at http://clu-in.org/asr.

2.1.2 Use of S/S in the EPA Superfund Program

The USEPA Superfund program has issued a number of excellent Superfund remedy reports, and other documents, describing treatment technologies for the remediation of the NPL sites under this program. Many of these documents can be accessed through the website http://clu-in.org/remediation.

Remediation of hazardous waste sites in the USEPA Superfund program is based upon a risk management approach and compliance with applicable, or relevant, and appropriate requirements (ARARs); meaning the site clean-up and remediation is managed in such a manner, that residual risk is reduced to an acceptable level and meets ARARs. Thus, since S/S generally does not remove or destroy the COCs (contaminants of concern), S/S accomplishes its risk reduction objective by blocking the pathway between the COC and the receptor (human or the environment). S/S does this by either reducing the solubility of the COC (stabilisation) or by containing the COC in a low permeability matrix (solidification), or by using both processes (solidification/stabilisation).

The Superfund Remedy Report, Thirteenth Edition, remarks 'solidification/stabilisation continues to be the most frequently selected *ex-situ* source treatment technology' (EPA 2010 page 9). In this same report, S/S is listed as the second most frequently selected *in-situ* source treatment technology from 2005-2008, exceeded only by in-situ soil vapour extraction.

The most recent Superfund Remedy Report, Fourteenth Edition, reported that for FY 2005-2008, and FY 2009-2011, ex-situ S/S was selected for 19 % and 13 % of the source treatments respectively, and "is still the second most commonly chosen ex-situ remedial technology for sources" (USEPA 2013, page 9). This same report indicates that in-situ S/S was selected for 9 % of the source treatments for each of the periods FY 2005-2008 and FY 2009-2011. Also of interest is that during FY 2009-2011, S/S was selected in 11 % of the 56 decision documents expressing remedy components for sediments (USEPA 2013, page 11, Table 2).

Appendix B of this report, (EPA 2013), provides a table showing treatment technologies selected for each fiscal year from 1982 thru 2011. Based upon this data, Figure 2.1 was prepared and displays the most frequently selected source treatment technologies over this 30 year period. Figure 2.1 shows S/S as the second most frequently selected technology at 22 % of remedy documents selecting source treatment, slightly behind soil vapour extraction at 24 %. Figure 2.2 uses this same data base to graph the frequency of selection for S/S verses all other source technologies over this 30 year period. Figure 2.2 shows that S/S (combined ex-situ and in-situ) has consistently been selected in about 20 % of all remedy documents selecting source treatment under Superfund.

In the absence of more recent data, Figure 2.3 gives the generalised contamination found at sites for which S/S was selected under the Superfund Program. The literature is lacking a study, evaluating over time, trends in the frequency of selection of S/S for treatment of sites with organics. However, in the author`s experience, there seems to be increasing acceptance that S/S can effectively treat many non-volatile organics, alone or mixed with toxic metals.

In-situ S/S seems to be increasing modestly as a remedial technology, slightly displacing ex-situ S/S. Figure 2.4 compares the relative frequency of selection of in-situ S/S with selection of ex-situ S/S. Although selection of S/S overall, combined in-situ and ex-situ has remained fairly constant since about FY 2000, in-situ S/S appears to be increasing as a proportion of all S/S treatment selections. Perhaps this is due in part to the ability of in-situ S/S to treat wastes to substantial depth, including below the water table, without the need to de-water.

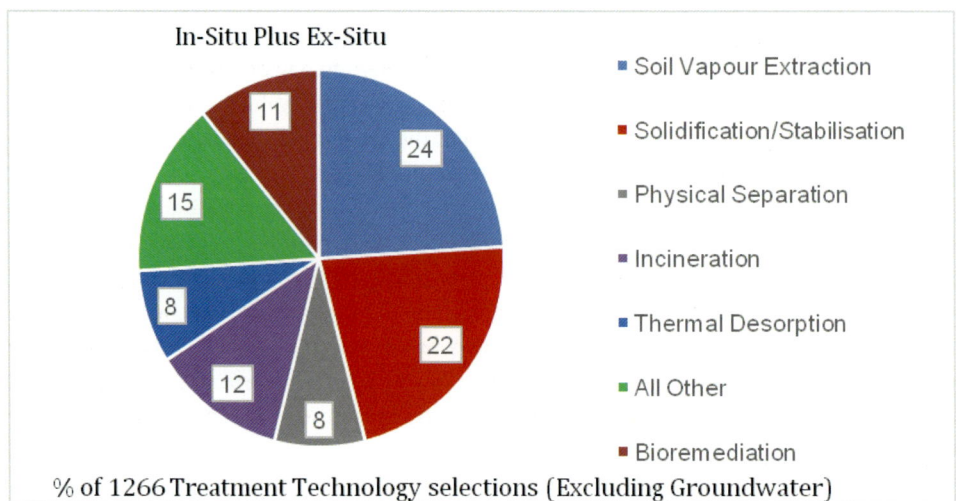

Figure 2.1: Cumulative source control technology selection from 1982-2011 (After EPA-542-R-13-016, Appendix B)

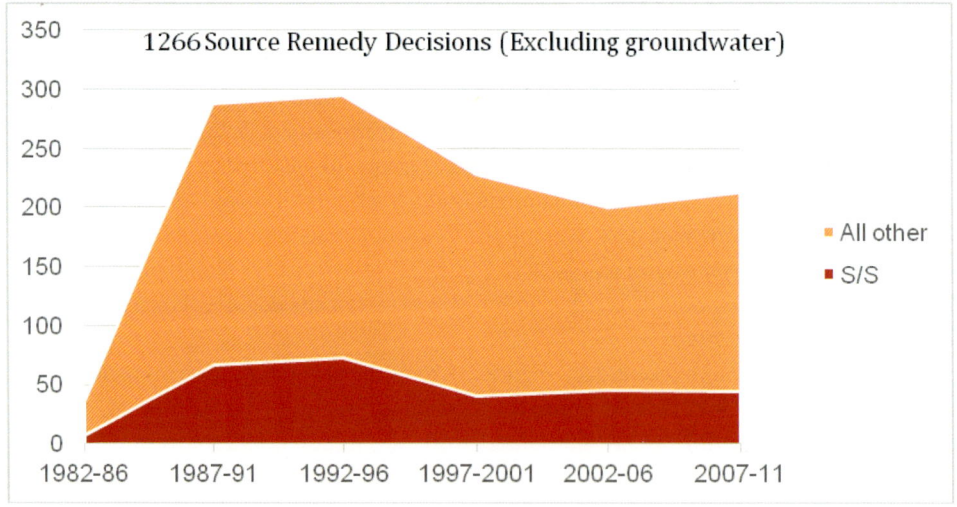

Figure 2.2: Selection of S/S vs all other technologies 1982-2011 (After: USEPA 542-R-13-016, Appendix B)

2.1.3 Conclusions

Decades of experience in the application of S/S technology to a wide variety of contaminant and soil types makes the USEPA Superfund Program experience a valuable resource for anyone involved with S/S, or any treatment technology. For the last 30 years, S/S has consistently been a technology of choice for the USEPA Superfund program, being selected consistently in about 20 % of all remedy documents with source area treatments. This trend continues today with a number of sites in active remediation employing S/S in 2014. The USEPA RCRA Program has also designated S/S as a BDAT technology for many waste types containing metals.

In view of the large number of successful applications over extended time within the USA, it is surprising that S/S has not been more widely used in other countries. This is even more remarkable considering that in the USA, for S/S treated sites, as with any site that leaves contaminants on the site, the Superfund program requires a review every 5 years, to assure that the remedy is performing as required.

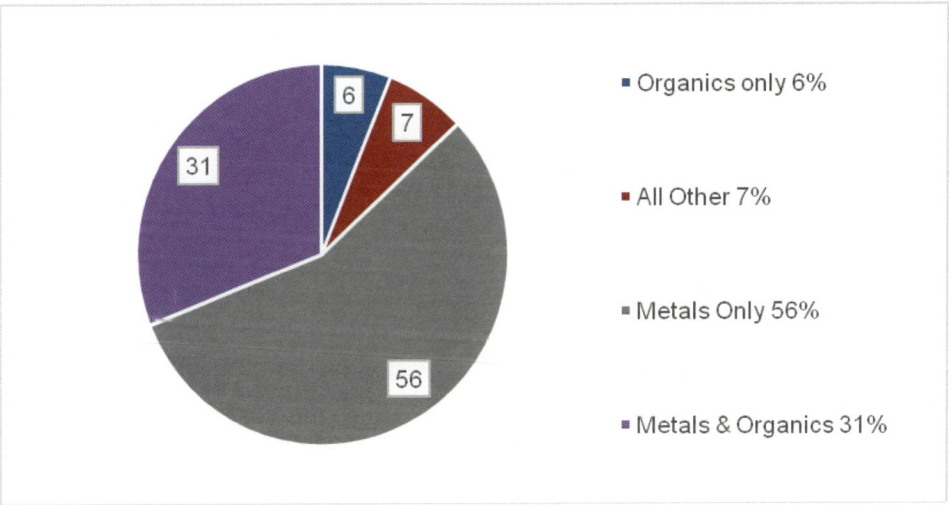

Figure 2.3: Waste types selected for S/S (EPA-542-R-00-010)

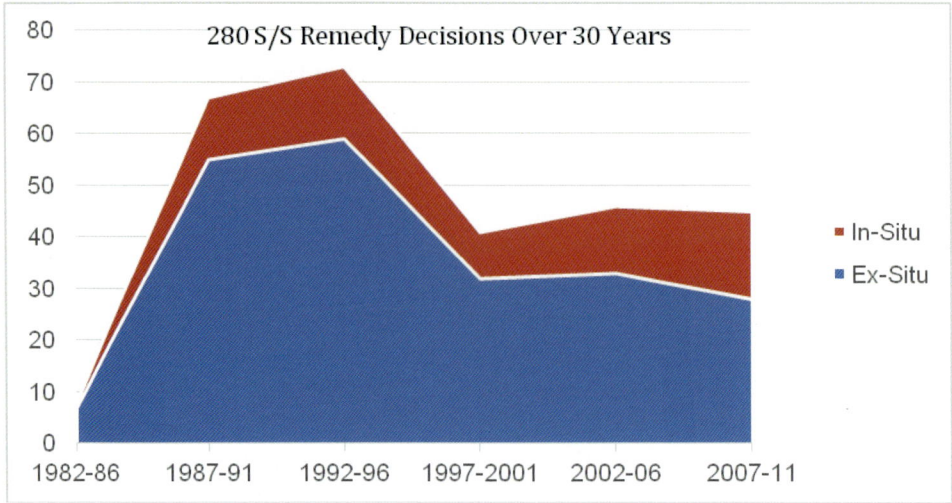

Figure 2.4: S/S selection of in-situ vs ex-situ 1982-2011 (After USEPA 542-R-13-016, Appendix B)

2.2 Use of S/S in the UK and Europe

Although the use of S/S can be traced back more than 50 years in North America, uptake of this technology (for treating contaminated soil) has been much slower in Europe. This primarily arose because of the historical low cost and plentiful supply of landfill space, and for much of this time, a pervasive immature legislative environment.

The use of S/S in the UK can be traced back to the early 1970's for the treatment of mixed wastes prior to co-disposal with domestic waste, in unlined quarry workings. Two centralised S/S treatment sites operated in the UK, one in the English Midlands, and the other East of London.

2.2.1 Treatment of waste and soil by S/S

By the late 1950's S/S was being used in France to treat low-solids containing waste (Conner, 1990) and by 1978 was routinely used to treat hazardous waste (Environment Agency, 2004a), having been originally developed for cementing radioactive residues. The focus on wastes remained and by the early 2000's, S/S was being used for waste treatment in France, Austria, The Netherlands and in Portugal.

In France, in 2004, there were 12 central processing facilities employing S/S to treat 400,000 tonnes of waste per year (Pojasek, 1978).

In the English Midlands, a waste treatment plant started operation in the 1970's primarily processing metal plating residues by a process called Sealosafe (Conner, 1990). Over 1 Million tonnes of materials were treated, before the operation was closed amid concerns over product quality (ENDS, 1988). The operating company was eventually successfully prosecuted in 1990 under Trades Descriptions legislation, although it should be said that doubts on the efficacy of the S/S product were reported as early as 1983 (ENDS, 1983). It is a widely held view that the reported failures at this commercial plant held back the uptake and acceptance of S/S in the UK for a number of years. Nevertheless, the second site, East of London, successfully employed the 'Stablex' process between 1978 and 1996, and up to 400,000 tonnes of hazardous wastes were treated each year and placed in a sanitary landfill (Conner, 1990).

The first widely reported example of contaminated soil treated by S/S in the UK took place at the former ICI explosives plant at Ardeer in Scotland in 1995. Here, 10,000m^3 of soil were treated by in-situ mixing using a lime, cement and slag-based binder system (Wheeler, 1995).

2.2.2 Legislative background

In the UK, the Public Health Act of 1848 recognised 'an accumulation or deposit' that was prejudicial to health or a nuisance. Attempts to regulate industrial waste (with reference to waste produced from alkali manufacture) were made in 1863 (Kiefer, 2012). However, since 1973, the direction of environmental protection across the European Community area has been articulated in the European environmental action programmes, which were first implemented in that year.

European environmental law, through single-issue 'Directives', and more recently 'Frameworks', shapes legislation in individual countries including the UK. This facilitates an integrated approach to environmental law making across the Community, but allows local variations in the way the laws are enacted.

Although the origins of European Environmental Law date back to the 1970's with air and water directives, the 1987 European Treaty (The Single Environment Act, 1987) was the first to include 'a policy in the sphere of the environment'. The Polluter Pays principle was subsequently established (The Sixth Environment Action programme of the European Community 2002-2012) introducing liability and an incentive to incorporate environmental considerations into the design of products or processes.

In the UK, the Control of Pollution Act (1974) dealt with a number of environmental issues including waste disposal, atmospheric pollution and public health. The COPA, as it is known, was augmented by the separate 'Special Waste' regulations in 1980, and amended in 1989 (Control of Pollution (Amendment) Act (1989)). The COPA was also re-enacted by the 1990 Environment Protection Act, allowing for improved control over emissions to land from industrial and other processes, and the introduction of 'risk' into the assessment of contaminated land and its condition.

In 1999, the Pollution Prevention and Control Act introduced the requirements of European Council Directive 96/61/EC on Integrated Pollution, Prevention and Control (IPPC) to ensure the best technical option is available to prevent emissions to air, water and ground. The strong emphasis on risk reduction and minimisation was a fundamental step-change over earlier legislation.

The European Council Directive 99/31/EC (*The Landfill Directive*) is also of great relevance to S/S as it categorised waste on harmfulness and the division of landfills into three classes:

- Landfills for hazardous waste
- Landfills for non-hazardous waste
- Landfills for inert waste

Standard waste acceptance procedures were introduced, and the pre-treatment of waste (prior to acceptance) was accompanied by a rigorous system of permitting. With respect to S/S, its use is facilitated by:

- The option to pre-treat waste/soil by S/S prior to landfilling
- The accepted use of appropriate risk management option to prevent waste going to landfill
- A substantial increase in the costs of landfill disposal particularly in the UK with the introduction of the Landfill Tax, promoting waste (and soil) recycling and re-use

2.2.3 Public and private initiatives

With an emphasis on risk management and the diversion of waste from landfill there was an increased interest in the use of S/S throughout the 1990's. In 2000, national guidance (England and Wales) was being developed in a partnership between industry and government. This project, called CASSST (Codes and Standards for Stabilisation/Solidification Technology), was initiated at the University of Greenwich, and resulted in the publication of national guidance on S/S by the Environment Agency (2004b).

In the first of its kind, this EA guidance was supported by an exhaustive review of the science behind S/S (Environment Agency, 2004a), to allow UK stakeholders to be conversant with the potential strengths and weaknesses of the technology, and to facilitate the formulation of the best design solutions employing S/S.

Other UK initiatives in the early 2000's included a research council sponsored academic-led network (STARNET, 2004), industry-led guidance BCA (2004) and an EU-supported project on S/S waste form performance (PASSIFY 2010). The latter project, called PASSIFY (Performance Assessment of Stabilised/Solidified Waste Forms) also involved the EA, the EPA, academics and stakeholders in the USA, UK and France, including ADEME. Under this initiative a number of S/S remedial operations were examined in detail in each country, including several Superfund sites.

In the years following the publication of Environment Agency Guidance, the uptake of S/S in the UK has increased and example remedial operations are cited in the literature (STARNET, 2004; Environment Agency, 2004b). The use of S/S is now firmly established as a credible risk management option for both soil and waste, as exemplified by its use at the Olympic Park in London (Penseart, 2008).

Currently, in Europe no comparable national (or Community) guidance on S/S exists and applications of the technology to contaminated soil remain second to those of waste treatment. In many EU member countries, landfill still remains a relatively cheap option, and until the cost and availability of disposal become prohibitive, landfill will remain as an attractive available alternative to risk management options such as S/S.

It is known that in France, guidance on S/S is being considered and reports indicate that within a few years ADEME may publish in this respect (Chateau 2012, pers. com).

2.3 Use of S/S in Canada

The use of S/S in Canada has been primarily focussed on the treatment of contaminated soil and management of mining spoil.

Examples of Canadian S/S projects since 1990 included remote military bases, former industrial facilities, urban waterfront redevelopments, and mining sites. Example remedial actions are documented by, for example, the Cement Association of Canada (www.cement.ca), and sites treated by S/S include:

- The Dockside Green redevelopment project in Victoria BC, for lead in soil to meet provincial hazardous waste leachate standards
- Western Steel Mill, Vancouver BC, for cadmium, lead, and zinc waste
- A rifle range in Burnaby, BC for lead, zinc, copper and antimony to meet the provincial hazardous waste targets
- Glacier National Park, BC, for lead in soil
- False Creek, Vancouver (Winter Olympic project) for zinc and pH in soil for a residential property redevelopment
- Canmore, Alberta, for mine waste recycling as on-site structural fill for a mixed-use residential development
- Swan Hills, Alberta, for spray dryer salts and bag house fly ash for disposal
- St. Catherines, Ontario, for lead and PAH in soil to meet the Ontario Land Disposal Regulations at a 5 ha (50,000m^2) property development
- Sydney, NS, for petroleum hydrocarbons in sediments at a former steel mill, to meet strength and permeability criteria

2.3.1 Performance (risk)-based versus prescriptive regulations

A "prescriptive" approach may call for a compound concentration in soil or water to be reduced to a specific value, so that the resulting risk or hazard by some assumed exposure conditions are acceptable. Prescriptive based environmental project goals were almost universally used in Canadian jurisdictions in the early 1990's.

A "performance" approach involves more site specific information, and may call for changes in compound concentration, migrating characteristics, exposure route conditions, receptor characteristics, any other variable, combination thereof such that the target acceptable resulting risk or hazard is not exceeded.

While an option for defining site specific performance goals existed in some Canadian jurisdictions in the early 1990's, Nova Scotia was the first province where performance goals were being consistently specified and used for remedial operations. Within 5 years of the 1996 Nova Scotia Environment's Guidelines for the Management of Contaminated Sites, the other Atlantic Canadian provinces (New Brunswick, Prince Edward Island, and Newfoundland and Labrador) revised their own regulations.

In 2007, new guidelines were issued in Alberta to encourage the use of performance-based environmental goals, followed by the same in British Columbia. Subsequently, the Federal Government of Canada released revised guidance for sites under federal jurisdiction. Nunavut, the Yukon Territory and the Northwest Territories now incorporate federal guidance for the use of S/S within their own jurisdiction.

Since 2005, S/S has been used in mining project developments in Canada to reduce the production of waste material and to protect ground and surface water resources. The result from the S/S of mine tailings is their accepted use as a sustainable construction material for use as structural backfill.

2.3.2 Relative use of S/S in Canadian provinces

Canadian jurisdictions with the greatest experience in using performance (risk)-based objectives tend to also be those where the most significant S/S projects have occurred (including Nova Scotia). This was noted in an earlier review (Ells, 2010) of the relative regulatory receptiveness among Canadian jurisdictions to S/S activities as:

- Relatively receptive (e.g. Nova Scotia, New Brunswick, Newfoundland and Labrador, Alberta, and British Columbia)
- Initially Receptive (e.g. Prince Edward Island, Ontario, Quebec, Government of Canada, and Manitoba)
- Not Yet Receptive (e.g. Northwest Territories, Nunavut, Yukon Territory, and Saskatchewan)

2.3.3 Sydney Tar Ponds and coke ovens project – Nova Scotia

When it was completed in 2013, the Sydney Tar Ponds project was the largest S/S remedial operation in the world, and the most publically and politically prominent contaminated site remediation project in Canada. At Sydney, over a million tonnes of sediments and soil contaminated with oily residues and metals from the production of steel and coke were treated using S/S.

The performance specification used at Sydney included hydraulic conductivity (permeability), SPLP (leachability) and unconfined compressive strength. Treatment protects the surface waters, the important fishery in Sydney Sound and provides for land for recreational and light industrial activities. Because of the prominence of this remedial action, a huge amount of technical information is publicly available. A case study on this project is included in Appendix B.

2.4 Risk management by S/S

The principles of risk-based remediation are now embedded in contaminated land legislation across the US and Europe. A "suitable for use" approach to contaminated land management is now widely adopted, implementing remedial action where it is demonstrated that unacceptable risks are being caused (or could be potentially caused) by contamination on a site for a designated end use or for potentially affected human or environmental receptors. This section will explain the basic principles of risk management, within which S/S is applied.

2.4.1 Definition of risk

Risk can be defined as a combination of the probability, or frequency, or occurrence of a defined hazard and the magnitude of the consequences of the occurrence. In the context of land contamination, this relates to the potential for contaminants to harm human health,

impact water resources and ecological receptors, and damage buildings and infrastructure.
Assessment of risk is founded on the "source-pathway-receptor" concept, identifying:

- Source - the pollutant hazards associated with the site
- Receptor - possible receptors at risk from the identified hazards (e.g. human, water resource, flora and fauna, buildings and infrastructure)
- Pathway - a route or means by which a source can impact a receptor

For risks to be present at a site, all three elements (source-pathway-receptor) of a plausible pollutant linkage must be present.

2.4.2 Risk management frameworks

Whilst factors such as acceptable levels of exposure and exposure models vary with individual nations' policies and legislation, most countries share a common risk-management framework structure based on site characterisation, environmental risk assessment, and evaluation of remedial measures, remediation implementation and validation (e.g. see Rudland *et al.*, 2001).

A tiered approach to risk assessment is adopted, with assessments increasing in detail where risks are potentially unacceptable:

Preliminary risk assessment: a desk based, qualitative assessment of potential risks. Information on the site history, ground conditions and environmental setting is collated and used to develop the conceptual model for the site.

Generic quantitative risk assessment: a comparison of contaminant concentrations with generic assessment criteria. Intrusive ground investigations are undertaken to obtain representative soil, water and soil-gas data to develop the conceptual site model. Generic assessment criteria are conservative, applying to limited defined land end uses and pollutant pathways.

Detailed quantitative risk assessment: the derivation of detailed assessment criteria using site-specific data. Comparison of soil, water and gas data site-specific assessment criteria or target levels to determine if unacceptable risks are present.

Phasing the risk-assessment process permits action to be taken rapidly to resolve obvious problems, whilst more detailed assessment in other scenarios may demonstrate no unacceptable risks are posed. A phased risk-based approach allows financial resources to be allocated most effectively.

Remediation using a risk-based methodology is an alternative approach to the total clean-up of a contaminated site. A risk-based approach can be used to establish that land use is neither technically or financially feasible, nor sustainable, or to establish that land use is possible and that the costs of treatment are acceptable.

However, by adopting a risk-based approach the ground conditions, environmental setting, and current and proposed site usage can be considered. Furthermore, soil type, geology, hydrogeology, the water environment, local ecosystems all impact on the risk posed by contamination. Similarly, the form, intensity and frequency of a receptor's exposure to contaminated soil, groundwater and soil gas are dependent on the end use of the site. As an example, as residential end use is more sensitive than an industrial end use, a greater degree of remediation would be required.

Where the level of risk justifies corrective action, risk-management can take the form of managing the receptor, breaking the pathway or reducing or removing the source material.

S/S does not remove or destroy contaminants, but immobilises them by a cementitious reaction involving the soil matrix, contaminants and binder reagents to promote sorption, precipitation or incorporation into crystal lattices, and/or by physically encapsulating the contaminants.

2.4.3 Summary

Treatment by S/S therefore breaks the pollutant linkages by both breaking the pathway and reducing the source material. However, the release of contaminants will still occur from treated material, and the objective of the design process is to produce a durable waste form from which release is controlled, to levels that do not pose an unacceptable risk to receptors.

It should be noted that dermal and volatile exposure pathways are not necessarily broken by S/S. Further thought as to the use of the technique in conjunction with others is needed, or possibly the careful reuse of treated materials as part of an engineered solution. These concepts are discussed in more detail in the following section.

2.5 S/S and the 'Risk Framework'

To understand how S/S fits into the risk framework, it is important to consider how S/S breaks pollutant linkages to different receptors. By assessing different potential receptors and pathways in turn, it is possible to identify issues specific to S/S that must be considered as part of the environmental risk assessment process and stabilisation design.

2.5.1 Water environment

The chemistry involved in S/S is complex, but the mechanisms considered to reduce the mobility of contaminants include:

- Sorption to soil and binder/additive materials
- Precipitation as a result of pH modification, reducing solubility
- Incorporation within the crystal lattice
- Encapsulation within the CSH gel formation

Infiltration and leaching pathways are reduced as a result of decreased permeability and reduced mobility of the contaminants. Mass flux of contaminants to groundwater and surface water receptors is therefore also reduced.

A robust hydrogeological model is an essential part of understanding the potential risks and the applicability of S/S. It is important not only to consider potential pollutant linkages, but also to be able to anticipate the groundwater flow regime and chemistry. No mass reduction in contaminants occurs; rather the risks are managed via a controlled very slow release of contaminants. Factors that are needed to develop a conceptual site model are discussed in more detail in Section 7.2. However, in the context of the risk-management framework, factors that can affect the potential durability and long-term leaching performance of treated materials require early consideration, to ensure they are adequately addressed throughout the design and approval process.

Considering the risks to water environment receptors (surface waters and groundwater) must first involve understanding the likely leaching behaviour from the S/S treated material. Key to this is the development of an appropriate testing regime representative of the likely end-use scenario. Potential exposure to aggressive (e.g. acidic and sulphate rich soils and groundwaters) or saline ground conditions or environments where a high degree of carbonation could be envisaged should be identified at the conceptual model development stage. The identification of aggressive conditions may require additional performance testing where necessary, and modified standard testing procedures to represent the site-specific conditions highlighted.

Following this, a comparison of likely release levels with remediation targets is then needed.

Although remediation targets are usually established by risk assessment, and are site-specific, they can be based on drinking water standards or health advisory levels for contaminants of concern. As such, the appropriate compliance points are selected at which environmental standards must be met to protect the receptors. These may be the aquifer or surface waters or some point nearer to the source. Target concentrations will reflect background concentrations, current and future use of the water resource and environmental standards.

A tiered approach to assessment is then usually undertaken, with an initial assessment determining whether the target concentration is exceeded at the compliance point, then subsequent tiers of assessment consider dilution, dispersion, retardation and degradation by biotic or abiotic processes at increasing levels of complexity. The level of assessment undertaken is usually dependent on the level of risk posed, or the relative sensitivity of the receptor.

When comparing leaching test results with remediation targets, care needs to be applied to ensure that the comparison is relevant. Groundwater models can require comparison with pore water concentrations, and consideration needs to be given to exactly what data is being compared to what target, and if a direct comparison is appropriate. As discussed in later sections, a wide array of leaching test methods is available, all reflecting different leaching behaviour of the treated materials. An understanding of what the leaching data is telling you is needed and the risk assessment and testing regime should ideally be developed in conjunction.

Potential surface water run-off pathways also need consideration. On a short-term basis, the potential risks to surface water receptors during construction (from material storage or partially treated materials) will require addressing during development of the environmental mitigation measure strategy. However, on a longer timeframe, the environment of service of the waste form, and what environmental loads may impact upon it (if any) will need to be assessed so there are no unforeseen adverse impacts resulting from the remedial action.

2.5.2 Human health

The key potential exposure pathways to human health receptors can be summarised as:

- Ingestion of soil, dust, home grown produce and waters
- Dermal contact with soil, dust and waters
- Inhalation of soil, dust and vapours from soil and waters

Whilst S/S can feasibly reduce generation of dust via production of a monolith, other pathways are not addressed by S/S treatment alone. Treatment may not reduce exposure to volatile contaminants, although the resulting reduction in permeability may restrict vapour movement to some extent. Treated material will also have a high pH, which would cause problems via the dermal pathway. Considerable attention therefore needs to be given to other potential pollutant linkages, and whether a treatment-train approach needs to be adopted.

The principal of a treatment-train is simple – the use of multiple treatment methods used either sequentially or simultaneously to enable the treatment objectives to be met. Common techniques may include bioremediation, chemical oxidation, soil vapour extraction and even incineration, depending on the range of contaminants present. Where other techniques are to be used in conjunction with S/S, the treatability trials must be designed to reflect this, to ensure testing is appropriate.

Another option is to use a low permeability-capping layer (Section 9). The placing of treated material at depth on a site can sometimes be sufficient by itself, supported by a detailed human health risk assessment to demonstrate that the risk from volatile contaminants is acceptable.

A final consideration with respect to human health receptors is whether the treatment process itself can increase the volatilisation of contaminants. The heat of hydration from binder components, particularly with a lime binder, may result in vapour emissions. This should be considered in the environmental risk assessment, so that appropriate mitigation or control measures can be put in place if needed.

2.5.3 Other receptors

The potential for adverse impacts on other receptors such as vegetation, ecological receptors and buried services must also be considered as part of a risk-based remediation solution. The reduced permeability, high pH and high contaminant concentrations in treated material could have an adverse impact on plant growth and wildlife, and reuse of stabilised materials within soft landscaping areas should not be recommended. Mitigation measures such as a capillary break layer may be needed to prevent alkaline groundwater causing problems.

With respect to buried services, considerations are more practical, such as whether the placement of the hardened treated material restricts access to services, and if it could be necessary to dig through it as part of later maintenance activities. Mitigation typically comprises a thorough validation of the extent and location of treated material, and inclusion within operation and maintenance plans to ensure these risks are communicated to future maintenance workers so that appropriate precautions can be taken. Sometimes a "clean Corridor" is constructed through the treated material and utilities are placed within this clean corridor.

2.5.4 Durability and integrity

In understanding the environment of service of a waste form, it may be necessary to incorporate additional protection measures. These may include a capping layer to, for example, restrict infiltration, or measures to mitigate the effects of freeze-thaw. The thorough investigation of likely environmental and other impacts after treatment is a key part of the development of the conceptual model and ensures an appropriate testing regime is adopted to test the S/S formulation adopted to provide confidence in future behaviour.

In ensuring the durability and integrity of the waste form, S/S can be part of a risk-based remediation framework. A good environmental risk assessment will consider the risk factors for degradation of S/S materials as the conceptual site model is developed, and in this respect a comprehensive review of relevant factors is provided by the Environment Agency (EA 2004b) and (Hills *et al.*, 2013).

2.6 Future trends

As was discussed in Section 2.1, S/S has been a workhorse technology for remediation of hazardous waste sites in the United States since at least 1982. S/S has been selected in ex-situ and in-situ applications in nearly one quarter of Superfund remedy documents that selected source treatment. Figures 2.1 and 2.2 (Section 2.1) indicated that the selection of S/S has remained nearly constant over 30 years from 1982-2011.

The specific reasons for the popularity of S/S are, in the authors' opinion, a combination of the following factors, as S/S:

- Is applicable to most inorganic and a wide selection of organic contaminants (see Section 3.1)
- Is competitive on cost for many applications
- Can be deployed using readily available equipment (and materials) with a minimum set-up time
- Can be completed faster than most other technologies
- Has an extensive tract record of success on a wide range of site and contaminant types

2.6.1 USA

Given the long history of selection and successful use, the authors expect that S/S will continue to be a popular risk-management strategy available well into the future. To this end in the USA, the 2011 publication of guidance on S/S by the Interstate Technology Regulatory Council (ITRC 2011) provides authoritative reference material to the regulatory community, and this will enable greater acceptance and use of this technology.

However, the authors believe there will be changes going forward, with S/S being employed more frequently at sites containing non-volatile organic contaminants, either alone or in combination with inorganic contaminants. This expanded application of S/S is supported by the increased ability/availability of experienced S/S experts in developing treatment formulations that successfully immobilise organic contaminants, especially non-volatile organics. The successful treatment formulations have included, for example, organophilic clays, activated carbon, and ground blast furnace slag, and will enable more sites (containing a wider variety of organic contaminants) to be successfully treated.

Historically S/S has been employed far more often in ex-situ than the in-situ applications. More recently, however, in-situ applications are increasing in frequency, as illustrated in Figure 2.5, which is based upon a recent USEPA Superfund remedy selection report (EPA, 2013). This trend may reflect the increased depth of treatment possible with higher power in-situ augers and the use of longer Kelly bars. It could also be due to greater availability of in-situ treatment equipment other than augers (described in Section 5).

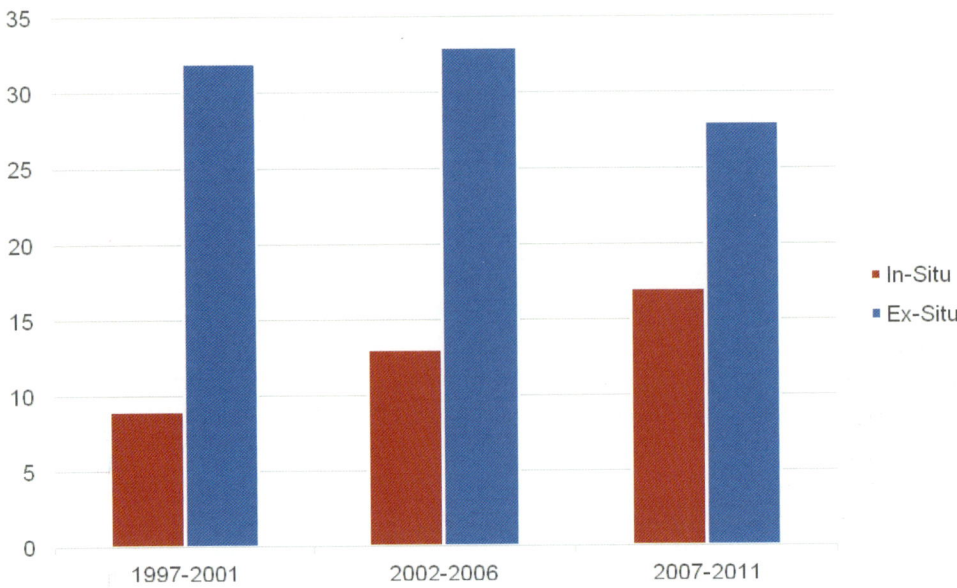

Figure 2.5: Relative increase in selection of in-situ vs ex-situ S/S (USEPA 542-R-13-016, Appendix B)

2.6.2 UK and Europe

In the UK and Europe one of the key drivers for using S/S is the availability and cost of landfill space, and the restrictions placed upon the 'digging and dumping' of waste and soil through the EU Landfill Directive (Directive 99/31/ED) and associated legislation, in different EU member countries. As options for managing soil and waste become more expensive and restricted in nature, S/S can be expected to become more widely appreciated and credible as a low-cost management option.

In the Authors' experience, the reluctance to treat organic or mixed contamination is becoming less and, by way of example, gas-works sites (MGP) are now routinely treated by S/S in England and Wales.

Development work also continues on the use of novel binding systems employing, for example, geo-polymers (Fernández Pereira *et al.* 2009; Guo and Shi, 2012)), waste materials and industrial by-products (Leonard and EA; 2004; Stegemann, 2010) and the use of CO_2 activated systems (Lange *et al.* 1996, 1997; Gunning, 2011) to produce carbonate-cemented S/S products.

2.6.3 Conclusions

For over two decades S/S has consistently been a favoured technology for use by the USEPA Superfund program. S/S has been selected consistently for over 20 % of all sites treated, and this trend continues today, not only within the Superfund but also with sites remediated under different jurisdictions.

The authors' believe that there will be an increase in use of S/S for sites with organic contaminants as the effectiveness of applying S/S to organic contaminants has improved and acceptance by the remediation industry and regulatory authorities for treating organics with S/S has increased. There is also an increasing trend for applying in-situ S/S, as available equipment continues to improve and its capability of working at greater depths, including below the water table and without having to de-water.

PART THREE

Applicability of S/S

One of the strengths of S/S is that it can be applied to metals, other inorganic compounds, and most organic compounds. This section briefly discusses the many inorganic and organic contaminants for which S/S has been successfully applied. However, for a detailed review of the chemistry of individual inorganic and organic compounds, and how these interact (with the binder) in an S/S system the reader is directed to the informative book written by Conner (1990).

Further discussion of the mechanisms involved with specific compounds can be found in the reference material identified in **Section 1.7**, and in the accompanying science volume of this work.

3.1 Metals

Early in the development of S/S, the treatment technology was primarily applied to metals, as the chemistry utilised to remove metals from solution was already established and practiced for water treatment. Since metals cannot be destroyed, S/S treatment focused on producing a less mobile and less toxic form of the metals in waste materials.

Though any metal can be chemically immobilised by S/S, the primary metals of environmental concern are antimony (Sb), arsenic (As), barium (Ba), beryllium (Be), cadmium (Cd), cobalt (Co), copper (Cu), chromium (Cr), lead (Pb), mercury (Hg), nickel (Ni), selenium (Se), silver (Ag), Thallium (Tl) and zinc (Zn). These metals are those listed in the EPA's Universal Treatment Standards (40CFR 268.48). Figure 3.1 and 3.2 shows typical metals-contaminated waste requiring S/S treatment.

3.1.1 Chemistry

The chemistry involved in S/S is complex, and still remains poorly understood for some contaminants. For both metals and (some) organic contaminants, sorption processes, precipitation and incorporation into solid cementitious phases may occur during S/S. In order to present key aspects of the nature of metal contaminants, the properties of important metals are discussed below.

Antimony, Sb, is a Group V element, has valence states of +3, +5, and -3, which complicates its chemistry. Antimony has definite cationic chemistry, but only in its trivalent state. It forms both oxides and sulfides readily in its trivalent state. Reduction of antimony to the trivalent state often simplifies its chemical fixation.

Arsenic, As, is a Group V element with principal valence states of +3, +5, and -3. The valence state can be changed readily and reversibly. It combines with other metals to form arsenides, and forms both oxides and sulfides. Arsenic chemistry is complicated by its variety of valence states and ability to be present in both cationic and anionic species in solution. Often the redox potential must be adjusted to cycle the arsenic into only one valence state prior to chemical immobilisation.

Figure 3.1: Lead-contaminated soil and battery casings from a former battery-recycling site, subsequently chemically stabilised

Beryllium, Be, is a Group II element with a predominantly covalent chemistry, though it forms a cation with a valence state of +2. Beryllium forms a low solubility hydroxide at near normal pH values. Beryllium hydroxide is amphoteric, creating a "beryllate" ($Be(H_2O)_4^{2-}$) anion in alkaline solution.

Cadmium, Cd, is a Group IIB element, with one valence state of +2. Cadmium forms low solubility compounds with carbonate and hydroxide, and can be chemically fixated by alkaline precipitation. Cadmium also forms a low solubility phosphate compound. Because cadmium forms stable complexes with ammonium, cyanide, and halides, it will not precipitate in the presence of these complexing agents.

Cobalt, Co, is a Group VIII transition metal with primary valence states of +2 and +3. Cobalt forms low solubility phosphate, sulfide, and arsenosulfer compounds. Cobalt forms stable hydroxides, which are amphoteric at high pH.

Copper, Cu, is a Group IB metal, with a primary valence state of +2. Copper forms low solubility carbonate, hydroxide, phosphate, sulfur, and arsenosulfur compounds.

Chromium, Cr, is a Group IVB metal. It has three valence states (+2, +3, +6), but +3 and +6 are the most prevalent. While Cr^{+3} is poisonous, Cr^{+6} is highly toxic and carcinogenic. In the +3 valence state, chromium is cationic and forms low solubility carbonate, hydroxide, phosphate, and sulfide compounds. In the +6 valence state, chromium is predominantly present as an anion (CrO_4^{-2} or $Cr_2O_7^{-2}$), and can form insoluble compounds with barium or lead. Typically, Cr6+ compounds are reduced to their trivalent state and then immobilised, reducing both the toxicity and solubility.

Figure 3.2: Chromium paint pigment-contaminated soil being chemically stabilised in-situ

Lead, Pb, is a Group IVA element, with a primary valence state of +2. Lead forms low solubility carbonate, chromate, halide, phosphate, sulfate, and sulfur compounds. Lead forms stable hydroxides, which are amphoteric at high pH.

Mercury, Hg, is a Group IIB metal, existing mainly in three valence states: 0, +1, and +2. While Hg^0, elemental mercury, is insoluble, it often has a layer of HgO on its surface giving it an appearance of solubility. Therefore, the surface of Hg^0 is often reacted with either sulfides to form insoluble mercuric sulfides or metals (e.g. copper) to form amalgams. Cationic mercury species form low solubility halide, hydroxide, phosphate, and sulfides.

Nickel, Ni, is a Group VIII transition metal, with a primary valence state of +2, though valence states of -1, 0, +1, +3, and +4 are also known. Cationic nickel species form low solubility carbonate, hydroxide, phosphate, and sulfide compounds.

Selenium, Se, is a Group VIB element, with principal valence states of -2, 0, +4, and +6. The valence state can be changed readily and reversibly. It combines with other metals to form selenides, and forms both oxides and sulfides. Like arsenic, selenium chemistry is complicated by its variety of valence states and ability to be present in both cationic and anionic species in solution. Adjustment of the redox potential can be used to adjust selenium into only one valence state prior to chemical immobilisation.

Silver, Ag, is a Group IB metal with a primary valence state of +1. Silver forms low solubility halide, hydroxide, and sulfur compounds.

Thallium, Tl, is a Group III metal, with valence states of +1 and +3, though it is typically found in the +1 valence state. As a cation, thallium forms low solubility compounds with halides (except F), hydroxide, and sulfide.

Zinc, Zn, is a Group IIB metal, with a primary valence state of +2. Zinc forms low solubility compounds with carbonate, hydroxide, phosphate, and sulfide.

3.2 Other inorganic compounds

As with metals, many inorganic compounds cannot be destroyed, or are recalcitrant in nature. Treatment of metals and other inorganic compounds by S/S is focussed on producing a less mobile, less toxic form of the compounds involved, or through physical entrapment in the cementing phase or pore structure of the waste form.

Anions that are often treated by S/S include: fluoride (F^-), cyanide (CN^-), nitrate (NO_3^-), phosphate (PO_4^{-3}), and sulfide (S^{-2}); however, chloride can be readily incorporated in a hardening cement-based system, but that exceeding the binding capacity of the cement involved will be readily lost by dissolution/diffusion on exposure to a hydraulic gradient.

3.2.1 Chemistry

Fluoride, F^-, is easily immobilised as calcium fluoride.

Cyanide (soluble) CN^-. Soluble cyanides, including loosely-complexed cyanides such as nickel or cadmium cyanides, readily form insoluble compounds with iron.

Nitrate, $NO3^-$. All nitrate compounds are readily soluble. Therefore, S/S treatment of nitrate in waste materials is often futile.

Phosphate, $PO4^{-3}$. Soluble phosphates form insoluble compounds with many cations, including calcium, iron, and lead.

Sulfide, S^{-2}. Soluble sulfides form insoluble compounds with most metal cations, particularly iron. This property makes sulfides a valuable reagent for the S/S treatment of metals, particularly as a sulphide-based system is a reducing system and can be used to manage hexavalent Cr, by reducing it to the trivalent species.

3.3 Organics

Traditionally, wastes contaminated with non-volatile and semi-volatile organics such as dioxins, explosives, lube oil range (>C28) and asphaltic petroleum hydrocarbons, pesticides, polyaromatic hydrocarbons (PAHs) polychlorinated biphenyls (PCBs), were treated via on-site thermal treatment technologies. Problems with the public and regulatory acceptance of thermal treatment technologies have limited the application of these technologies on such wastes.

S/S has become an acceptable low cost treatment alternative to reduce the mobility, and thereby manage the risk by minimising exposure pathways, of these recalcitrant organics. S/S is typically not applied to volatile organics.

Figure 3.3 shows wastewater sludge being excavated at a former automobile plant. The sludge contained an organo-nickel complex from plating operations at the plant. The sludge was chemically stabilised for disposal in an on-site RCRA landfill.

Figure 3.4 shows a S/S treatment system set up at an acid sludge tar pond.

3.3.1 Chemistry and physics of organic stabilisation

The S/S of organic-contaminated waste may involve the alteration/transformation of the organic compounds themselves, or their participation in physical processes, such as adsorption and encapsulation. The S/S of organic compounds has recently been reviewed (Hills et al., 2010).

The outcome of S/S is to retard the movement of the hazardous constituents within prescribed safe limits, defined by leachate quality. However, the mechanisms of organic stabilisation involved are very difficult to distinguish apart, and this tends to reflect the fact that S/S of organic compounds rarely involves the formation of solid organic salts.

The immobilisation assumed to be due to adsorption might involve some sort of chemical bonding or change in the speciation of a compound, or the formation of degradation products. The pervasive chemical environment within S/S treated hazardous waste is highly alkaline with a high number of soluble components, and may promote both short and longer-term changes in the nature of certain organic contaminant species.

In practice, however, inorganic stabilisation systems operating at ambient temperatures and pressures in non-exotic aqueous environments can induce stabilisation through hydrolysis, oxidation, reduction and compound formation.

Hydrolysis refers to the reaction of a compound with water. This usually results in the exchange of a hydroxyl group (-OH) for another functional group at the reaction centre.

Hydrolysis may be catalysed by acids or bases (e.g. H^+, OH^-, or H_3O^+) and many involve intermediate species (Dragun, 1988). Metal ions such as copper and calcium may act as catalysts for certain organic structures, and adsorption on surfaces such as clay and activated carbon may also facilitate reactions involving organic molecules.

Many organic compounds are resistant to hydrolysis, including some esters, many amides, all nitriles, some carbamates, and alkyl halides. Those less resistant to hydrolysis include alkyl and benzyl halides, poly-methanes, substituted epoxides, aliphatic acid esters, chlorinated acetamides, and some phosphoric acid compounds.

Figure 3.3: A contaminated wastewater sludge being excavated

Figure 3.4: Treatment facility to chemically S/S acid sludge tar

Oxidation and hydrolysis are the most common pathways for the reaction of organics in stabilisation systems. Oxidation of organics occurs via two pathways (Dragun and Heller, 1985). In one, an electrophilic agent attacks an organic molecule and removes an electron pair; in the other, only one electron is removed, forming a free radical. The former is heterolytic, the latter homophilic.

Free radial formation reactions have lower energy barriers than the oxidation of a polar compound or cleavage of a covalent bond. It is worthy of note that organic oxidation reactions in the chemical industries are typically catalysed by crystalline aluminosilicates at elevated temperatures and pressures. It has been recognised that this also occurs at ambient temperatures with clay and soils, not only in oxidation, but in reduction, hydrolysis, and neutralisation reactions (Dragun, 1988). Iron, aluminium and trace metals within layered silicate minerals have been identified as specific catalysts (Garrido-Ramirez et al., 2010), though not all clays exhibit this property.

Based on work by Dragun and Heller (1985), two generalities can be made concerning the oxidation of organics by soils and clay minerals:

- Many substituted aromatics undergo free-radical oxidation, e.g. benzene, benzidine, ethyl benzene, naphthalene, phenol
- Chlorinated aromatics and polynuclear organics are unlikely to be oxidised

Water content may be one of the more important constituents of soils, and possibly also in wastes. Partially saturated systems are more likely to undergo oxidation reactions than saturated ones. The above comments apply to "natural" oxidation by reagents normally used in stabilisation systems, or characteristics of the waste itself. The deliberate addition of strong oxidisers such as potassium permanganate or hydrogen peroxide is a different matter. These reagents have been used in stabilisation for the oxidation of phenols and other organics, as well as inorganics and treatment of metal complexes.

An important consideration in all organic reaction schemes involving hazardous constituents is the product of the reaction, which may be equally or more hazardous than the original reactant. This is especially true when oxidation processes are employed, as their use in stabilisation systems, as in addition they may also create hazardous species from other organics in the waste which were previously non-hazardous or less hazardous.

The use of strong oxidants on wastes, which contain chromium, might result in the formation of Cr^{6+}, necessitating a subsequent reduction step to Cr^{3+}. Reduction of organic compounds may be defined as either:

- An increase in hydrogen content or a decrease in oxygen content, or
- A net gain in electrons

Reduction can occur in clay systems with the clay acting as a reducing agent. Reductive alteration of organic compounds in waste has been poorly studied and is not well understood.

Compound formation between organic compounds and metals or other cationic species can form less soluble and/or less toxic compounds. For example, the solubility of oxalic acid is 95,000 mg/L, compared with 6 mg/L for the calcium salt. Organic acids of environmental concern might be effectively immobilised in the calcium-rich environment of commercial stabilisation systems. In addition to the formation of salts, a number of other direct reactions are possible between organic contaminants and organic or inorganic reagents under ambient conditions.

Boyd and Mortland (1987) developed the concept of a two-step reaction scheme where the organic molecule is first adsorbed on a clay surface, then reacted via surface catalysis. They claimed to have achieved polymerisation and de-chlorination reactions in this manner, including detoxification of dioxin analogues. Ortego (1989) discussed the formation of stable organo-metallic complexes, such as those formed between Cu2+ and benzene, in the interlayer of laminar-structured clays. Zinc, Cd, and Hg co-ordination compounds with polyamines have been also reported, as has the decomposition of acid-sensitive organics by acid adsorbed water at the clay surfaces (Kostecki, 1992).

As a subset of compound formation, several investigators have postulated the formation of linkages between organic compounds and insoluble substrates such as stabilisation binders like cement (e.g. Montgomery et al., 1983). It seems likely that the instances of strong "sorption" at these surfaces may really be the result of exchange reactions, similar to those known to occur during the production of organo-clays from clay minerals by treatment with organic cations of the form $[(CH_3)NR]^+$.

3.3.2 Physical aspects of organic stabilisation

Work on "physical" immobilisation of organics has been primarily focused around several materials and mechanisms. Cote (1987) has shown that a variety of organics can be sorbed fairly effectively by cement-based stabilisation processes, incorporating activated carbon and bentonite additives, whereas fly ash and soluble silicates are less effective. Other conclusions from this work were:

- Volatile organics were not well immobilised
- Water soluble contaminants were not well immobilised
- Organics with low water solubility were well immobilised

Kyle *et al.* (1987) compared a number of lime, kiln-dust and fly ash-based mixtures with organic reagent (vinyl ester, acrylic, epoxy, polymer cement) on several industrial wastes spiked with various priority pollutant organics. They found that the organic reagents produced poorer results, as measured by total organic carbon (TOC) in the leachate, than did the inorganic reagents. The addition of activated carbon to lime/fly ash systems lowered the TOC. Work by Lear and Conner (1992) found a similar reduction in the leachability of organics when activated carbon and other adsorbents were added to Portland cement systems. The efficacy of oxidisers was also noted in that study.

Christenson and Wakamiya (1987) found that Kepone leaching was increased in highly alkaline systems such as cement/soluble silicate, but decreased by encapsulation in either an organic polymer or a proprietary molten-sulfur blended binder. Co-precipitation in ferric hydroxide precipitation systems was found to remove chlorendic acid, humic acid, PCBHs and other compounds from landfill leachate (Pojasek, 1980).

3.4 Geotechnical considerations

The suitability of soil for S/S is somewhat different than for typical geotechnical projects. There is little choice but to utilise the soil type that is contaminated at a given site. Given this inevitability, the type of soil presented for S/S and its spatial variability will have an influence on the choice of equipment used and the selection of the binder system/additives used for treatment.

It is therefore essential that site conditions are well understood to enable treatment to be applied in a thorough, timely and cost-effective manner. A proper geotechnical/hydrogeological site investigation will limit risks to project owners and consultants. Following is an introduction to soil classification, site characterisation, and site variability as it applies to suitable applications of S/S technologies.

3.4.1 Soil classification

The classification of soil can be a relatively easy procedure for experienced geotechnical lab and field personnel. However, a site investigation for an S/S project that uses incorrect or inconsistent soil classifications creates confusion for third party consultants, owner consultants, regulators and contractors and could potentially lead to litigation. There are many different forms of proper soil classification depending on the country where the work is being performed.

None of these classification systems is necessarily more correct than the others. However, what is important from a soil classification standpoint is that the site investigation report is systematic and clear and uses a recognised classification system. Depending on the location of the work, this decision will be governed by the state of practice in the geographical region. Figures 3.5, 3.6, and 3.7 show the classification system provided in BS 5930 (code of practice for site investigations). A Unified Soil Classification System (USCS) can also be found in ASTM D 2487.

Most soil classification systems provide an indication of particle size distribution, colour, organic content and the physical behaviour of the soil in the classification. These soil classifications found in borehole and test pit logs are often performed on a visual basis and laboratory basis using tests such as particle size and Atterberg Limits.

Although following the proper soil classification protocol will assist in ensuring clarity in S/S planning and design, site investigations incorporating soil classifications should also include indications of compactness/consistency, colour (e.g. Munsell system; Goddard, 1979), and the presence of materials such as debris, organic-based plant matter (i.e. peat), free-phase organic contaminants, etc. Also, man-made fill material should be distinguished from naturally occurring soil deposits.

Where debris is present, efforts should be made to provide as much description as possible as related to the nature of the debris (brick, concrete, steel rebar etc.) as well as the size of this debris, if known.

Particle size information will be required to address requirements of pre-processing of the soil prior to S/S activities. Depending on the size of the debris, the amount of cobbles or boulders present, or the density/consistency of the material on a potential S/S site, the type of construction method used to provide S/S treatment may vary. For example, it may be necessary to screen out oversize material prior to processing through a pug-mill, excavate debris prior to treating soil with an in-situ auger, etc.

In some cases conventional ex-situ and in-situ S/S approaches may not be practical, given the presence of a large amount of boulders. In this case a less conventional approach, such as in-situ jet grouting, may be selected. As a worst-case scenario, portions of the site may be non-treatable with soil stabilisation if excessive debris or large quantities of cobbles and boulders are present.

APPLICABILITY OF S/S | 45

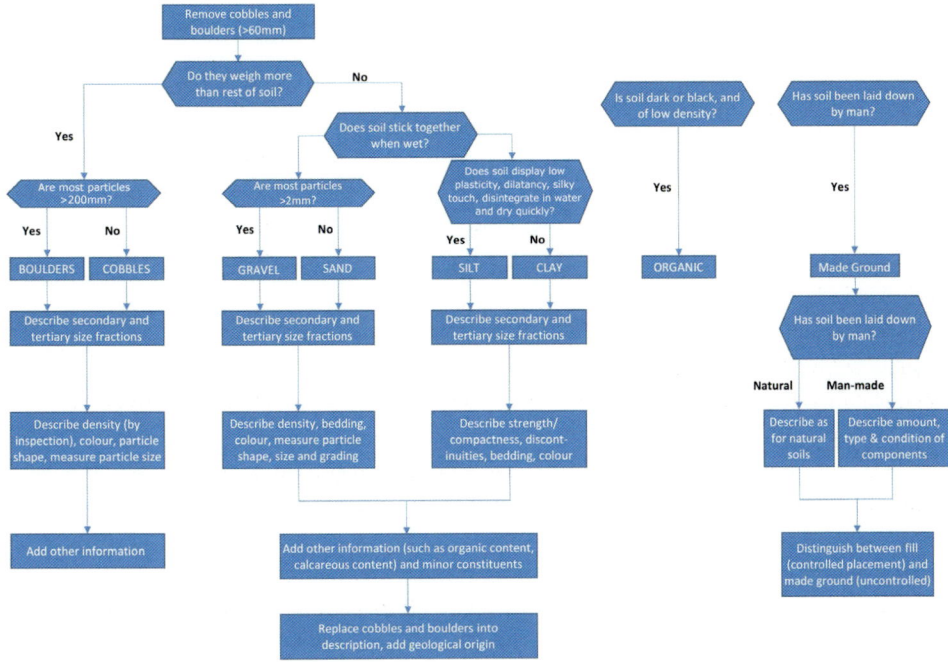

Figure 3.5: General identification and description of soils (modified from British Standard, BS 5930)

PRINCIPAL SOIL TYPE	VISUAL INDENTIFICATION	MINOR CONSTITUENTS	STRATUM NAME	EXAMPLE DESCRIPTIONS
BOULDERS	Only seen complete in pits or exposures			
COBBLES	Often difficult to recover whole from boreholes			
GRAVEL	Easily visible to naked eye; particle shape can be described; grading can be described	Shell fragments, pockets of peat, gypsum crystals, flint gravel, fragments of brick, rootlets, plastic bags, etc		Loose brown very sandy sub-angular fine to coarse flint GRAVEL with small pockets (up to 30 mm) of clay (TERRACE GRAVELS)
SAND	Visible to naked eye; no cohesion when dry; grading can be described			
SILT	Only course silt visible with hand lens; exhibits little plasticity and market dilatancy; slightly granular or silty to the touch; disintegrates with water; lumps dry quickly; possesses cohesion but can be powered easily between fingers	Using terms such as: with rare with occasional with abundant/ frequent/numerous	RECENT DEPOSITS, ALLUVIUM, WEATHERED, BRACKLESHAM CLAY, LIAS CLAY, EMBANKMENT FILL, TOPSOIL, MADE GROUND, OR GLACIAL DEPOSITS?	Medium dense light brown gravelly clayey fine SAND. Gravel is fine (GLACIAL DEPOSITS)
	Intermediate in behavior between clay and silt. Slightly dilatent	% defined on a site or material specific basis or subjective		Stiff very closely sheared orange mottled brown slightly gravelly CLAY. Gravel is fine and medium of rounded quartzite. (REWORKED WEATHERED LONDON CLAY)
CLAY	Dry lumps can be broken but not powdered between the fingers; they also disintegrate under water but more slowly than silt; smooth to the touch; exhibits plasticity but no dilatancy; sticks to the fingers and dries slowly; shrinks appreciably on drying usually showing cracks			Firm thinly laminated grey CLAY with closely spaced thick laminae of sand. (ALLUVIUM) Plastic brown clayey amorphous PEAT. (RECENT DEPOSITS)

Figure 3.6: Identification and description of soils (modified from British Standard, BS 5930)

Soil Group	Density/Compactness/Strength		Discontinuities		Bedding		Colour	Composite Soil Types (mixtures of basic soil types)		Particle Shape	Particle Size	PRINCIPAL SOIL TYPE
	Term	Field Test	Scale of spacing of discontinuities		Scale of bedding thickness			Term	Approx % secondary			
Very Coarse Soils	Loose	By inspection of voids and particle packing	Term	Mean spacing mm	Term	Mean thickness mm				Angular	– 200	BOULDERS
	Dense									Sub-angular	– 60	COBBLES
Coarse Soils (over about 65% sand and gravel sizes)		Borehole with SPT N-Value	Very widely	Over 2000	Very thickly bedded	Over 2000	Red	Slightly (sandy)	<5	Sub-rounded	Coarse – 20	
	Very loose	0-4	Widely	2000 to 600	Thickly bedded	2000 to 600	Orange			Rounded	Medium	GRAVEL
	Loose	4-10	Medium	600 to 200	Medium bedded	600 to 200	Yellow	(sandy)	5 to 20	Flat	– 6	
	Medium dense	10-30	Closely	200 to 60	Thinly bedded	200 to 60	Brown			Tabular	Fine	
	Dense	30-50	Very closely	60 to 20	Very thinly bedded	60 to 20	Green	Very (sandy)	>20	Elongated	– 2	
	Very dense	>50	Extremely closely	Under 20	Thickly laminated	20 to 6	Blue			Minor constituent type	Coarse – 0.6	
					Thinly laminated	Under 6	White				Medium	SAND
	Slightly cemented	Visual examination: pick removes soil in lumps that can be abraded	Fissured	Breaks into blocks along unpolished discontinuities			Cream	SAND AND GRAVEL	about 50	Calcareous, shelly, glauconitic, micaceous etc. using terms such as	– 0.2	
							Grey				Fine	
							Black				– 0.06	
Fine Soil (over about 35% silt and clay sizes)	Un-compact	Easily moulded or crushed in the fingers	Sheared	Breaks into blocks along polished discontinuities	Inter-bedded	Alternating layers of different types Prequalified by thickness term if in equal proportions. Otherwise thickness of and spacing between subordinate layers defined		Term	Approx % secondary		Coarse – 0.02	SILT
	Compact	Can be moulded or crushed by strong pressure in the fingers						Slightly (sandy)	<35	slightly calcareous calcareous	Medium – 0.006	
	Very Soft 0-20	Finger easily pushed in up to 25mm			Inter-laminated					% defined on a site or material specific basis or subjective	Fine – 0.002	CLAY/SILT
	Soft 20-40	Finger pushed in up to 10mm										
	Firm 40-75	Thumb makes impression easily	Spacing terms also used for distance between partings, isolated beds or laminae, dessication cracks, rootlets etc				Light	(sandy)	35 to 65			CLAY
	Stiff 75-150	Can be indented slightly by thumb					Dark					
	Very stiff 150-300	Can be indented by thumbnail					Mottled	Very (sandy)	>65			
	Hard (or very weak mudstone) Cu > 300kPa	Can be scratched by thumbnail										
Organic Soils	Firm	Fibres already compressed together	Fibrous		Plant remains recognisable and retains some strength		Transported mixtures	Colour		Contains finely divided or discrete particles of organic matter, often with distinctive smell, may oxidise rapidly. Describe as for inorganic soils using terminology above		
							Slightly organic clay or silt	Grey as mineral				
							Slightly organic sand	Dark grey				
							Organic clay or silt	Dark grey				
	Spongy	Very compressible and open structure	Pseudo-fibrous		Plant remains recognisable, strength lost		Organic sand	Black				
							Very organic clay or silt	Black				
							Very organic sand					
	Plastic	Can be moulded in hand and smears fingers	Amorphous		Recognisable plant remains absent		Accumulated in situ			Predominantly plant remains, usually dark brown or black in colour, distinctive smell, low bulk density. Can contain disseminated or discrete mineral soils		
							Peat					

Figure 3.7: Identification and description of soils (modified from British Standard, BS 5930)

3.4.2 Site Characterization of Soil and Groundwater

If possible, previous knowledge of the history of the site development (e.g. past land use, original topography prior to site development) should be used during the desk study phase to inform and develop a conceptual "model" of the soil, rock and groundwater conditions (see Section 7.2). Often this understanding of site conditions needs to extend to neighbouring properties, especially if the site boundaries are close.

The nature of soil, rock and groundwater conditions must be delineated vertically and laterally with test pits and/or boreholes, in the contaminated area and outside of its borders. For large projects, test pits and boreholes are usually both performed to take advantage of both types of investigations and reveal the most information about the site. The type of investigation (i.e. borehole versus test pit) will be governed by many different factors but below is a list of some of the considerations in selecting the type of investigation for potential S/S sites.

	Test Pits	Boreholes
Advantages	• The presence of debris can be seen • Bulk samples for testing of binder formulation are available • Excavation-related issues can be quickly identified • Construction is relatively quick and inexpensive to perform over a wide range of geographical/geological environments)	• Enables sampling at greater depth compared to test pits • Allows in-situ testing (SPT, CPT) to obtain quantitative data • Allows installation of monitoring wells and results from pump tests, k-tests • Not as intrusive or disruptive as test pits • Appropriate for a variety of groundwater conditions
Disadvantages	• Limited by excavator reach (15-30 ft; 5-30 m), may be shallower for dense layers at depth • Can cause significant site disturbance, especially for deep test pits • Large excavations may require remedial actions prior to development • Difficult to obtain useful information for non-cohesive soils below the water table • In-situ testing (e.g. Standard Penetration Testing (SPT), Cone Penetration Testing (CPT)) is not performed due to sample disturbance (i.e. limitations are placed on the amount of quantitative information available) • Does not provide information for seismic design • Monitoring wells cannot be properly installed where a test pit is constructed	• Sophisticated interpretation of ground conditions needed to inform site conceptual model • More expensive, especially for remote sites • Small sample sizes may not allow binder-formulation trials • Identification of extent of buried debris and oversize material not possible

Table 3.1: Considerations for choosing test pits versus boreholes

As for ground engineering projects, prior to remediation, a site investigation (SI) is required to examine the lateral and vertical properties of soil, rock and groundwater across a site. The SI is also used to obtain samples to enable the nature and extent of contamination to be characterised, however there is normally a trade-off between the amount of material collected during the SI and the cost. If ground conditions prove difficult, or the extent and nature of contamination is unforeseen, then more detailed and focussed investigative work is carried out at a later stage.

Thus, these studies are performed early in the project-development process and are used to inform the decision-making process leading to the choice of remedial design. The SI supplies fundamental information on groundwater flow regimes, the nature of contamination (distribution, type, level) and the host soil/rock characteristics. Information obtained from the SI is also used in the design of any dewatering scheme, the choice of complementary treatment technologies, including as part of a treatment train, as well as preliminary selection of S/S equipment that will be most suitable for use at a particular remedial site.

However, generally if S/S is selected as the method to manage contamination at a site, additional investigations may have to be performed to increase the amount (and reliability) of data available on:

- The variability of contaminants present
- Soil and rock conditions
- The geochemical properties of the soil and groundwater, as pertains to S/S treatment (i.e. pH, sulphate, etc.)

Additional factors such as the presence of free-phase organics may need to be established to avoid problems with respect to their interaction with available binder/additives. The appointed remediation contractor may often undertake these additional investigations.

For smaller projects, the costs associated with more detailed phases of investigation may be prohibitive. However, with larger projects, the acquisition of key data that can facilitate a higher degree of certainty in the remedial design reduces the residual risks to both owners and contractors. With larger projects, a phased programme of ground investigation is often undertaken to better understand soil, rock and groundwater conditions, allowing the sequential development of a site conceptual model, enabling refining both the model and remedial options to be employed at the contaminated site.

3.4.3 Approaches to site variability

Ground conditions at any site will be variable; the spatial variation of contaminants is no exception. The approach taken will assist in reducing the risk-potential to acceptable levels. As previously mentioned (Sections 2.4 & 2.5) it is impossible to eliminate risk entirely.

In increasing the intensity of data acquisition from the site investigation (i.e. increasing the number of test pits, boreholes etc.) it will be possible to reduce the risks associated with ground variability to an acceptable level.

However, given that ground investigative techniques are expensive, as the intensity of the site investigation increases so too will the total cost of the investigation. It is a judgment-based decision (by the regulator, consultant or site owner) as to what constitutes an adequate amount of intensity, and the level of risk that is acceptable.

For large projects however, the increased cost of the SI can save money in the construction phase of the project. For smaller projects, the added cost of an enhanced SI may outweigh the cost-benefits associated with better defining site variability. This paradox means that contractors enter a bidding process with more uncertainty that can ultimately lead to cautious (i.e. expensive) bids for the project.

For S/S projects, the level of variability that is present on a site can influence the type of equipment selected for the remedial project and often results in a binder formulation that is conservative. By way of example, if ground and other site conditions show that 25 % of a site can be treated with 8 % cement addition, and the remainder with 10 % cement addition, it is likely that a contractor will apply 10 % to the whole site.

Industrial sites that present predominately man made fill or 'made-ground' have potentially large variability in soil properties over small distances and hence conservative binder formulations are chosen. This conservatism may inevitably add additional costs to the project, but allows a consistent approach to the 'whole' site, limiting potential mistakes and reducing the chance of failing performance specifications, which can ultimately delay the project, diminish contractor confidence, and escalate the final cost of the remedial operation.

3.5 Ex-situ and in-situ application of S/S

There are two major methods of applying S/S treatment to contaminated soil: ex-situ and in-situ.

The ex-situ method of applying S/S involves the excavation of contaminated media (soil, sludge, sediment, etc.) and its transfer to another location, either on- or off-site for subsequent mixing with reagents/binders. In-situ treatment by S/S involves mixing binders/reagents into the contaminated media in the place where it is found and thus does not involve excavation and transport to another location.

Although the definitions given above are quite distinct and are used in this document, the reader should be aware that there are a number of other definitions of a regulatory nature, and these may be different.

3.5.1 Technical considerations

The major technical advantage to ex-situ S/S treatment is that the properties of the media (e.g. moisture content, texture, contamination level, degree of aggregation/agglomeration, amount and type of debris) are directly observable in the media to be treated, and this facilitates the production of a quality 'controlled' product.

If the properties of the contaminated media deviate from those used in the design criteria the process can be quickly adjusted to compensate, and to maintain treatment goals. This is particularly important where, for example, drum carcasses are encountered where not expected, or there are localised and discontinuous (but significant) 'hot-spots' of contamination, not foreseen in the original SI.

Ex-situ S/S has the advantage that both liquid and solid reagents can be used to treat the contaminated media, using commonly available mixing equipment. Furthermore, water, which is often critical for thorough mixing, can be easily added where necessary to maintain mix properties.

An additional advantage is that samples of the treated material are also easily obtained during processing, such as at the mixer discharge point (see Figure 3.8), enabling rapid evaluation of S/S product as it is being formed.

Furthermore, the application of ex-situ S/S involves more commonly available equipment, and there are more 'vendors' available that can apply this method of S/S. In addition, it is possible to identify more quickly any inadequate treatment and to re-route this material back into the 'system' to ensure the maintenance of treatment goals, or disposal by another method.

Ex-situ treatment also has the advantage of being suitable for shallow soils, located above the water table, and this facilitates selective materials removal, and a lower-cost treatment option.

The major technical disadvantages for ex-situ S/S include primarily space requirements and ease of materials handling. However, there must be space available to stockpile materials to be treated and to apply S/S and to hold the treated product until verification of performance criteria is received. It should be noted that ex-situ S/S involves significant logistical issues, related to the transport of the contaminated media to the site of S/S treatment and then onto the location of final placement (which may be off or on-site). Figures 3.9 and 3.10 are examples of typical ex-situ S/S treatment-systems.

If the final deposition of the treated material is on-site, the placement and compaction of the treated material becomes another requirement for ex-situ S/S treatment. Another technical disadvantage of the ex-situ S/S is the increased design required for deep (greater than 15 feet or 4.5 metres) contamination. Excavations this deep often involve sheeting and shoring, increasing the complexity of the excavation. Also, excavation below the water table will require management of the groundwater that may infiltrate the excavation.

The major technical advantage of in-situ S/S treatment is that the contaminated media is

Figure 3.8: Ex-situ treatment in a roll-off box

treated in place (Figure 3.11). There is no need for a separate on or off-site treatment area and no need to consider the logistics of transport prior to treatment, and transport only becomes a concern if the final treated material is disposed of off-site.

Other major advantages, depending on the site, are that treatment below the water table is feasible without dewatering and emissions from in-situ treatment are far less than from ex-situ excavation and treatment.

The major technical disadvantage for in-situ S/S is that properties of the media must be inferred from or interpreted from how the in-situ S/S equipment reacts during treatment. Skilled operators are often required to effectively operate the equipment. In-situ S/S treatment often requires the treated material to develop enough compressive strength to support the equipment as treatment progresses. Figure 3.12 shows in-situ S/S equipment operating on previously treated material.

On some sites, it may be possible to sequence the S/S work so that the in-situ S/S equipment never has to operate on top of treated material. However, many sites will require the in-situ S/S equipment to operate on top of treated material.

In-situ S/S ancillary and equipment involves track-mounted cranes or carriers (see Figure 3.13 and Section 5), which can require treated material to attain compressive strengths up to 50 psi or 0.35MPa before the equipment can safely operate on it.

Many of the in-situ S/S rotary mixers and augers require that S/S reagents be liquid or slurry,

Figure 3.9: Typical ex-situ S/S equipment, including pug-mill, silos, reagent tank, water tank, stacker and haulage

Figure 3.10: Ex-situ S/S treatment enclosure

Figure 3.11: In-situ S/S treatment of oily sludge

Figure 3.12: In-situ mixing equipment operating on top of S/S treated material

Figure 3.13: Typical in-situ S/S equipment, batch plant with reagent silos and mix tanks: crane, Kelly bar, and auger

Figure 3.14: A 3 ft limestone boulder excavated to facilitate in-situ treatment

as these equipment types use the reagent slurry as drilling fluid to ease the movement of the equipment into the subsurface and to mix the reagents into the media.

Figure 3.13 shows a batch plant for slurried S/S reagents. Reagents that are difficult to slurry, such as zero-valent iron, can pose a problem for in-situ S/S. Specialised sampling devices are required to collect samples of the in-situ S/S treated material and the sampling is often complicated by the viscosity of the treated material.

3.5.2 Physical material handling

One of the advantages of in-situ S/S treatment is the limited material handling required to prepare the media for treatment. The presence of debris (foundations/footers, drum carcasses, piping, etc.) in the subsurface can be detrimental to the in-situ S/S equipment (see Section 5), causing damage to the augers and/or penetration refusal at depths less than the desired S/S treatment depth. Figure 3.14 shows a 3 ft (0.92 m) limestone boulder excavated in order to facilitate in-situ treatment. Dense, compacted lithology, such as layers of boulders, compacted gravel, or glacial till, may also hinder penetration rate and/or cause penetration refusal at depths less than the desired treatment depth.

If the site information includes boring logs with n-values from standard penetration testing, n-values of greater than 50 for a 6" depth (15 cm) may be indicative of a dense compacted lithology. If these impediments are known or considered likely to occur at a site, material handling in the form of pre-excavation of the media is often employed to remove or loosen the impediments. If the lithology and depth requires this pre-excavation under a bentonite-slurry, the added bentonite often improves the chemical and physical properties of the final treated material.

Ex-situ S/S mixers (see Section 4) require debris such as rock and brick to be removed or separated from the media and the media size-reduced if necessary before it can be introduced into the mixer. This handling is necessary to prevent pinching of the media between the mixing paddle, auger or screw and the side of the mixer vessel. This not only can damage the paddle or screw, but also impacts the gear drive and motor which turn the paddle or screw. This is an easily avoidable issue for ex-situ S/S treatment.

The major types of material handling equipment commonly utilised for the preparation and/or separation of media for ex-situ mixers are: screening equipment, crushing equipment, shredding equipment, and magnetic separation equipment (see Section 4).

Mixing energy

The mixing energy imparted to the media during S/S treatment varies depending on S/S equipment, whether in-situ or ex-situ. Typically, the ex-situ S/S mixers, such as a pug-mill, exert the highest mixing energy on the materials to be treated. Hydraulically driven in-situ S/S rotary mixers, many of which are mounted on excavators, are close to ex-situ S/S mixers in mixing energy, but impart less mixing energy to the media as it is treated. The in-situ S/S auger mixers, driven by large hydraulic turntables or gearboxes, are moderate in their mixing energy. Excavators, whether used for in-situ or ex-situ mixing, impart the lowest mixing energy to the media during treatment.

Mixing energy is an important factor in the production of a consistent, reproducible, homogenous treated material. But it is not mixing energy alone, but the combination of mixing energy, mixing speed, and mixing time that is required to produce a homogeneous treated material. This combination can be adjusted or tailored for the specific media being treated, regardless of the specific mixing equipment selected for S/S treatment, to produce a consistent, reproducible, homogenous treated material. Therefore no type of S/S mixing equipment, whether in-situ or ex-situ, can be deemed superior to the others on the basis of mixing energy.

Site considerations

Treatment depths of less than 15 feet or 4.5 metres are often an advantage for ex-situ S/S over in-situ, as these shallower treatment depths are typically amenable to excavation without the need for sheeting or shoring. Deeper treatment depths are advantageous for in-situ S/S treatment.

The presence of the water table within the treatment depth is advantageous to in-situ S/S. Not only does one avoid the need to remove and manage the water, but the water reduces the amount of reagent grout needed to act as a drill fluid while it increases the penetration rate. The excess moisture in the media complicates the excavation of the media, and its material handling, for ex-situ S/S.

3.5.3 Economics

Mobilisation/demobilisation

The in-situ rotary mixers and excavators are also less expensive to mobilise and set up. The mobilisation typically involves just excavator, mixing heads, and a reagent batch plant. Only the batch mixing plant requires set-up, which can typically be done in less than a day.

Though more complicated, the ex-situ mixing and material handling equipment is typically only slightly more expensive to mobilise and set-up. Most of the ex-situ S/S equipment is transportable and can be off-loaded and set up for operation in less than a day.

The in-situ S/S auger is the most expensive to mobilise and set-up. The cranes, though transportable when disassembled, require re-assembly on-site. The gearboxes or turntables then need to be attached, followed by the auger. Typically this takes three to four days to complete.

Equipment

The daily equipment costs for the ex-situ mixing equipment are typically on the order of $1000/day and are the least expensive when compared to the in-situ mixing equipment. The daily costs for the excavator-mounted in-situ S/S equipment is only slightly more expensive. However, the crane-mounted in-situ S/S equipment is significantly more expensive, on the order of $3,000/day.

The daily cost for most of the ex-situ material handling equipment is typically less than $1,000/day. Ex-situ S/S also requires excavators or loaders to feed the media to the mixing or material handling equipment and then remove and stockpile the treated material. In-situ S/S also requires an excavator to handle swell from the S/S treatment.

Overall, the total daily equipment costs for ex-situ S/S are typically less than those of in-situ S/S. However, daily equipment cost for certain in-situ S/S configurations (e.g. excavator-mounted mixers with the ability to accept dry reagent addition) can be comparable to those for ex-situ S/S. In-situ S/S using crane-mounted mixers and batch plants have the highest daily equipment costs.

Labour requirements

Most in-situ mixing systems with dry reagent addition require two workers to manage the reagent storage and delivery and operate an excavator-mounted in-situ mixer. Ex-situ S/S systems typically require three to four workers to feed the system, operate the mixer (and material handling equipment, if necessary), manage the reagent storage and delivery, and stage the treated material. In-situ S/S involving crane mounted equipment mixers can require five to six workers to handle the reagent storage and grout preparation, operate the mixer, and manage swell.

Reagent storage and delivery

Silos for reagent storage typically cost on the order of $500/day, whether used to supply ex-situ mixers or in-situ mixing equipment set up for dry reagent addition or batch plants. Batch plants typically run about $2,000/day, including the cost for all of the tanks, mixers, pumps, and hoses needed to prepare and deliver the reagent grout to the ex-situ or in-situ mixer.

Dry reagent addition, whether for ex-situ or in-situ S/S, typically requires only a reagent silo to store the reagent on-site and deliver that reagent to the ex-situ or in-situ mixing equipment. This is the least expensive reagent storage and delivery option. Reagent delivery in the form of a reagent grout typically requires silos to store the dry reagents and batch plants (Figure 3.13) to prepare the reagent grout, making this a more expensive option for reagent storage and delivery.

For the in-situ mixing equipment that can receive dry reagent feed, the reagent is delivered pneumatically, which makes it difficult to consistently and accurately deliver the reagents.

For the in-situ mixing equipment, reagent grout addition, though more expensive, results in more accurate, consistent, and reliable reagent delivery.

Production

Ex-situ S/S systems typically treat 350-1,000 yd^3 (270-760 m^3) of media per day, per mixer. Site-specific production rates vary based on media properties (e.g. moisture content, texture, plasticity), reagent addition level (i.e. productivity decreases with increasing reagent addition), and site considerations (e.g. space available for equipment or to stockpile untreated and treated materials).

In-situ S/S systems typically treat 400-1,200 yd^3 (300-920 m^3) per day, per mixer. Site-specific production rates vary based on media properties (e.g. moisture content, texture, plasticity), reagent addition level (i.e. productivity decreases with increasing reagent addition), and site considerations (e.g. depth of treatment, depth to water table).

Overall costs

Typically the cost per cubic yard for ex-situ S/S processing, excluding reagent costs, ranges from $15 to $30/yd^3. This cost does not include the cost to excavate, transport, and stockpile the untreated media or the cost to stockpile, transport, and place the treated material.

For in-situ S/S, the processing costs, excluding reagent costs, can range from around $20 /yd^3 (0.8 m^3) for using excavator-mounted equipment with dry reagent addition to $50 /yd^3 (0.8 m^3) for crane-mounted equipment with reagent grout addition.

In general, the costs for ex-situ S/S and in-situ S/S processing cover a similar range. However, site-specific factors (media properties, depth of treatment, depth of water table, and site space considerations) may give one method an advantage over the other.

PART FOUR

Ex-situ S/S equipment and application

The nature of a soil or waste and the reagents to be mixed with it is an important factor on the choice of mixing equipment to be used on site. Site characteristics such as its layout, the space available, size and geographical location, are also important. There are various types of ex-situ mixing equipment including pug-mills, excavators, screw mixers, ribbon blenders, tillers, asphalt mixers, and paddle aerators which are discussed below. In addition, design and application of mixing pits is also presented. Figure 4.1 outlines the processes involved in ex-situ application.

4.1 Mixing Chambers

Mixing chambers are designed to mix, or blend different materials to produce a homogenous product. The body or casing of the mixer contains or restricts the movement of the materials being combined during the blending process. Within the mixing body, rotating paddles, screws, ploughs, or ribbons lift, throw, fold, or knead the material. The mixing action and energy input (into the forming mixture) from the rotating paddles, screws, ploughs, or ribbons influence:

- What materials can be treated/mixed
- How long it takes to mix the materials being presented
- The thoroughness of the mixing action (i.e. how homogeneous is the product)
- The cost of the mixing process

The arrangement of the mixing chamber can be horizontal or vertical. Typically, horizontal mixing chambers are used for continuous mixing, while vertical mixing chambers are used for batch mixing. Continuous mixing chambers receive the material to be mixed at one end and discharge the mixed product at the other.

Batch mixing chambers have their feed materials introduced before mixing begins, and when mixing is completed the product is then removed (from the mixing chamber).

Continuous mixing allows for high materials throughput. Mixing time is set by the rate at which material is fed into the mixer, its volume and mixing action. Batch mixing has the advantage that the mixing time can be controlled, but processing capacity is related to the size of the mixing body and its cycle time, including loading of materials and unloading of product.

During batch mixing, the final product is usually discharged (by dumping) requiring a homogeneous product that is free flowing or fluid enough to flow out of the mixer by gravity.

PART FOUR

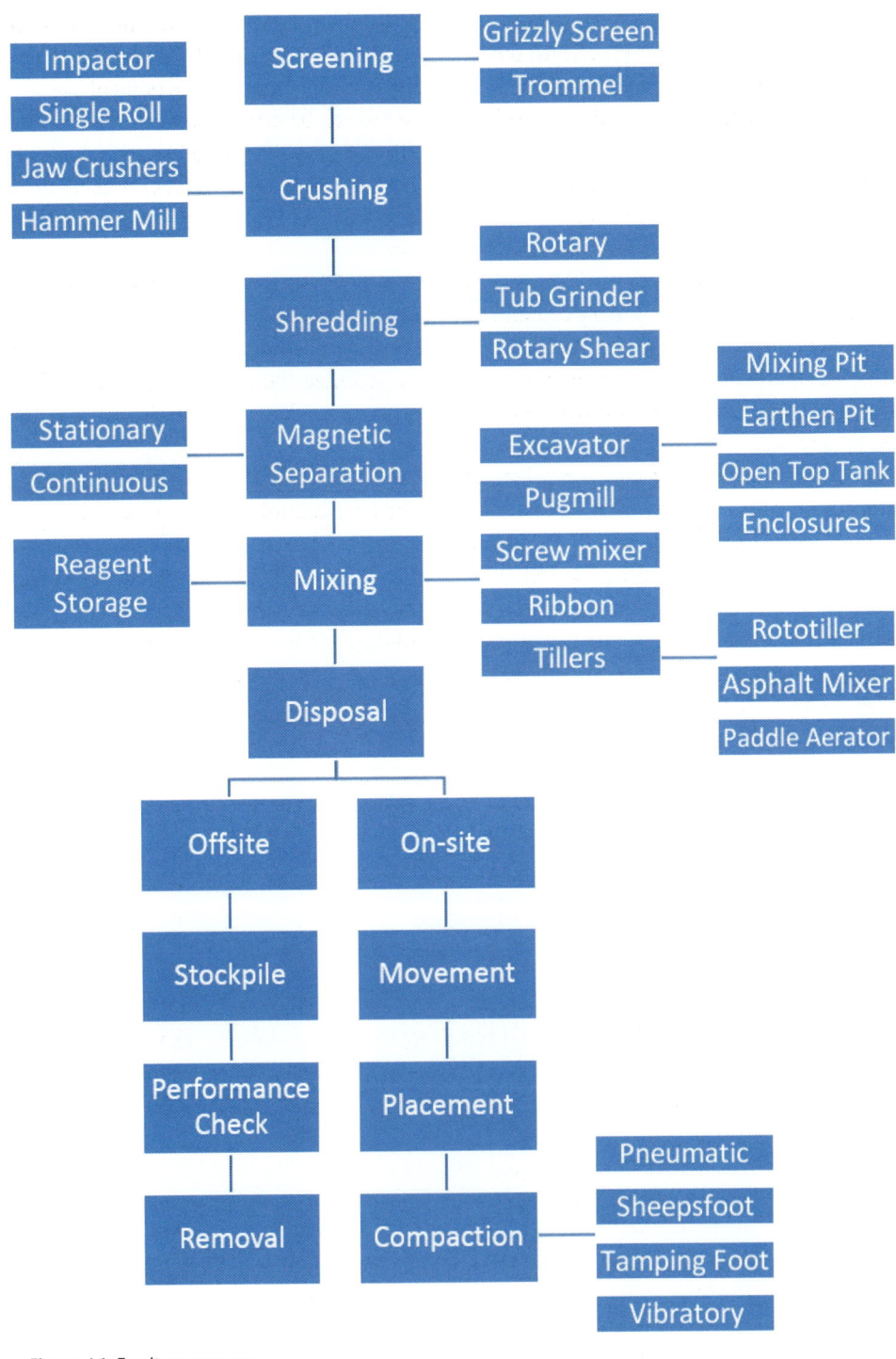

Figure 4.1: Ex-situ processes

Many continuous mixers use conveyors to move the discharged product, which must be viscous enough to be conveyed. Thus, the rheology of the freshly mixed S/S product is very important, and needs controlling for effective mixing and conveyance to the point of use.

Mixers can receive both liquids and solids. The solids can be dry powders, thin slurries, or pastes. However, the types of solids to be processed will need to be mixed in such a way as to produce a homogeneous product with a minimum of processing time. As an example, ribbon blenders and screw mixers cannot be used effectively with cohesive soil or waste and are not appropriate for blending these materials with binders.

There are environmental considerations for mixing chambers. Since materials are conveyed into and out of mixers, secondary containment is required to collect and contain any spillage, especially when one of the materials being introduced is a waste material; this is typically achieved by using a bermed or lined mixing platform. Often the materials to be mixed are dusty in nature and require effective dust control measures, such as a water-mist/spray within the mixer body.

However, a shroud with dust control measures may be required under some circumstances. This may include dust filtering or a bag-house, to effectively control any dust emissions. This is especially important when operations are within a populated urban area and where sensitive receptors are within close proximity (tens of meters) of the site. Similarly odour control during mixing can often be addressed by supplying a mist or spray, or an odour suppressant polymer or foam within the mixer body. However, mixer shrouding and use of an activated carbon filter or a reactive scrubber may be suitable mitigating measures when sensitive receptors may be close to the site of operation.

4.1.1 Pug-mill mixers

Pug-mill mixers are horizontal mixers that can be used for the continuous mixing of dry free-flowing powders, thin slurries, and thick pastes. A typical pug-mill consists of a horizontal box-like or trough mixing chamber with a top inlet on one end and a bottom discharge at the other end, 2 shafts with opposing mixing paddles, and a drive assembly to rotate the shafts.

A continuous pug-mill is an effective mixing platform that can achieve a thoroughly mixed, homogeneous product within a few seconds. The action of the pitched paddles moves the material from the bottom of the trough, up each side, and forces the material back down between the shafts. This both kneads and folds the S/S mixture. Mixing efficiency and product residence time are influenced by the size of the paddles, paddle swing arc, overlap of left and right swing arc and the dimensions (and volume) of the mixing chamber, and the nature of the material being mixed.

Pug-mills are often supplied as part of a pug-mill mixing 'system' including a feed hopper and feed conveyor to ensure a consistent supply of materials. The feed hopper can employ either belts or augers to control the transfer rate of the waste material onto the transfer conveyor, which often includes a belt-scale to allow the operator to accurately determine feed rate and quantity of material to be treated. Since the reagents are typically added on a weight-for weight basis, the waste feed rate from the belt-scale can be used to control the reagent addition rate.

A discharge conveyor to stack the treated product exiting the pug-mill can be included, but this is often a separate piece of equipment. Similarly, the reagent silos and feed-system can also be part ancillary equipment supplied. Figures 4.2 and 4.3 show a typical pug-mill system with support equipment. Figure 4.4 shows the trough of a pug-mill and its paddle design. The advantages and disadvantages of pug-mill mixers are listed in Table 4.1.

A list of pug-mill suppliers by country is shown in Table 4.8.

Figure 4.2: A pug-mill with its ancillary equipment

Figure 4.3: Ancillary equipment of a pug-mill

Advantages	Disadvantages
• Pug-mills are high capacity (200 tonnes of materials per hour) • High shear/mixing-energy mixers. The twin shafts in a pug-mill can rotate at up to 500rpm • Very homogeneous and well-mixed S/S product • Pug-mills are very tolerant of debris and will typically tolerate small stones and pieces of bricks, concrete, wood, and metal debris • Can effectively process soil or waste of a widely variable consistency, handling effectively viscous fluids, free-flowing powders, and plastic and/or sticky solids • Transportable and often mounted on a wheeled frame for road transportation • Can be ready for operation in less than half a day • Are relatively inexpensive to mobilise, set-up, operate and demobilise	• Throughput is dependent on the characteristics of the waste media (moisture content, plasticity, particle size distribution) • The waste/soil to be treated must be screened to remove debris greater than 2in (50mm). This may require additional handling/processing steps, including the drying of waste materials with moisture contents above their plastic limit, as is needed to allow effective screening • For waste volumes less than 2,000 m^3 costs can significantly impact the economics of on-site treatment

Table 4.1: Advantages and disadvantages of pug-mill mixers

Figure 4.4: Paddles inside a typical pug-mill

4.1.2 Screw Mixers

Screw mixers are horizontal or near-vertical mixers for the continuous mixing of dry free-flowing powders and solids, with moisture contents below their plastic limit. A screw mixer consists of a circular-shaped mixing body (usually at least ten times longer than it is wide), a screw auger mounted on a shaft, and a drive assembly to rotate the shaft.

A screw mixer is usually loaded at one end (typically the 'lower' end when used in a near vertical configuration). Mixed material is typically discharged from the other end. The rotation of the screw lifts and moves the material to be blended, creating a folding effect as it moves up the auger flight. The screw is set at a fairly close clearance to the mixer body so no material remains along the mixer body. The pitch, rotational speed, and length of the screw can be used to set the mixing time.

Screw mixers cannot be used for sticky or plastic solids, as they tend to 'ball up' and bind in the mixer. Figure 4.5 shows a screw mixer with its lid removed. The advantages and disadvantages of screw mixers are listed in Table 4.2.

A list of screw mixer suppliers by country is shown in Table 4.8.

Advantages	Disadvantages
• The high rotational speeds thoroughly blend waste media with reagents • Screw mixers can process in excess of 100 tonnes per hour • Screw mixers are simple to operate and have few moving parts to maintain	• Throughput is dependent on the waste materials characteristics (moisture content, plasticity, particle size distribution) and screw design (which controls the residence time in the mixer) • The waste material must be screened to remove debris larger than 1 inch, as screw mixers are not tolerant of hard debris, e.g. concrete, bricks, metal • Other handling/processing steps may be required to prepare the waste media, including, for example, drying of waste materials with moisture contents above their plastic limit (to allow screening and subsequent processing) • Are not capable of handling waste media that are above their plastic limit or are sticky or tacky in nature • Generally not suited for mixing liquids • Screw mixers are typically designed for fixed facilities

Table 4.2: Advantages and disadvantages of screw mixers

4.1.3 Ribbon Blenders

Ribbon mixers are also horizontal mixers for the continuous mixing of dry free-flowing powders, thin slurries, and solids with moisture content below their plastic limit.

A typical ribbon blender has a trough-shaped mixing body (usually two to three times longer than it is wide) with a semi-circular bottom, a horizontal longitudinal shaft upon which are mounted arms supporting a combination of ribbon blades and paddles. A drive assembly is used to rotate the shaft.

Loading a ribbon mixer with materials is usually done at (the top of) one end while discharging is done at the bottom of the other end. A long and complex single paddle is mounted axially on the shaft and is used to disperse the material to be mixed to the outer ribbon. Compared to a pug-mill, the paddles on a ribbon blender move relatively slowly, feeding the material to be blended into the ribbon.

Figure 4.5: Screw mixer

The most effective ribbon design is the double spiral in which the outer ribbon moves the product in one direction and the inner ribbon moves it in the opposite direction. This opposed movement creates axial flow of the material though the mixer and prevents build-up of materials in the mixer. The pitch of the ribbons are designed to mix material slowly, resulting in long residence times.

To improve the dispersion of materials in ribbon mixers, the ribbon blades are built to be close to the cylindrical wall to provide high shear mixing. Rubber wipers can also be fitted on the ribbons to lift material from near the wall into the middle of the ribbon. Figure 4.6 shows a ribbon mixer with its cover removed to display its internals. The advantages and disadvantages of ribbon blenders are listed in Table 4.3.

A list of ribbon blender suppliers by country is shown in Table 4.8.

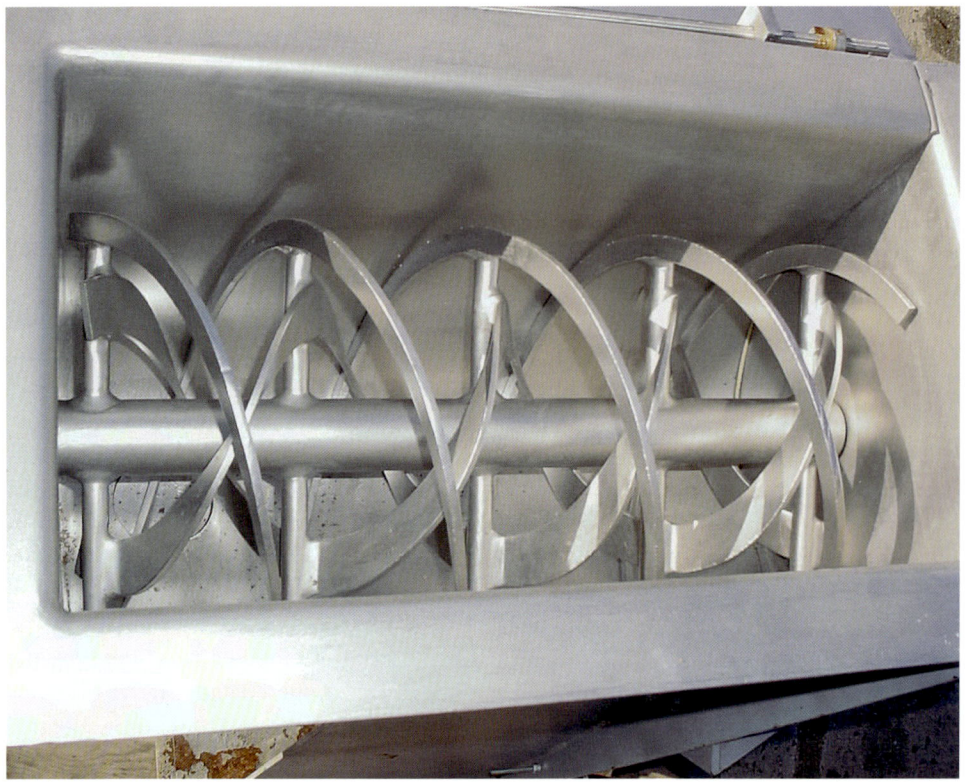

Figure 4.6: Ribbon mixer

Advantages	Disadvantages
• The high rotational speeds thoroughly blend waste media with reagents • Ribbon mixers can process in excess of 20 tonnes per hour • Simple to operate and have few moving parts to maintain • Many fixed facilities that treat sludges and liquids utilise ribbon mixers due to their simplicity of operation and thoroughness of mixing	• Throughput is dependent on the waste materials characteristics (moisture content, plasticity, particle size distribution) and ribbon design (which controls the residence time in the mixer) • The waste material must often be screened to remove debris larger than this offset, which is typically less than 1 cm • Extremely intolerant of debris, especially hard debris • Not capable of handling plastic, sticky or tacky waste media • Are typically designed for fixed facilities

Table 4.3: Advantages and disadvantages of ribbon blenders

4.1.4 Quality control using mixing chambers

Typically, the S/S mix design determined during the treatability study (see Section 8) is expressed on a weight/weight ratio, or the percentage of reagent to waste material to be treated. For example, a 0.05 mix ratio or 5 % additive addition rate, both equate to 5 tonnes (5.1 tons) of reagent added to every 100 tonnes (102 tons) of waste treated.

For pug-mills and screw mixers, the quality control revolves around calibrating the waste and reagent feed-rates to the mixer. The waste feed rate for pug-mills or screw mixers is typically determined by a belt-scale on the feed conveyor belt to the mixing box. This belt-scale should be calibrated at least weekly, preferably daily, using the manufacturer's recommended calibration procedure. Figure 4.7 shows the control panel for a belt-scale.

The reagent feed can be conveyed to the pug-mill mixing box either by a conveyor belt or screw auger. If the reagent feed is conveyed by a belt, a belt-scale can be used to measure the feed rate and this belt-scale can be calibrated at the same time as the waste feed belt-scale. If a screw auger is used to convey the reagents to the mixing box, a calibration between the screw auger rotation speed and the weight of reagent delivered needs to be completed. This calibration should be verified at least weekly, preferably daily. Figure 4.8 shows a pug-mill crew calibrating the reagent feed from a screw auger.

The waste and reagent feed rates are typically manually controlled in older pug-mills, but are automated in newer pug-mills. Figure 4.9 shows the control screen from an automated pug-mill.

Figure 4.7: Control panel for waste feed conveyor

Figure 4.8: Calibrating reagent feed from a screw auger

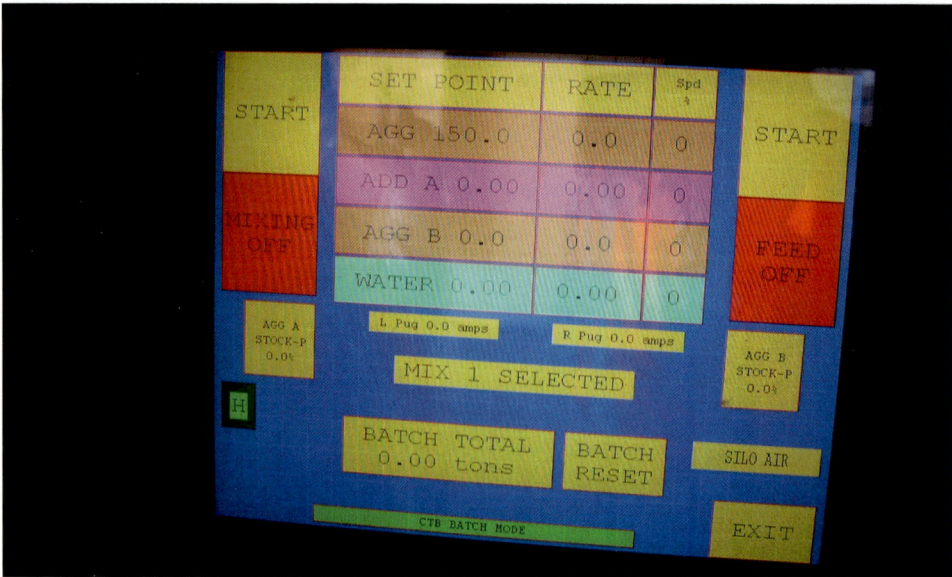

Figure 4.9: Display panel for an automated pug-mill

Figure 4.10: Modified excavator bucket for improved mixing

Similar to pug-mills and screw mixers, the quality control of ribbon blenders involves calibrating the waste and reagent feed rates to the mixer. However, most ribbon blenders are operated in a batch (rather than continuous) mode. Ribbon blenders are typically mounted on load cells, which should be periodically calibrated on the manufacturer's recommended schedule. During operation, a known weight of waste along with a known weight of reagents are charged into the ribbon blender and then mixed.

4.2 Excavator bucket mixing

Standard excavators can be used to lift, sift, turn, and mix materials. A conventional bucket, a bucket modified to improve efficacy with sifting and mixing (Figure 4.10), or a specialised mixing head (Figures 4.11, 4.12, and 4.13) can be used for 'excavator' mixing.

Excavators can be used to treat waste materials either in-situ or ex-situ. For in-situ S/S with an excavator, the waste to be treated is divided into treatment cells. The dry or liquid reagents are spread over the surface of the treatment cell, and the excavator operator uses the bucket or mixing head attached to the boom and stick to mix the reagents into the waste, until homogeneous. The treated waste is then either left in place or removed and stockpiled for transportation to the disposal location.

Advantages	Disadvantages
• Common construction equipment and requires minimal training for operation • Conventional or modified buckets are typically unaffected by the presence of debris (wood, concrete, or metal) in the waste • Up to 1,200 yd^3 (920 m^3) per day can be treated using an excavator with a conventional bucket	• It must be possible to spread the reagents on top the waste to be treated • Specialised mixing attachments are not tolerant of debris and debris should be removed prior to treatment • Excavator buckets are typically not designed for thorough mixing of wastes and reagents and the treated material may not be thoroughly homogeneous unless sufficient mixing time is employed • Plastic soils are not adequately mixed • Specialty mixing attachments can be expensive, may not be available locally, require more training to operate

Table 4.4: Advantages and disadvantages of excavator bucket mixing

Figure 4.11: Allu PMX Power Mixer™ attachment

Figure 4.12: Rotating mixing head attachment

Figure 4.13: Rotating rake mixing attachment

For ex-situ S/S with an excavator, the waste to be treated is laid out on a concrete pad or lined area (see Section 4.4) and the dry or liquid reagents are spread over the surface of the waste. The excavator operator uses the bucket or mixing head attached to the boom and stick to mix the reagents into the waste, until homogeneous.

The treated waste is then removed and stockpiled for transportation either to its original, or an alternate, on-site location for placement, or to an off-site disposal location. The next batch of waste is then placed on the treatment area and amended with reagent for treatment. The advantages and disadvantages of excavator bucket mixing are listed in Table 4.4.

A list of excavator bucket suppliers by country is shown in Table 4.8.

4.2.1 Quality Control using excavator bucket mixing

Excavator mixing (either ex-situ or in-situ) is a batch process, so the quality control revolves around determining the weight of waste and reagent added during the process. For all in-situ excavator mixing and much ex-situ mixing, the weight of the waste is calculated using the in-place volume and density of the waste material. The in-place volume is that of the treatment cell during in-situ excavator mixing and is typically the excavated volume for ex-situ excavator mixing. Occasionally, the waste material is transported over truck scales to obtain the waste weight for excavator mixing.

For ex-situ and in-situ excavator mixing, the reagent is typically conveyed either by conveyor belt or screw auger or pneumatically. If the reagent feed is by belt, a belt-scale can measure the weight of reagent conveyed to the treatment cell or treatment pad. If a screw auger is used to convey the reagents, the screw auger rotation speed and duration and the weight of reagent delivered needs to be calibrated.

Alternatively, reagents are pneumatically conveyed to the treatment cell. In this case, the weight of waste in the treatment cell or on the treatment pad is managed in 'full truck loads' of reagent, for convenience.

4.3 Tillers and Other Ex-situ Mixers

Tillers employ rotating shafts with tines, teeth, or paddles designed to cut, fracture, fold, lift, and/or shear soils. Tillers are typically not stand-alone equipment but are coupled with other construction equipment such as excavators, front-end loaders, tractors, or skid steers. The motor driven rotating shaft is typically mounted on the tiller. The hydraulic power supply provided to the equipment is also attached to the tiller.

For ex-situ S/S, the waste media are placed in lifts, ranging from 1-3 ft (0.3-1 m) in depth in a treatment area, often a concrete pad or lined area (see Section 4.4). The reagents (dry or liquid) are spread on top of the soil lift; typically fertiliser spreading equipment is used for this task. The tilling equipment is then used to mix the soil and the reagents. Multiple (three to four) passes are typically used to mix the waste media with the reagents. When the waste and reagents are thoroughly mixed, the treated material is then removed from the treatment area and stockpiled awaiting confirmation of successful treatment.

There are environmental considerations associated with tillers. The treatment area often used is a concrete pad or lined and berm-enclosed area designed to contain the waste media and any spillage. Reagents and waste materials are sometimes dusty in nature, and dust control, involving sprays or water mists, are often a consideration. However, shrouding the treatment area within a sprung structure or building and pulling air through the building and into a baghouse may also be necessary to control dust emissions, especially when the mixing will occur in an urban area, where sensitive receptors may be close (tens of meters) to the site boundary.

Similarly, odour control during mixing can often be addressed by supplying a mist or spray of an odour suppressing polymer or foam within and/or over the treatment area. However, enclosing the treatment area within a sprung structure or building and pulling air through the building and into an activated carbon canister or a reactive scrubber may be necessary. Again, this level of odour control is likely only when the mixing will occur in a populated area where sensitive receptors may be close to the site boundary.

4.3.1 Rototillers

Rototillers use a rotating shaft with curved tines designed to cut, fracture, loosen, and lift soil as the equipment rolls over the surface. Rototillers are typically mounted on the three-point hitch on the back of backhoes, front-end loaders, or tractors and use the hydraulic system of the host equipment to power its motor and rotate the shaft. Figure 4.14 shows a rototiller attachment for a small tractor.

Rototillers can be used to treat soil either in-situ or ex-situ. For in-situ S/S using a rototiller, the dry or liquid reagents are spread over the surface of the waste to be treated. The rototiller is then driven over the waste, mixing the reagents into the top 8 to 12 in (20 to 30 cm) of the waste. The treated waste is then removed and stockpiled for on-site placement after all of the waste is treated, or for transportation to an off-site disposal location. The next 8 to 12 in (20 to 30 cm) lift of waste is then amended with reagent and treated, until the depth of contamination is reached.

Figure 4.14: Rototiller attachment

For ex-situ S/S with a rototiller, the waste to be treated is laid out in 8 to 12 in (20 to 30 cm) lifts on a concrete pad or lined area (see Section 4.4) and the dry or liquid reagents are spread over the surface of the waste. The rototiller is then driven over the waste, mixing the reagents into the waste. The treated waste is then removed and stockpiled for transportation either to its original, or alternate, on-site location for placement or to an off-site disposal location. The next 'lift' of waste is then placed in the treatment area and amended with reagent for treatment. The advantages and disadvantages of rototillers are listed in Table 4.5. A list of rototiller suppliers by country is shown in Table 4.8.

Advantages	Disadvantages
• Can easily treat up to 1 ft (30 cm) thick lifts of soils, either ex-situ or in-situ • For in-situ S/S, up to 2,500 yd^3 (1,912 m^3) per day can be treated using a rototiller	• It must be possible to spread the reagents on top of the waste to be treated • The spreading operation involved in ex-situ S/S limits the daily treatment rate to closer to a maximum of 1,000 yd^3 (765 m^3) per day • They are typically not designed for thorough mixing of soil and reagents and the treated material may not be thoroughly homogeneous unless multiple passes are employed • Plastic soils may not be adequately mixed • Not amenable to brick, concrete, or metal debris greater than 2 in (5 cm) in diameter • Have difficulty operating on un-even soil surfaces • Are not standard roadway construction equipment and may not be available locally

Table 4.5: Advantages and disadvantages of rototillers

4.3.2 Asphalt millers

Asphalt grinders, cold planers, or millers employ a rotating drum with cutter teeth to loosen and remove worn asphalt as the equipment rolls over the asphalt surface. Bomag supplies asphalt millers, which can scarify to a depth of up to 12 in (30 cm) over a treatment width from 39-78 in (1-2 m). This equipment can be used to mix soil with reagents for S/S. Figure 4.15 shows a Bomag asphalt miller.

Asphalt mixers can be used to treat waste materials either in-situ or ex-situ. For in-situ S/S with an asphalt miller, the dry or liquid reagents are spread over the surface of the waste to be treated. The asphalt miller is then driven over the soil, mixing the reagents into the top 12 in (30 cm) of the waste, removing the treated waste, and conveying the treated waste to dump trucks. The treated waste is then stockpiled for on-site placement after all of the waste is treated, or for transportation to an off-site disposal location. The next 12 in (30 cm) lift of waste is then amended with reagent and treated, until the depth of contamination is reached.

For ex-situ S/S with an asphalt miller, the waste to be treated is laid out in a 12 in (30 cm) lift on a concrete pad or lined area (see Section 4.4) and the dry or liquid reagents are spread over the surface of the waste. The asphalt miller is then driven over the waste, mixing the reagents into the waste, removing the treated waste, and conveying the treated waste to dump trucks. The treated soil is then stockpiled for transportation to its original location for placement, or to an off-site location. The next lift of waste is then placed and amended with reagent for treatment. The advantages and disadvantages of asphalt millers are listed in Table 4.6. A list of asphalt mixer suppliers by country is shown in Table 4.8.

Advantages	Disadvantages
• Can easily treat up to 1 ft (30 cm) thick lifts of soils, either ex-situ or in-situ • Are standard roadway construction equipment and can typically be found worldwide • Little technical expertise is required to operate the equipment • Capable of handling non-metallic debris • For in-situ S/S, up to 3,500 yd^3 (2,680 m^3) per day can be treated	• It must be possible to spread the reagents on top the waste to be treated • The spreading operation involved in ex-situ S/S limits the daily treatment rate to closer to a maximum of 1,000 yd^3 (765 m^3) per day • They are typically not designed for thorough mixing of soil and reagents and the treated material may not be thoroughly homogeneous • Non-cohesive or plastic soils may not be adequately moved onto the conveyor • Have difficulty operating on soil surfaces

Table 4.6: Advantages and disadvantages of asphalt millers

Figure 4.15: Asphalt miller

4.3.3 Paddle aerators/compost turners

Paddle aerators consist of triangular paddles mounted on a rotating shaft. Paddle aerators can be mounted in the front of front-end loaders, bulldozers, tractors or skid steers, using the equipment's hydraulic system to power the motor, which turns the rotating shaft. The paddles are curved to lift and turn material. As the equipment moves forward, the rotating paddles lift and turn the material, providing a mixing action. Paddle aerators vary in width from 6-11 ft (1.8-3.4 m) and can mix material to a depth of 5 ft (1.5 m). Figure 4.16 shows a paddle aerator mixing soil in a windrow.

Paddle aerators can be used to treat waste either in-situ or ex-situ. For in-situ S/S with a paddle aerator, the dry or liquid reagents are spread over the surface of the waste to be treated. The paddle mixer is then driven over the waste, mixing the reagents into the waste, to a depth of up to 3 ft (0.9 m). If greater treatment depths are required, the treated waste can be removed and stockpiled for later on-site placement, or for transportation elsewhere. The next layer of waste is then amended with reagent and treated, until the depth of contamination is reached.

For ex-situ S/S with a paddle aerator, the waste to be treated is laid out in a lift of up to 3 ft (0.9 m) deep on a concrete pad or lined area (See Section 4.4) and the dry or liquid reagents are spread over the surface of the waste. The paddle aerator is then driven over the waste, mixing the reagents into the waste. The treated waste is then removed and stockpiled for transportation either to its original location for placement or to an off-site disposal location. The next 'lift' is then placed and amended with reagent for treatment. The advantages and disadvantages of paddle aerators are listed in table 4.7.

A list of paddle aerator suppliers by country is shown in Table 4.8.

4.3.4 Quality control using tillers and paddle mixers

Tiller mixing, either ex-situ or in-situ, is a batch process, so the quality control revolves around determining the weight of waste and reagent added during the process. For all in-situ rototiller mixing and during ex-situ mixing, the weight of the waste is calculated based on the in-place volume and density of the waste material.

The in-place volume is that of the treatment cell during in-situ rototiller mixing and is typically the excavated volume for ex-situ rototiller mixing. Occasionally, the waste material is transported over truck scales to obtain the waste weight for rototiller mixing.

For ex-situ and in-situ rototiller mixing, the reagent is typically conveyed by conveyor belt, screw auger or pneumatically. If the reagent feed is conveyed by belt, a belt-scale can be used to measure the weight of reagent delivery to the treatment cell or treatment pad. If a screw auger is used to convey the reagents, the screw auger rotation speed (and duration) and the weight of reagent delivered needs to be calibrated. Alternatively, reagents can be pneumatically conveyed to the treatment cell, and in this case the weight of waste (in the treatment cell or on the treatment pad) is often controlled so as to require full truckloads of reagent to simplify reagent delivery.

Advantages	Disadvantages
• Can treat lifts of soils up to 2-5 ft (0.6-1.5 m) in depth either ex-situ or in-situ • For in-situ S/S, up to 2,500 yd^3 (1900 m^3) per day can be treated using a paddle aerator	• It must be possible to spread the reagents on top of the waste to be treated • The spreading operation involved in ex-situ S/S limits the daily treatment rate to closer to a maximum of 1,000 yd^3 (765 m^3) per day • Cohesive and plastic materials may not be thoroughly mixed • Not amenable to brick, concrete, or metal debris greater than 4 in (10 cm) in diameter • Paddle aerators are specialised equipment produced by a limited number of suppliers and may not be available locally

Table 4.7: Advantages and disadvantages of paddle aerators

Figure 4.16: Paddle aerator mixing windrowed soil

78 | PART FOUR

Table 4.8: mixer manufacturers

Location	Paddle Aerator	Asphalt Miller	Rototiller	Ribbon Blender	Screw Mixers	Pug-mill
Australia				• Andritz Group		• Aran International Pty
Austria					• Andritz Group	
Belgium	• Menart Sprl					
China	• Shandong Sunco Agricultural Equipment Technology • Zhenzhou Repale Machinery Co.	• Weifang General Machinery • Wuhan Kudat Industry and Trade	• Shandong Yuntai Machinery Co. • Yucheng Dadi Machinery Co. • Zhenzhou Whirlson Trade Co. Ltd	• Shanghai Senfan Machinery Co. • Yangzhou Nouya Machinery Co.	• Yangzhou Nouya Machinery Company • Jinan Xucheng Co. • Double Crain Machinery Manufacture (Leling) Co., Ltd.	
Canada				• Steelcraft, Inc Engineered Products	• UniTrak	
Denmark			• Baltic Korn A/S			
Finland	• Allu Group					
Germany	• Backhus Kompost-Technologie • Doppstadt	• Wirtgen BmbH	• Backhus Kompost-Technologie		• J. Engelsmann AG • CATS GmbH • Doppstadt • Putzmeister	
India	• Ambica Engineering Works	• Shitla Road Equipment • Shiv Shakti Road Equipments		• Krishna Engineering • Rana Perforators	• Rana Perforators	• Leo Road Pvt Ltd • Atlas Industries • S.P. Enterprises • Shitla Road Equipment
Italy					• WAMGROUP • IMER Group	
Switzerland					• Gericke AG	
UK		• Roadtec • Maddock Equipment • Terex Cedarapids • Bomag America	• BEFCO • Rotomec • Servis-Rhino	• Winkworth Mixer Co. • Aaron Process Equipment • Applied Chemical Technology • National Bulk Equipment • Jaygo	• US Air Filtration • Auger Manufacturing Specialists • Acrison	• BG Europa Ltd
United States	• Midwest Bio-Systems • Brown Bear					• DustMASTER Enviro Systems • Eagle Iron Works • Excel Machinery • JW Jones • Kohlberg/Pioneer • Maxon Ind. • Peerless Conveyor and Manufacturing Co. • McLanahan Corp. • Pug-mill Systems, Inc • Rapid International, Inc • Terex Cedarapids

4.4 Mixing pits

Mixing pits are earthen pits, open-topped tanks of metal or concrete, or three-side concrete (though can also be metal or wood) enclosures in which ex-situ S/S treatment can take place.

The waste media are placed into the mixing tanks, and the dry reagents are added pneumatically or via super-sacks (e.g. 1 to 2 ton bags or 0.9-1.8 tonnes); liquid reagents are pumped into the pit. Typically an excavator is used to mix the waste media with the reagents, though tillers are sometimes employed. When the waste and reagents are thoroughly mixed, the treated material is then removed from the mixing pits and may be stockpiled, to await confirmation of successful treatment.

There are environmental considerations for mixing pits. Since materials are conveyed into and out of mixers, secondary containment is typically required to collect and contain any spillage. For tanks and enclosures, secondary containment may involve double-walled structures, or placement within a bermed and lined area. Often the reagents, and sometimes the waste materials, to be mixed are dusty in nature and dust control via water mist or spray within and/or over the mixing pit may be necessary. However, shrouding the mixing pits within a sprung structure or building and pulling air through the building and into a baghouse may be necessary. Similarly, odour control during mixing may be addressed by an odour suppressant polymer mist or spray or foam within and/or over the mixing pit.

However, enclosing the mixing pit within a sprung structure or building and pulling air through the building and into an activated carbon canister or a reactive scrubber may be necessary. This level of odour control is likely only when the mixing will occur close to a populated area or sensitive receptors. Figure 4.17 shows an excavator with a standard bucket mixing Portland cement into lead contaminated soil and debris on a concrete pad.

Figure 4.17: Excavator mixing of Portland cement into Pb-contaminated soil and debris

4.4.1 Earthen Pits

Earthen mixing pits may be excavated into contaminated soil or established within impoundments and lagoons. Waste media are transferred into the pits using excavators or loaders, and the reagents (wet or dry) are added to the pit. An excavator is then used to mix the waste and the reagents until a visually homogeneous product is formed.

The treated waste is sampled to verify successful treatment, and then may be removed and stockpiled for off-site disposal or later placement back on-site; alternatively, the product can be left in-place, on-site. Figure 4.18 shows an excavator with a standard bucket mixing hydrated lime into an oily waste in an earthen mix pit. The advantages and disadvantages of earthen pits are listed in Table 4.9.

Advantages	Disadvantages
• Simplicity and flexibility	• Limited by the reach of the excavator arm
• The size can be varied as necessary to contain the mixing volume, which can range from 20 to 1,000 m^3 (26-1300 yd^3)	• Excavators or loaders may need to track onto the treated material to access other pit areas
• No need for secondary containment	• Sprung structures or buildings are required to contain dust or odour controls - these structures would need to be mobile if the treated material is left in place within the mixing pit
• Many simple options to place the waste media into the pits	
• No feed preparation is required	
• Typically, only an equipment operator and perhaps a labourer are required to run a mixing pit and only an excavator is needed for the mixing operation	

Table 4.9: Advantages and disadvantages of earthen pits

Figure 4.18: Bucket mixing hydrated lime and oily waste in earthen pit

4.4.2 Open-top tanks

Open-top tanks are metal or concrete structures in which the waste/soil are placed for mixing. The waste material is conveyed to the open-top tank by conveyor, excavator, or loader and the reagents, whether dry or wet, are added directly to the tank.

An excavator is then used to mix the waste and the reagents until the product is visually homogeneous. The treated waste is then sampled to verify treatment, and the product is removed and stockpiled for off-site disposal or placement on-site. Figure 4.19 shows an excavator bucket mixing reagents into hexavalent chromium-contaminated soil in a roll-off box. The advantages and disadvantages of open-top tanks are listed in Table 4.10.

Advantages	Disadvantages
• Simple metal containers such as roll-off boxes or inverted sea-land containers (with the bottoms removed) can be used for tanks. Tanks can range in size from 10 to 1,000 yd^3 (7.7-765 m^3) • Many simple options to place waste into the tank, and little to no feed preparation is required. • Minimal staff required	• The lateral extent of a mixing tank is only limited by the reach of the arm of the mixing excavator • The metal bottoms of the tank require careful consideration by the excavator operator to avoid puncturing with the excavator bucket during mixing

Table 4.10: Advantages and disadvantages of open-top tanks

Figure 4.19: Stabilisation of hexavalent chromium-contaminated soil in a roll-off box

4.4.3 Enclosures

Enclosures often have three concrete sides and a concrete bottom. The concrete bottom of the enclosure is sloped to allow the enclosure to drain away from the open side.

The waste media can be transferred into an enclosure using conveyers, excavators or loaders. Reagents are added directly to the tank. An excavator, loader, or bulldozer is then used to mix the waste and the reagents thoroughly, until visually homogeneous. The treated waste is then sampled to verify successful treatment has been carried out. The treated waste is often removed and stockpiled for off-site disposal or placement on-site. The advantages and disadvantages of enclosures are listed in Table 4.11.

Advantages	Disadvantages
• Enclosure may comprise a concrete slab with the sides made of portable concrete barriers (e.g. K-rails or Jersey barriers) • The size of the enclosure can range from 20 to 1,000 yd^3 (14.4-765 m^3) • Little to no feed preparation is required prior to waste placement • personnel requirements are generally limited to an equipment operator and labourer; an excavator is needed for mixing	• The footprint of an enclosure is generally only limited by the reach of the excavators mixing arm, which must be able to treat all of the waste material presented • Liquid or self-levelling wastes and treated materials are unsuitable for treatment in mixing enclosures as even with a profiled/sloping base (away from the opening) the enclosure will not often adequately contain these types of materials

Table 4.11: Advantages and disadvantages of enclosures

4.5 Ancillary equipment for ex-situ mixing

Except when excavator mixing is employed, the wastes often require pre-processing to remove debris and to reduce their particle size to enable adequate blending with the S/S reagent to be achieved. This section will discuss some of the material handling equipment requirements when preparation of waste material is needed before ex-situ mixing.

4.5.1 Screening equipment

Screening equipment is often employed to classify or separate waste material by size. Screening equipment may vary on the size classification (or "cuts") required, the dimensional opening of the screen material, and the material throughput.

Grid screeners or grizzlies consist of a parallel grid of inclined iron or steel bars. Hydraulic vibration or an oscillation of the grid is often employed to improve the separation efficiency and throughput of the grid. Grizzly screens (Figure 4.20) are often used to remove large material, greater than 6 in (15 cm) in diameter from the waste material.

Screening plants employ inclined vibrating screen-decks for size separation. Screen decks are 4 ft wide by 8 ft long (1.23 x 2.46 m), with screen openings ranging from 4 in x 4 in (10 x 10 cm) down to 100 mesh (149 μm), and a screen plant can have one or multiple screen decks.

The screen decks are vibrated to provide maximum screening efficiency and the speed of the vibration can be adjusted.

Trommels are horizontal rotating cylindrical screens with up to 500 ft² (46.5 m²) of screening area (Figure 4.21). Their large screening area produces highly efficient action and high throughput rate. The screen dimension is commonly 1 in x 1 in (2.5 x 2.5 cm), but can be varied as necessary. The pitch or incline of the trommel can also be varied to improve screening efficiency.

For all of the screening equipment, the undersized material passes through the screen, while the oversized material moves across the screen and is discharged off the end of the screen. The oversized materials from multiple screen decks can be combined or kept separate, depending on the screen plant design. For most ex-situ S/S mixers, the final screened material is often needed to be less than 1 in (2.5 cm) in diameter.

Grizzly screens are often used to separate large material (greater than 6 in (15 cm) in diameter) from the media, while screen plants or trommels are used to provide a less than 1 in (2.5 cm) in diameter material for S/S treatment (Figure 4.21).

Figure 4.20: Grizzly screen for separation of large debris, with over-sized material, to the left and processed soil for ex-situ mixing, right

Figure 4.21: Grizzly screen (foreground) combined with a trommel to remove battery casings and debris from excavated material

4.5.2 Crushing equipment

Crushing equipment is employed to size-reduce hard waste material and debris, often in combination with screening equipment. Crushing equipment typically relies on impact (collision of the material with moving surfaces), attrition (intra-aggregate abrasion between two hard, moving surfaces), shear (cleaving action which occurs as aggregates are pinched between two surfaces), and compression (crushing material between two surfaces) to fragment the waste material.

Impactors use fixed hammers attached to a rotating axis and breaker bar lining the internal radial surface, while hammer mills employ free-swing hammers attached to a rotating axis with screening bars located along the radius of the hammers.

The offset of the breaker bars from the hammer radius determines the size of the crushed material. The screen bars in the hammer mill also produce attrition of the material before it exits the crusher. Single roll crushers employ breaker bars attached to a rotating drum above an impact plate.

A combination of shear, impact, and compression fragments the material as it passes between the breaker bars and the impact plate. The offset distance between the radius of the breaker bars and the impact plate determines the size of the crushed material. Jaw crushers use a swinging jaw to compress material against a fixed jaw. Impactors handle wet, sticky material best, as they are open-bottomed. Hammer mills and single roll crushers handle hard materials, but hammer mills are not tolerant of plastic or sticky materials as they blind the screen bars.

Jaw crushers are best at handling very hard materials, but cannot be used with sticky or abrasive materials. Jaw and single roll crushers produce coarse (greater than 1.5 in (3.75 cm) in diameter) crushed material, while impact hammers and hammer mills can produce a smaller material, down to less than ½ inch (1.25 cm) in diameter, depending on the offset.

4.5.3 Shredding equipment

Shredding equipment is often employed to size-reduce soft shear-able waste material and debris, including wood, vegetation, plastics, and some metals. Shredding is often used in conjunction with screening equipment, and size-reduction.

Shredding equipment typically relies on shear to fragment waste material and debris. Shredding equipment varies in the mechanics of the shredding and the size of the shredded product.

Rotary shear shredders (Figure 4.22) employ counter-rotating shafts or cutter blades. Material is shredded as it is drawn between the interfaces of the two counter-rotating cutter blades. The close tolerance of the cutter blades performs the shearing action.

Tub grinders use an inclined tub or rotary screen to feed a rotor or hammer mill. The tumbling action of the tub screens the material prior to shredding.

Rotary shear shredders can process wet and plastic material. Tub grinders are more suited to handle hard and abrasive materials, due to their efficient use of impact forces. Rotary shear shredders produce a coarser shredded product than tub grinders, since the tolerance of the cutter blades is greater.

Figure 4.22: A rotary shear shredder processing scrap metal

4.5.4 Magnetic separation equipment

Magnetic separators are used to remove ferrous metal debris from waste or soil. After removal, the metal is often taken off-site for recycling. Additionally, separating the ferrous metal prior to further processing (such as crushing or shredding operations) protects equipment and reduces maintenance. Magnetic separators are typically suspended over the body or head pulley of a conveyor.

Stationary overhead magnets can be suspended using a cable sling-support and periodically must be swung to the side for the ferrous metal to be removed from the surface of the magnet. Continuous, self-cleaning magnets involve a belt with a movement that is used to "clean" the separated ferrous metal from the belt.

There are two configurations for magnetic separators: in-line or cross belt. In-line continuous, self-cleaning magnets are installed with the cleaning belt running parallel to the conveyor movement. Therefore, the on-line configuration is only available above the head pulley. Cross-belt continuous, self-cleaning magnets are installed with a cleaning belt running perpendicular to the direction of conveyor movement, and can be placed over the body or head pulley of the conveyor. The 'In-line' configuration requires less magnetic force than the 'cross belt', but the discharge of the cross belt is easier to implement.

4.6 On-site placement of ex-situ treated material

Waste materials treated by ex-situ mixing are often either disposed of in an off-site disposal facility (e.g. landfill) or are disposed of on-site. When the material is disposed of off-site, it is typically stockpiled until it is confirmed to have met the required performance criteria (discussed in Section 7) and then loaded into trucks, or railcars, for transport to the off-site disposal facility. When the material is disposed of on-site, it often must be properly placed, spread, and compacted. This section will discuss the procedures for placing and compacting the treated material, and the material properties that facilitate the placement and compaction.

4.6.1 Timing

When the performance criteria for the ex-situ S/S treated material include specified properties related to durability, UCS strength and/or permeability, the treated material must be placed and compacted within 24 hours, (preferably 12 hours), of treatment.

As the development of physical performance criteria relies on cementitious or pozzolanic reactions, initial set generally occurs within the first 24 to 48 hours. It is therefore necessary to have the treated material placed and compacted before the initial set has occurred. If the initial set is disrupted or placement and compaction occurs after the initial set, the treated material will become impaired and unlikely to achieve the required performance goals. Therefore, ex-situ S/S treated material should be placed and compacted on-site within the same day as it is treated.

It is important to note that setting of S/S materials is generally much slower than that of normal concrete on account of the nature of the materials being treated and the interaction of waste and soil matrices, and contaminants with cementitious reactions. That said, the rate of setting is an important parameter to S/S as it is to concrete, and should be monitored to ensure compliance with performance targets established by bench-scale testing.

If the performance criteria for ex-situ S/S treated material does not include the development of physical performance criteria, the timing of the on-site placement and compaction is not as critical. Under these circumstances, the treated material can be stockpiled after mixing until it is operationally advantageous for its placement and subsequent compaction.

4.6.2 Placement

Typically, ex-situ S/S treated material is transported on-site in dump trucks or off-road dump trailers. The treated material is transported to a placement area, dumped and spread using a dozer in loose lifts that tend not to exceed 18 in (45 cm) in depth. Figure 4.23 shows the transport and spreading activities for pug-mill-treated S/S material, involving a dozer making multiple passes over the loose 'lift' to remove any large voids in the placed material. Figure 4.24 shows the surface of material after multiple passes of the dozer.

For ex-situ S/S treated material with physical performance criteria, the treated material needs to have sufficient moisture to allow cementitious and pozzolanic reactions to occur. Therefore, the treated material may have to be placed with a higher-moisture content than the optimum required for maximum density.

Additional water may also be applied to such treated material during placement to ensure that it does not dry out to the point where cementitious reactions cannot proceed. The avoidance of water 'starvation' involves careful consideration by the field engineer overseeing materials placement.

4.6.3 Compaction

Compaction of treated, placed material involves maximising the density of the product and thereby minimisation of its volume.

Compaction requirements

Compaction can be based on achieving greater than specified maximum dry density, as defined by its moisture-density relationship, or degree of 'effort' needed in the field.

For the first case, the compaction requirement is based on attaining a greater than a specified percentage (typically 90 % or 95 %) of the maximum dry density from Proctor compaction testing. The Proctor method (ASTM D698 or D1557) is a laboratory determination of the moisture content-density relationship for a compacted soil, or soil-like material, and moisture content at which the material has its maximum compacted density.

In Figure 4.25, the moisture content range for compaction to 95 % of the maximum Proctor density is shown. Compaction of placed ex-situ S/S treated material with moisture content between 6.8 % and 22.3 % should result in a dry density of greater than 103.4 pcf (0.016 tonnes/m^3) (95 % of the maximum dry density). Ex-situ S/S treated material with a moisture content below 6.8 % would have to be wetted or a moisture content above 22.3 % would, depending on the material, need to be either wetted or dried before it could be compacted.

Figure 4.25 shows the moisture/density relationship from the Proctor compaction testing of an ex-situ S/S treated material.

This first case is only applicable when the performance criteria for the ex-situ S/S treated material do not include the 'physical' performance criteria (durability, strength and/or permeability).

As discussed (Section 4.6.1), physical performance criteria rely on cementing reactions within the treated material, with placement and compaction being completed before the initial set is complete. Changes to the moisture content of the placed material will disrupt the initial set leading to failure to meet performance criteria. Therefore, the compaction requirement for ex-situ S/S treated material that is needed to achieve physical performance criteria, is typically expressed as a specified number (typically at least two) of passes with specified compaction equipment (see below).

Compacting equipment

The following discussion on compaction equipment is based on the Caterpillar Compaction Manual (Caterpillar Tractor Company, 1989).

Pneumatic tire compactors: Pneumatic tire compactors are used on small to medium sized compaction jobs, primarily on bladed, granular base materials. Pneumatic tire compactors are not suited for high production, thick lift compaction projects. The compaction forces (pressure and manipulation) generated by the rubber tires work from the top of the lift down to produce density. The amount of compaction force can be varied by altering the tire pressure (the normal method) or by changing the weight of the ballast (done less frequently). The kneading action caused by the staggered tire pattern helps seal the surface.

One advantage of pneumatic compactors is that there is little bridging effect between the tires, thus they seek out the soft spots in the fill and compact them. For this reason, they are sometimes referred to as "proof" rollers.

Sheeps-foot roller: Sheeps-foot rollers are named after Roman road builders, who used to herd sheep back and forth over base material until the road was compacted. The sheeps-foot roller (Figure 4.26) consists of pads attached to a cylindrical drum. These pads penetrate through the top lift and actually compact the lift below. When a pad comes out of the soil, it 'fluffs' the material at the surface, resulting is a loose-layer of surface material, and when the next lift is placed this becomes compacted, and the next layer is 'fluffed' up.

Because the top lift of soil is always being fluffed, the process helps aerate and dry out treated materials with excessive moisture contents. However, the top lift at the end of each day is loose and not compacted, and this has the disadvantage of acting like a sponge, should it rain, slowing the compaction process.

By their very nature, sheeps-foot rollers only exert pressure to effect compaction – they do not provide impaction or vibration. As such, multiple-passes (six to ten in total) may be necessary to achieve the required density in 8 to 12 in (20-30 cm) lifts.

Tamping foot compactors: Tamping foot compactors are high speed, self-propelled, non-vibratory rollers. They usually have four steel padded wheels and are equipped with a dozer blade. Their pads are tapered with an oval or rectangular face.

Like the sheeps-foot, a tamping foot compactor compacts from the bottom of the lift to the top. But because the pads are tapered, the pads can walk out of the lift without fluffing the soil.

Figure 4.23: Transport and placement of ex-situ S/S material

Figure 4.24: Placed ex-situ S/S-treated material after multiple passes of the dozer

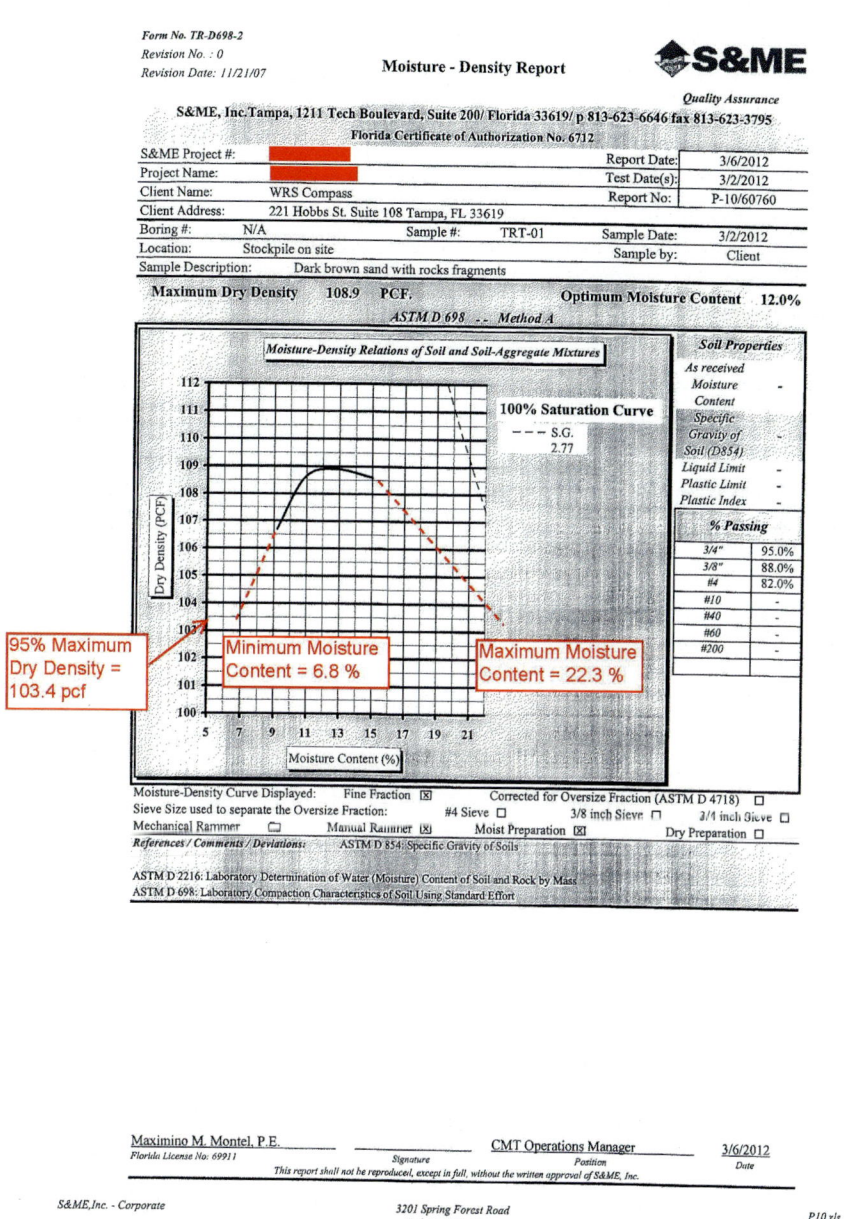

Figure 4.25: Proctor compaction-testing results for ex-situ S/S treated material. Note the moisture content range defining greater than 95 % maximum dry density. (pcf = pounds per cubic foot)

Figure 4.26: Sheeps-foot roller compacting S/S treated material (note the oval pads on the roller)

Therefore, the top of the lift is compacted and the (top lift) surface is relatively smooth and 'sealed'. Because tamping foot compactors are capable of speeds near 15-20 mph (24-32 kmh), they develop and exert all four forces in compaction: pressure, impaction, vibration and manipulation to the lift of S/S material. This not only increases their effective compaction (of a lift) but the production rate is relatively much higher as well. Generally two to three passes will achieve the desired density in 8 to 12 in (20-30 cm) lifts, though four passes may be needed in materials that are plastic in nature.

The main limitation to the use of tamping foot compactors is that they are best suited for large projects, as long uninterrupted passes at a high speed are key to a high production rate. Tamping foot compactors are considerably more expensive than vibratory compactors.

Vibratory compactors: Vibratory compactors work on the principle of the rearrangement of S/S particles to decrease voids and increase density. Vibratory compactors come in two types: smooth drum and padded drum varieties. Smooth drum vibratory compactors generate three compaction forces: pressure, impaction and vibration. Padded drum units also generate manipulative force. Compaction is generally assumed to be uniform throughout the lift subject to vibratory compaction.

The forces generated by a vibrating drum hitting the ground are large, resulting in a densification of the lift. The amount of compaction achieved is a function of the frequency of these blows, the force of the blows and the time period over which the blows are applied. This frequency/time relationship accounts for slower working speeds and a speed of 2 to 4 mph (3.2 to 6.4 kph) provides the best results.

Smooth drum vibratory compactors (Figure 4.27) were introduced first and are most often used on granular materials, with particle size ranging from large rocks to fine sand.

They are also used on semi-cohesive soils with up to 10 % cohesive soil content. The thicknesses of a lift will vary according to the size of the compactor but, generally, the lift thickness of granular material should not exceed 24 in (60 cm).

When padded drum machines were introduced, the materials that could be treated were extended to include soils with up to 50 % cohesive material and a greater percentage of fines. When the pad penetrates the top of the lift it breaks the bonds between the particles within the cohesive soil and improves compaction results. This is a result of the geometry of the 'pads' which are involute (i.e. curled inward), self-cleaning, and able to "walk" out of the lift without fluffing the surface. The typical lift thickness for padded drum units on cohesive soil ranges from 12 to 18 in (30-45 cm).

Figure 4.27: Smooth roller compacting ex-situ S/S treated material

PART FIVE

In-situ S/S equipment and its application

5.1 In-situ auger mixing

Auger mixing is the most widely used technique of *in-situ* soil mixing and is especially useful for deeper applications, greater than 15 feet (4.6 m).

Auger mixing is suitable for a wide range of soil types to depths in excess of 100 ft (30.8 m), although specialised methods are necessary for depths beyond 60 ft (18.5 m) below ground surface (BGS).

5.1.1 Equipment

Generally the equipment used for in-situ soil mixing consists of augers and their support carriers to conduct the injection and mixing, a batch plant to prepare the reagent slurry and pump it to the auger assembly, and miscellaneous support equipment.

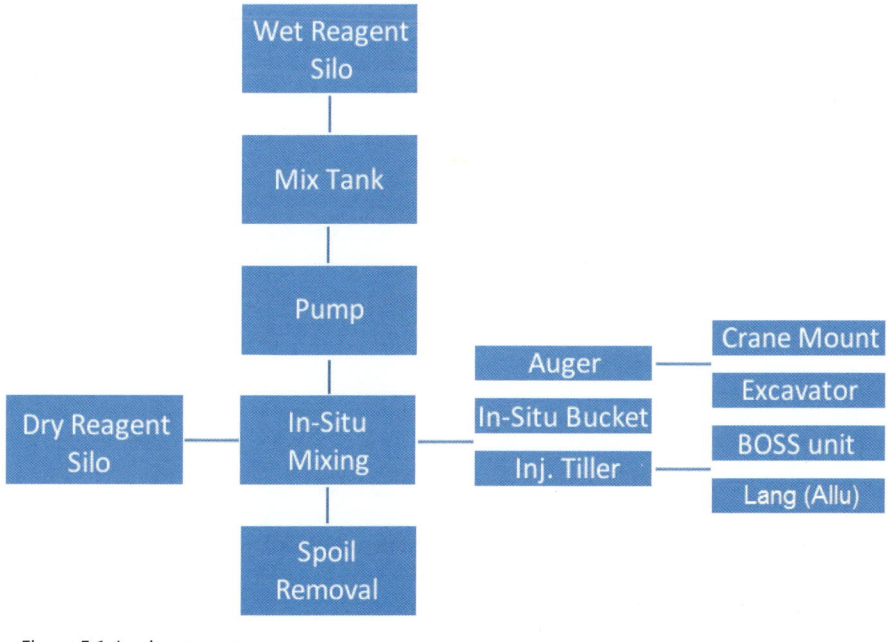

Figure 5.1: In-situ processes

Augers and their support carriers

There are two main types of drilling rigs used for in-situ soil mixing using an auger: crane- and excavator-mounted units. Crane-mounted units consist of mechanically driven rotary drilling heads fitted on a crawler crane. Excavator-mounted units consist of hydraulically driven, rotary drilling heads fixed to an excavator base.

Crane-mounted units can generally accept slightly larger diameter augers (10-12 ft or 3-3.6 m) and, in theory, have deeper mixing depth capabilities (by increasing the amount of boom on the crane and installing a longer Kelly bar). On the other hand, excavator-mounted rigs are generally best suited to smaller diameter augers (9-10 ft or 2.8-3.1 m) and have stroke lengths limited to the mast height of the equipment (generally less than ~60 ft or 18.5 m). However, recent advances in excavator-mounted equipment have made these systems comparable and (in many ways) more advantageous to use, than crane-mounted systems. This results from their superior mobility, higher operating torque and lower mobilisation costs. Examples of crane-mounted rigs are shown in Figures 5.2 and 5.3, while excavator mounted rigs are shown in Figures 5.4 and 5.5.

The mixing head of the crane-mounted system is mechanically powered by an engine/transmission combination, whereas the mixing head of the excavator-mounted system is hydraulically powered using hydraulic pumps running off the engine system of the host machine. Crane mounted systems generally have maximum torque outputs (typically 250,000–350,000 ft.lbs or 34.56-48.39 kgf-m) that far exceed the listed maximum torques of their excavator mounted counterparts (typically 100,000–250,000 ft.lbs or 13.83-34.56 kgf-m). However, the operating torque of crane-mounted systems is much less than the maximum torque, which is only achieved during small portions of the engine power curve. Higher torque requirements result in lower rotation speeds and therefore less thorough mixing. The operating torque of excavator-mounted systems is very close to maximum available torque and often exceeds the operating torque of much larger crane-mounted systems. This allows higher rotation speeds at higher torque, resulting in a better mixing.

In both the crane and excavator-mounted systems, reagents are normally pumped through a hollow Kelly bar and out of the auger ports. Reagents are most commonly added in a liquid or grout form that acts as both the drilling lubricant and the final stabilisation reagent, but reagents can also be added in a dry-powder form. Occasionally, for shallow/smaller applications, reagents are added at the surface, by spreading.

Auger mixing provides the highest quality in-situ mixing available. The high torque available in auger-mounted rigs makes them ideal for mixing dense sands and stiff clays. A number of manufacturers produce drill rigs that can be used for soil mixing, but most commercially available rigs require modification through additional specialty equipment to be fully utilised in this application. Auger mixing involving both crane and excavator mounted systems has been in use for 25–30 years, however, in recent years the remediation industry has increasingly tended to use excavator-based systems, due to their high level of mobility and consistent torque output.

Figure 5.7 illustrates the auger often used today. In this case it is a 10 ft diameter auger and one can see the cutting teeth on the forward side of the blade and the reagent injection ports on the trailing edge of the blade. Below the blade is a projection, commonly called a "stinger" which helps guide the auger into the soil.

IN-SITU S/S EQUIPMENT AND ITS APPLICATION | 95

Figure 5.2: A crane-mounted soil mixing rig

Figure 5.3: A crane-mounted soil mixing rig

Batch plant equipment

Batching plants consisting of silos, pumps, and mixing tanks are critical to the success of most in-situ soil mixing projects, relying on reagent delivery through ports located on the mixing head of an auger. Batching plant configurations vary widely, as their make-up/configuration includes the practitioner's preference, reagent type and quantity, the number of reagents to be used and the reagent pumping distance, and site–related constraints.

Figure 5.4: An excavator-mounted rig

For applications where the reagent(s) is being added as a grout, the batch plant can become quite extensive. An example of an automated batch plant is shown in Figure 5.8.

The majority (>95 %) of in-situ soil mixing projects use pre-mixed reagents in a fluid grout or slurry form prior to injection. With the increased acceptance of this S/S technology, it is not uncommon to blend two or three dry reagents in a batching plant that is automated.

Automated plants use weigh-scales to accurately measure mix components in a process that improves efficiency and quality control.

For applications where the reagent is being delivered from dry, pressurised storage tanks, pneumatic pumps are required. An example of a self-propelled dry pneumatic reagent hopper is shown in Figure 5.9.

Ancillary equipment

In addition to the batching plant, a variety of ancillary support equipment is necessary for the successful completion of an auger-based soil mixing project. The support equipment may include excavators, dozers, loaders, forklifts, man-lifts, pumping systems, hoses, survey equipment and data loggers.

5.1.2 Staffing requirements

Typically an auger-based soil mixing project requires a supervisor, drilling rig operator and support labour, a batch plant operator and support labour, and QC/engineering staff. However, staff requirements vary from project to project, depending on ancillary work required and chosen batch plant configuration.

Figure 5.5: An excavator-mounted soil mixing rig

Figure 5.6: Auger used at the USX Site, Duluth, MN

Figure 5.7: Auger used at a coal gas plant site in FL

Figure 5.8: An automated batch plant

Figure 5.9: Self-propelled dry storage silo

5.1.3 Treatment metrics and considerations

The maximum treatment depth, the optimum auger diameter and reagent addition methods and production rate are highly variable, depending on site-specific conditions and the equipment employed. Obstructions in the sub-surface, such as concrete slabs, pipes, rocks, and disused cables can substantially slow production, break augers, and significantly increase cost. Such sub-surface objects should be removed prior to starting auger-treatment to avoid potential costly delays. Additional treatment metrics and considerations follow below.

Depth of treatment

The depth of treatment is dependent on a number of factors, including soil-type and relative density, the auger diameter and its configuration, the torque available to the Kelly-bar, mast length and downward force capability (drill crowd and tool weight).

As previously mentioned, auger systems are generally limited to depths less than 60ft (18.46 m), unless specialty high-torque power units are used with smaller diameter augers. Crane-mounted systems tend to have slightly deeper maximum treatment depths, but mast extensions for excavator-mounted rigs are available to match the depth capabilities of almost any crane-mounted unit.

Most of the excavator-based systems currently in use are limited to maximum depths in the 45–55 ft (14-17 m) range. It is important therefore, that practitioners should comment on the depth limitations of their respective equipment in the soil conditions anticipated at the project site. The effective maximum working depth can be derived from geotechnical data, such as from CPTs, SPTs, and other common site investigation-derived data. A decrease in the auger diameter (other factors being equal) will generally allow for treatment at greater depth, however smaller augers have a major adverse impact on production rates and cost. Table 5.1 lists some remedial operations employing in-situ auger mixing, including the depth treated and auger diameter employed.

Anticipated production rates

Generally, auger-based systems are capable of treating between 200-600 yd^3 (153-460 m^3) of soil or sludge per working day (based on an 8 hour shift). However, this is somewhat dependent on a variety of factors including relative soil geotechnical properties, maximum treatment depth and reagent dosage, etc. At the optimum production rate, with treatment depths of 10-40 ft and 8-10 ft diameter augers (3.1-12.3 m and 2.5-3.1 m respectively), it is not unusual to treat over 800 yd^3 (612 m^3) per auger in an 8-10 hour shift. However, due to maintenance requirements and the occasional (expected) equipment breakdown, a lower average production rate should be anticipated.

Reagent addition methods

Reagent addition is normally supplied by pumping/injecting through a wet Kelly-bar/ auger, or by adding the reagents at the soil surface (and then mixing in). However, adding the reagents at the surface significantly limits the effective treatment depth that can be achieved.

Reagent addition through the mixing tool provides improved delivery distribution and therefore improved quality control by comparison. Depending on the equipment and batch plant make-up, both dry and wet reagent addition are possible, with the former being advantageous on projects with very high moisture content soils. Wet reagent delivery is however, better for an even vertical reagent distribution within each S/S column. Bench (Section 8.3) and pilot scale (Section 8.4) treatability tests are critical for determining the appropriate reagent formula and field application methods.

5.1.4 Treatment plan

A site-specific and detailed treatment plan (otherwise known as a work-plan) is produced by the contractor, prior to starting the remedial operation/treatment.

The treatment plan should include the specific equipment to be used, staffing, proposed work schedule, reagent addition mixing and dosage rates and the plan for sample collection/curing/testing, site safety and reporting requirements.

For in-situ auger mixing, a critical component of the plan is the precise layout and planned depth of each column, so as to achieve the desired area of treatment with overlapping columns. By employing overlapping columns, complete treatment is achieved without leaving any untreated 'void' spaces.

The degree of overlap will vary depending on project-specific needs. Overlapping columns will cause some portions of the soil to be mixed and treated two or possibly three times. Increased overlapping will provide greater assurance that all the soil is being treated, but in a slower production rate and at a higher treatment cost.

Figure 5.10 illustrates a portion of a typical column layout showing overlapping of adjacent columns.

For the constructor, over-lapped columns requires careful planning of the sequence of column construction, as cutting into previously treated and solidified columns can be difficult, if too much time has passed since initial treatment. A pragmatic approach involves treating every other column, then returning and 'cutting' the skipped overlapping columns on the second or third day. Figure 5.11 shows S/S columns that have been excavated, illustrating that the overlapping of the columns facilitates the complete treatment of the target interval.

5.1.5 Quality control

The level of quality control available for auger systems is very high in comparison to other in-situ soil mixing methods. The quality control procedures available vary based on the equipment being used, the reagent type (liquid or dry), and the preference of the practitioner.

Table 5.2 gives key features of a typical quality control program from an auger-based S/S mixing application.

5.1.6 Operational Issues

Equipment

The essential maintenance of equipment is an important part of the management of any construction project, including S/S by in-situ soil mixing. Production can be adversely affected by equipment that is poorly maintained and where wearable parts are not regularly inspected and kept in an operational condition. Both the grout and soils being treated are abrasive, and can have adverse effects on exposed portions of the equipment, due to wear and chemical degradation.

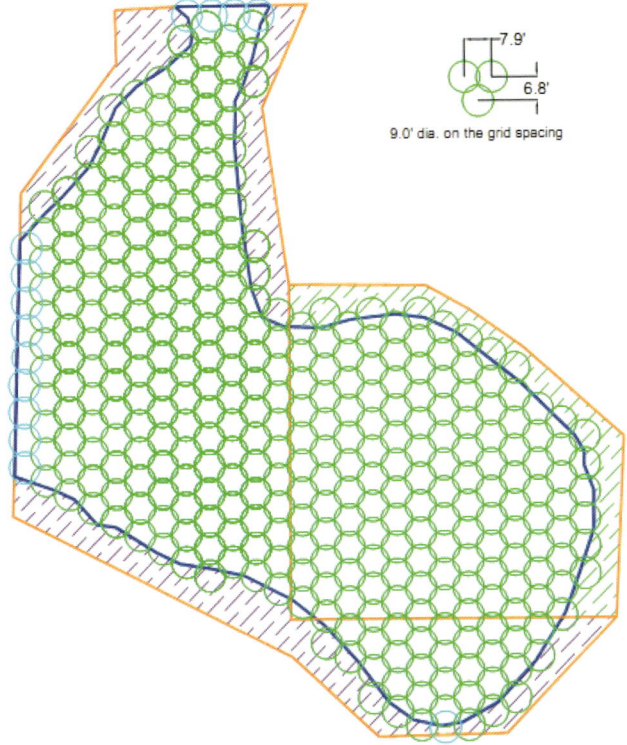

Figure 5.10: Typical column layout showing overlapping columns to achieve 100 % coverage

Figure 5.11: Showing typical excavated overlapping columns

Site Name	Constructor	Date Completed	Auger Diameter, Ft (m)	Max. Depth, Ft (m)	Volume Treated, cy (m³)	Considerations
Confidential	Geo-Solutions	Fall 2011 – Spring 2012	10 (3)	41 (12.5)	58,000 (44,340)	Dense glacial till with cobbles, high torque drill
Inner Slip Site Remediation	Geo-Solutions	Fall 2011	3 and 8 (0.9 and 2.4)	32 (9.7)	6,500 (4,970)	Loose dredge sediments, tight access and poor subgrade
SAR Levee Repair	Geo-Solutions	Fall 2010	9 (2.7)	54 (16.5)	5,500 (4,205)	High torque drill rig, dense sandy soils
MW-520 Site Remediation	Geo-Solutions	Spring 2010	9 (2.7)	20 (6.1)	15,200 (11,620)	Clayey silts
Ameren Site Remediation Pilot Study	Geo-Solutions	Fall – Winter 2010	5 (1.5)	37 (11.3)	500 (382)	Dense glacial tills, test program (small volume)
Joachim Creek South Alignment Bearing Capacity Improvement	Geo-Solutions	Winter – Spring 2010	9 (2.7)	40 (12.2)	2,500 (1,911)	High plasticity clay
OMC Plant 2 Site Remediation	Geo-Solutions	Fall – Winter 2011	9 (2.7)	25 (7.6)	8,900 (6,805)	Dense sand & gravel
Former Municipal Wastewater Treatment Lagoon Stabilisation	Geo-Solutions	Summer – Fall 2012	8 (2.4)	26 (7.9)	8,600 (6,575)	Very dense clays
P&G Site Remediation	Geo-Solutions	Spring 2012	9 (2.7)	30 (9.1)	19,500 (14,910)	Lagoon sediments
Front and T Street Site Remediation	Geo-Solutions	Summer – Fall 2012	10 (3)	33 (10)	40,000 (30,580)	High torque drill
Former Hanley Area Site Remediation	Geo-Solutions	Spring 2012	5 (1.5)	30 (9.1)	1,400 (1,070)	Tight access, small drill, dry mixing

Table 5.1: Examples of in-situ auger treatments including auger diameter and depth treated

Quality Control Plan	
Information	**Details**
Lines of communication	
Key personnel & responsibilities	
Methods and procedures for verifying reagent addition at depth	
Project staging	Column layout showing 100% coverage of the treatment area
Layout procedures	GPS, Total Station, Triangulation
Sampling procedures	
Non-conformance procedures	
Daily monitoring requirements	
Information	**Details**
Grout consistency (wet applications)	Density, Viscosity, Temperature, pH
Treated Columns	Dimensions (effective treated area & treated depth)
	Column centre-point locations
	Unique column identification
	Target reagent weight – based on effective treated area & treated depth
	Number of mixing strokes
Reagent Addition (dry and wet applications)	Total volume of grout added via flow meter (wet)
	Weight of reagent via weigh-scales (dry)
Mixing energy	Rotary head (RPM)
	Lift rate (if applicable)
	Grout pressure/flow rate
Sample collection and curing	
Information	**Details**
Molds, store, transport, and testing	
Completion of QC reports	
Information	**Details**
Daily Report	Daily report: site activities, problems, safety issues, progress map (what has been completed), total daily volume treated, cumulative volume
Reagent Usage	QC report: reagent usage – total and per treated volume, effective area calculations (treated volume calculations), start stop time
Quality assurance of operational reporting by an independent engineer	

Table 5.2: Quality control planning for auger-based S/S mixing

Obstructions
Auger-based mixing equipment is sensitive to the presence of large sub-surface or overhead obstructions. Given the deeper soil treatment depth-limitations of auger-based systems, some obstructions can be too deep and too costly to effectively remove.

Shallow obstructions are more easily dealt with when the treated soil is in a "liquid" state, such as immediately after mixing. Site constraints may limit the removal of obstructions during the mixing operation, and so the best approach might be to stabilise all the soils around the obstruction and then remove the obstruction (from between stabilised soils).

Spoil
In general, 15 % to 30 % of the volume of the treated soil becomes spoil, sometimes called 'swell', 'slop' or 'float'. This spoil is formed above ground, as treated material, which accumulates due to the mixing process and addition of reagents.

Above the water table some of the spoil becomes subsumed into the S/S column, filling the pore spaces between soil particles. However, below the water table spoil is displaced upwards to the surface of the column.

The spoil requires removal and/or disposal where possible. It generally has a high slump value, being composed of soil, liquid grout, and groundwater. Spoils can be moved and channelled in their liquid state or allowed to take an initial set, before selective removal and transport/placement for disposal.

5.1.7 Summary of limitations, advantages, and disadvantages

In comparison to the other types of in-situ soil mixing, auger-mixing systems have few limitations. All methods of in-situ soil mixing are heavily influenced by the presence of sub-surface obstructions, and auger mixing of soil is no different.

Occasionally, a skilled operator can navigate the auger around small obstructions, but in general, unforeseen sub-surface obstructions stop a soil mixing project in its tracks. Buried utility lines must be located and cleared if they are in the area to be treated.

A rare exception to this is shown in Figure 5.12, which shows a 10 ft (3.1 m) diameter in-situ auger treating soil immediately under an active fibre optic cable. The fibre optic cables are located within the PVC pipes (just to the right of the Kelly bar) and were temporarily exposed and supported while the auger bit was located underneath the lines and then into the soil below.

Figure 5.13 shows the mixing taking place in the location of the fibre optic cable mentioned above.

Auger mixing can be difficult in extremely dense soils or very 'fat' clays (liquid limit greater than 50) as dense soils tend to cause accentuated wear of the mixing equipment components and it's difficult to achieve a consistent mix in 'fat' clays.

The advantages and disadvantages of auger mixing are summarised in Table 5.3.

Figure 5.12: In-situ treatment of soil under a live fibre optic line

Figure 5.13: In-situ mixing under fibre optic line

Advantages	Disadvantages
• High production rate/reduced schedule time • Possible in difficult drilling conditions, stiff clays/dense sands • Deeper depth capabilities • Treat below water table without dewatering	• Sensitive to obstructions • Requires specialty expertise and equipment • Less efficient than other types of in-situ or ex-situ mixing for shallow depths (< 5 ft)

Table 5.3: Advantages and disadvantages of auger mixing

5.1.8 Costs

Typical costs for stabilisation using the auger mixing system are indicated below. However, it should be noted that the costs involved are very sensitive to the depth of treatment, the types of soils being treated, obstructions to mixing and the chemical reagents being used. In addition, the specifics of performance sampling requirements may also be important. As a rule of thumb, current prices for mobilisation are $75,000 to $250,000, whereas the application of treatment is in the range $30 to $60 /yd^3 (0.7 m^3) + reagent costs.

5.2 Injection tillers and rotary drum mixers

Injection tillers and rotary drum mixers are suitable for the in-situ mixing of a wide range of soil types to depths up to about 12 ft (3.7 m), although deeper mixing is sometimes achievable in certain sludge-like materials.

For the purpose of clarity, injection tillers and rotary drum mixers are considered to be an attachment to a standard excavator and can also be referred to as a Backhoe

Operated Soil Stabiliser, or BOSS unit. Quality control is limited in comparison to auger mixing, but is better than that achieved with bucket mixing.

5.2.1 Equipment

Mixing equipment generally consists of some form of rotating mixing head through which reagent is injected either dry, or more commonly, as grout slurry and mixed in place at the target depth. In addition a batch plant is required to prepare the reagent slurry according to the formula developed during bench scale treatability tests (Section 8.3) and refined during the field pilot test (Section 8.4). Ancillary equipment may include pumps, hoses, support excavators and/or front loaders, a dozer, and survey equipment.

Mixing head types

In-situ soil mixing with injection tillers and rotary drum mixers is typically accomplished using attachments to standard construction equipment, i.e. excavators, dozers, and front-end loaders. Excavator "arm" attachments have the greatest depth capabilities and are often referred to as BOSS systems for short. BOSS systems replace the digging bucket of an excavator assembly with a mixing arm that has a rotary drum mixer at the end. Figures 5.14 and 5.15 show commercially available BOSS systems used by Geo-Solutions.

With a BOSS system, the mixing head(s) is powered by the hydraulic system of the host machine, or a separate hydraulic power pack can be mounted on the host machine. Reagents may be pumped through the mixing arm or added at the surface.

Most of the BOSS units have the capability of pumping reagents through piping which discharges just above the mixing head and this is of critical importance if mixing to depths of more than 2-3 ft (0.6-0.9 m) is required.

If reagents are applied to the surface, it is very difficult to mix them uniformly to a depth of more than 2-3 ft (0.6-0.9 m). Thus, the BOSS systems that inject the reagents at the point of mixing have the capability to achieve thorough mixing to depths of 12-15 ft (3.6 to 4.6 m), sometimes more, depending on the length of the mixing arm and how this arm is articulated.

The BOSS-type unit works well in sludges and soft soil applications. If soils are stiff or of a clay type material, the mixing heads require a much higher torque, which is only available on a few of the currently available BOSS models.

The BOSS units, which are available in a variety of forms, have been available to the remediation industry for over 20 years. Commonly available commercial systems include the Lang and the Allu Mixer.

The Lang mixer (langtool.com) has frequently been used in the USA and is available in several models (e.g. Figure 5.16) with somewhat different depth limitations depending on the length and articulation of the arm.

Reagent is injected just above the mixing head, which is sealed and custom fitted to the excavator body, in one complete unit. This facilitates mixing in wet soil.

The Allu mixer (allu.net) has been less available in the USA. The Allu mixer is provided as an attachment to be placed on a standard excavator body. Several models with varying depth capability are available. However in the model recently used on a coal tar site in Florida, it was observed that the top of the Allu attachment was not sealed and thus could not be immersed in the treated soil. Figure 5.17 illustrates one model of the Allu mixer.

Batch plant equipment

Batch plants consisting of silos, pumps, and mixing tanks are critical to the success of most in-situ soil mixing projects, namely those relying on reagent delivery through ports at the mixing head such as the BOSS systems.

- Batch plant configurations vary widely, due to plant makeup and configuration and the practitioner's preference. The type and quantity of reagents, as well as the pumping distance and site constraints are also important. Where the reagent is being added as a grout, the batch plant can be extensive. An example of a large soil mixing grout production plant is shown in Figure 5.18
- When the reagent is being delivered as a dry powder, pressurised storage tanks and pneumatic conveyance pumps are required. An example of a self-propelled dry pneumatic hopper is provided in Figure 5.9

Figure 5.14: Backhoe (excavator) operated soil stabiliser

Figure 5.15: Backhoe (excavator) operated soil stabiliser

Figure 5.16 Lang mixer - excavator, arm, and mixing head as one unit

Ancillary equipment

In addition to the batch plant, a variety of support equipment is necessary for the successful completion of a soil mixing project using the BOSS system. Supporting ancillary equipment may include excavators, dozers, loaders, forklifts, man-lifts, hoses, pumps and surveying apparatus.

5.2.2 Staffing requirements

Typically a rotary drum mixing or injection tilling remedial project requires a supervisor, mixing apparatus operator, mixing apparatus support labourer, batch plant operator, batch plant support labourer, and QC/engineering staff. Labour requirements vary from project to project depending on ancillary work and batch plant configurations.

5.2.3 Treatment metrics and considerations

Some treatment metrics and considerations are briefly discussed below. However, in practice these are very site- and equipment-specific, and the reader should consult with an experienced practitioner regarding the application to any specific site.

Treatment Depth

As previously discussed, rotary drum mixers and BOSS systems are typically limited to depths shallower than about 12-15 ft (3.69-4.62 m). Applications of this technology to depths 15 ft (4.62 m) or deeper are possible, but mixing quality and quality control become limited at increased depth. Note however that equipment designs and capabilities are evolving and

Figure 5.17: Allu mixer head attached to a standard excavator

Figure 5.18: A batching plant for a large S/S project

newer equipment may achieve good mixing at greater depths. Practitioners should prepare and submit "digging" charts to illustrate the full extent of their equipment's capabilities, i.e. maximum treatment depth in relation to the machine body.

Due to the limited power and penetration capacity of these mixers at greater depths, auxiliary equipment may be necessary to pre-excavate or loosen the soils prior to treatment. In these cases, BOSS treatment may not be considered purely as an in-situ treatment.

Anticipated production rates
BOSS systems are capable of treating between 150-600 yd^3 (115-459 m^3) of soil or sludge per working day (assumed 8-hour shift). The production rate is highly dependent upon the site soils/wastes, formula addition rate, depth of treatment, and operator skill. The production rate will significantly slow for treatment depths over 10 ft (3.07 m).

Reagent addition methods
Reagent addition may be delivered by pumping through the mixing tool or by adding the reagents at the surface. Reagent addition through the tool provides improved mixing and quality control in comparison to reagent addition at the surface. Depending on the equipment and batch plant configuration, both dry and wet reagent addition are possible options. Dry reagent addition can be advantageous on projects with high moisture content soils.

Pre-construction investigations
Prior to commencement of field operations, it is important to thoroughly investigate and delineate the treatment area and depth of treatment. Bench scale studies (Section 8.3) are usually employed to determine efficient (or worst case) reagent application rates and the mode of reagent delivery (grout or dry powder). In addition, bench scale studies can be used to estimate spoil volume. Generally pilot-scale field tests are conducted to refine the reagent dosage and the planning for execution of the desired treatment option at full-scale.

5.2.4 Quality control

The level of quality control for BOSS systems is much less than that available on the auger-based soil mixing systems. The quality control available for BOSS systems varies based on the equipment used and the method of reagent delivery (wet or dry).

The components typical of a quality control program on a BOSS application are given in Table 5.4.

For all in-situ S/S treatments, performance samples are collected and cured in a similar manner to that described in Sections 6.2 and 6.3.

The correct care and attention should be given to sample storage and transportation of quality control samples, which should be stored in a temperature-controlled environment. During storage, samples should not be subjected to movement or vibration, particularly during the initial 24 to 72 hours when undergoing initial set. Samples should not be transported until they have initially set and achieved a reasonable strength, usually within the period 3 to 7 days. When being shipped by courier, samples should be properly packed to minimise movement and damage during transportation.

Quality Control Plan	
Information	**Details**
Lines of communication	
Key personnel & responsibilities	
Lines of communication	
Methods and procedures for verifying reagent addition at depth	If available, GPS can be used to assist the operator in verifying complete mixing of the entire treatment block
Project staging	Grid map or daily treatment area. Development of a grid to determine reagent application zones. Each zone should be sized for a proportion of the treatment area that can be completed in a few hours (certainly less than 1 day).
Layout procedures	GPS, Total Station, Triangulation
Sampling procedures	
Non-conformance procedures	
Daily monitoring requirements	
Information	**Details**
Grout consistency (wet applications)	Density, Viscosity, Temperature, pH
Treated panel or volume per 'stroke'	Dimensions (length, width, depth)
	Location
	Unique identification
	Target reagent weight
	Number of mixing strokes
Reagent Addition (dry and wet applications)	Total volume of grout added via flow meter (wet)
	Weight of reagent via weigh-scales (dry)
Mixing energy	Rotary drum (RPM)
	Lift rate (if applicable)
	Grout pressure/flow rate
Sample collection and curing	
Information	**Details**
Molds, store, transport, and testing	
Completion of QC reports	
Information	**Details**
Daily Reporting	Daily reporting of site activities, problems, safety issues, progress map (what has been completed), total and cumulative volume treated
Reagent Usage	QC reporting of reagent usage (total/per treated volume, effective area calculations (treated volume calculations), start stop times etc.
Quality assurance reporting by Independent engineer	

Table 5.4: Quality control planning for BOSS-based S/S mixing

5.2.5 Operational issues

Equipment

Equipment maintenance procedures are an important part of every construction project, but can be especially important on in-situ soil mixing projects. Rotary drum mixers have a limited mixing depth due to the equipment configuration, being constrained by the length of the mixing arm.

Mixing at extended depth requires that the entire mixing arm, machine joint, and machine boom are beneath the soil/sludge surface, and the grout and soil undergoing mixing can have adverse effects on mechanical joints leading to excessive equipment downtime and additional maintenance requirements.

Obstructions

The configuration of mixing arms allows them to more easily move around obstructions than an auger-based unit, but they are still limited by the presence of sub-surface and overhead obstructions. Given the shallow soil treatment limitations, most obstructions in these applications can be easily removed using a support excavator.

Obstructions are more easily removed when the treated soil is in a "liquid" state immediately after mixing. Site constraints may, however, limit removal of obstructions during the mixing operation. The best approach is then to stabilise all the soil around the known 'obstruction' and then remove this by excavation between stabilised soil-units.

Spoil

In general, 15 % to 30 % of the treated soil volume becomes spoil due to bulking by the addition of reagents and the mixing process. The spoil materials require handling and/or disposal by channelling whilst in their liquid state or after an initial set, when they can be treated and disposed of as a soil.

5.2.6 Advantages and disadvantages of rotary tillers and injection drum mixers

The advantages of rotary tillers and injection drum mixers are listed in Table 5.5.

5.2.7 Costs

The costs of treatment with the BOSS system are very sensitive to the depth of treatment desired, the soils being mixed and the reagents being used. As a guide, current prices for BOSS mobilisation are $50,000 to $150,000, whereas the application of treatment is in the range $15 to $30 /yd^3 (0.7 m^3), excluding reagent costs.

5.3 In-situ bucket mixing

In-situ bucket mixing (excavator mixing) refers to using a standard excavator that may have an extended reach, and an excavator bucket to mix reagents into the soil in place.

Reagents may be added dry or as a slurry, but the effective depth for mixing is however quite limited compared to other options. This approach to mixing is the simplest form (in terms of equipment) of in-situ S/S treatment and has been successfully used for decades.

Advantages	Disadvantages
• Lower cost than auger mixing • High production rate • Reduced schedule • Can readily utilise reagent in a wet (grout) or dry state • Reagent injection can be applied just above the mixing drum(s)	• Limited to a maximum depth of about 15 ft (4.62 m) below the working surface • Applicable only in sludge or soft soils • Reduced level of quality control in comparison with auger mixing • Obstructions require removal for complete mixing

Table 5.5: Advantages and disadvantages of rotary tillers and drum mixers

5.3.1 Equipment

Bucket mixing is generally carried out with excavators and standard excavation buckets or specialty buckets (see Figure 5.19) designed to facilitate high slump soil mixing. Shallow (1-2 ft or 0.3-0.6 m) mixing applications may be completed with bull-dozers or front-loaders, but this application is uncommon and is limited.

Although the equipment used is universally available, it should not be assumed that anyone with an excavator can successfully accomplish in-situ bucket treatment. The mixing techniques require experience and the quality control needs are rigorous.

Figure 5.19: Example of bucket mixing

Ancillary Equipment - A variety of support equipment is necessary for the successful completion of a soil mixing project utilising bucket mixing. Support equipment may include excavators, dozers, loaders, forklifts, man-lifts, etc. If reagents are added as slurry, then a batch plant like that used for in-situ auger or BOSS mixing will be required. Whether reagents are added dry or as slurry, equipment is required to accurately measure reagent addition per unit of soil.

5.3.2 Staffing requirements

Typically a soil mixing project completed using bucket mixing requires a supervisor, mixing excavator operator, mixing excavator support labour, and QC/engineering staff. Labour requirements vary from project to project depending on ancillary work and reagent addition procedures.

5.3.3 Treatment metrics and considerations

Treatment depth

Bucket mixing is limited to treatment depths less than 8 ft (2.5 m), as the uniform mixing of surface-added reagents beyond this then becomes progressively more difficult to achieve. However in some cases, bucket mixing may be extended to about 15-20 ft (4.6-6.2 m) if specific procedures are carefully followed:

The first 5-6 ft (1.5-1.8 m) are initially mixed uniformly with reagent slurry. It is necessary that the treated soil have a high slump. The excavator then carefully removes several buckets of soil from below the previously mixed and slurried soil, depositing the fresh soil above that previously mixed, and then proceeds to uniformly mix this soil whilst adding fresh reagent. The void created by excavating below the previously mixed soil is immediately filled by the mixed soil-slurry. This process is repeated as necessary. Due to restrictions on excavator reach, often a portion of the targeted cell is mixed to full depth, then the excavator re-positioned to mix another vertical slice.

This technique will not work with all soils/sludges and requires constant quality control monitoring to assure that the treated soils are homogenously mixed with reagent. Quality control is already challenging with bucket mixing and even more so at greater depth. Figure 5.20 depicts successful bucket mixing of coal tar contaminated soil to a depth of about 15 ft (4.6 m). Note that the picture shows clumps of deeper soil excavated to the surface, but not yet mixed so as to be homogenous. Part of the quality control process is to frequently extract an excavator bucket of mixed material from various depths, and visually check for clumps of unmixed soil.

Anticipated production rates

It's possible to treat 200-700 yd^3 (153-536 m^3) of soil per working day using in-situ bucket mixing (assumed 8-hour shift). However this is very site-specific depending on soil type, equipment, depth, and operator skill. Bucket mixing can often break up and treat clays and other hard soils that are difficult to treat with BOSS systems.

Reagent addition methods
Pumping through the mixing head is not possible in soil mixing completed using bucket mixing. Reagents are typically spread over the surface of the mixing area and mixed into the soils as the bucket mixes the soils. Reagents may be added dry or as a slurry.

Spoil
As with other in-situ treatment methods, a significant soil bulking may take place and a large amount of spoil created. The amount of spoil will vary to 15 % or more, depending on the reagent type and dosage, and whether the untreated soil/waste has air-filled pore space or is saturated. The spoil will have the same characteristics as the rest of the treated material.

Pre-construction investigations
Prior to starting the remedial operation, it is important to thoroughly delineate the treatment area and the depth of treatment. Bench-scale studies are used to determine most effective reagent type and its application. The bench-scale studies can also be utilised to estimate spoil volume.

5.3.4 Quality control

The level of quality control required for bucket mixing is considerably less than that available in auger or BOSS applications. The components typical of a quality control program on a bucket mixing application are given in Table 5.6. As mentioned, it is important to inspect material from various points and depths in the treatment cell to assure that no clumps of untreated soil remain. The maximum acceptable size of a clump of untreated soil should be declared in the S/S specifications. As an example, a 4 in (10 cm) size has been specified on several projects.

Figure 5.20: Bucket mixing of coal tar soils to depth of 15 ft (5 m)

For all in-situ S/S treatments, performance samples are collected and cured in a similar manner to that described in Sections 6.2 and 6.3. The correct care and attention should be given to storage and transportation of quality control samples, which should be stored in a temperature-controlled environment.

During storage, samples should not be subjected to movement or vibration, particularly during the initial 24 to 72 hours after manufacture, when undergoing initial set. Samples should not be transported until they have initially set and allowed to cure to a reasonable strength, usually within the period 3 to 7 days. When being shipped by courier, samples should be properly packed to minimise movement and damage during transportation.

5.3.5 Operational Issues

Equipment
The maintenance and operational requirements are similar to a normal excavation operation, except that the excavator arm and bucket are subjected to continuous contact with stabilisation agents and or contaminated groundwater. This contact can result in an increase in equipment maintenance and repair.

Obstructions
Small obstructions can be removed using the mixing excavator during the soil mixing. Larger obstructions may require a breaker or other means of sizing the obstructions prior to removal.

Spoil
In general, about 15 % to 30 % of the treated volume becomes spoil material requiring handling and disposal. Spoil generally displays a high slump as it is composed of a soil/grout/groundwater mixture. Spoil can be moved by channelling while liquid or allowed to take an initial set, after which it can be handled and disposed in a similar way to soil.

Quality Control Plan	
Information	**Details**
Lines of communication	
Key personnel & responsibilities	
Project staging	Grid map or daily treatment area. Development of a grid to determine reagent application zones. Each zone should be sized for a proportion of the treatment area that can be completed in a few hours (certainly less than 1 day's production).
Layout procedures	GPS, Total Station, Triangulation
Sampling procedures	
Non-conformance procedures	
Daily Monitoring Requirements	
Information	**Details**
Treated panel or volume per 'stroke'	Dimensions (length, width, depth)
	Location
	Unique identification
	Target reagent weight
	Number of mixing strokes
Reagent Addition (dry or wet applications)	Total volume of grout added via flow meter (wet)
	Weight of reagent via weigh-scales (dry)
Mixing energy	Rotary drum (RPM)
	Lift rate (if applicable)
	Grout pressure/flow rate
Sample Collection and Curing	
Information	**Details**
Molds, store, transport, and testing	
Completion of QC Reports	
Information	**Details**
Daily reporting	Daily reporting of site activities, problems, safety issues, progress map (what has been completed), total and cumulative volume treated
Reagent Usage	QC reporting of reagent usage (total/per treated volume, effective area calculations (treated volume calculations), start stop times etc.
Quality Assurance Reporting by Independent Engineer	

Table 5.6: Quality control planning for bucket-based S/S mixing

PART SIX
Quality assurance and quality control

6.1 Quality assurance and quality control during S/S

A great deal of general information on project quality assurance and quality control can be found on the USEPA website using the following link: http://www.epa.gov/quality/qapps.html. The discussion in this section is specific to the use of S/S technology as commonly practiced. Additional information relevant to QA/QC can be found in the following sections: sample collection, including frequency and number of replicates 6.2, sample preparation and curing 6.3, test methods 7.3, bench-scale treatability testing 8.3, pilot field tests 8.4, and selecting samples for testing 8.5.

6.1.1 Quality assurance and quality control during S/S

An important part of any remedial action involving S/S is the construction quality assurance (CQA) and construction quality control (CQC) monitoring activities. The CQA is the task of the responsible engineer and/or the site owner's representative, whereas the CQC is the primary responsibility of the construction contractor with oversight by the responsible engineer:

- CQA is the independent monitoring and verification by the responsible engineer to verify that the remedial works are meeting the agreed technical specification and performance objectives
- CQC is the system of measurement and monitoring activities conducted by the construction contractor to assure that construction will meet the agreed upon technical specifications

During an S/S project, CQC involves monitoring and documentation of all aspects of the S/S operation from preparation of the reagent through mixing of reagent with the target media (i.e. soil or sediment).

To maintain effective CQC the contractor establishes operational metrics (e.g. reagent addition/mixing times etc.) for each aspect of the S/S operation to ensure compliance with the agreed technical specification and performance objectives. The CQA process reviews the CQC to provide the assurance that the operation conforms to the technical specification. This process includes sampling and analysis by an independent laboratory to verify that the performance objectives (e.g. minimum strength, maximum permeability, and leaching parameters) are being met.

6.1.2 The importance of CQA in the S/S design and construction process

Verification of S/S performance involves a comprehensive CQA program that examines each phase of the S/S project, from bench-scale testing through to the full-scale field operation.

An effective CQA program will help control costs by highlighting failure to meet the agreed specification, especially when there is (as there often is) a time-lag between S/S sample collection and receipt of laboratory data. A key component, therefore, is the rapid assessment of construction performance criteria that can indicate problems early on (and which allow for timely field adjustments or corrections), thereby reducing potential costs and minimising potential risks.

The transition from bench to pilot-scale testing is a critical juncture for assessing the metrics from CQC and CQA effectiveness. The careful consideration of full-scale reagent preparation and application is critical in meeting the design objectives and for managing project costs. For example, during in-situ S/S, important reagent properties, such as accurate reagent densities, are critical for defining the appropriate water:reagent ratios, binder delivery rate and required mixing effort.

The effectiveness of *in-situ* mixing is dependent upon factors that include the mode of application, e.g. auger or bucket mixing, and operational parameters, such as the number of auger 'passes' through a column, the rate of auger rotation and duration of mixing required. For *ex-situ* operations, verification of accurate reagent addition rates calibrated to the unit weight of soil being treated and constant monitoring/adjustment of water addition rates to adjust for varying field moisture of soil being treated are necessary to achieve desired product properties.

6.1.3 Development of an effective CQA program

In general, a successful S/S operation is measured by its ability to meet the agreed technical specification. Furthermore, a consistent performance data set, received from both field and the laboratory testing showing achievement of treatment goals, will provide a high level of confidence in the work carried out, and facilitate public and regulatory acceptance of the treatment.

The monitoring of the complex performance factors involved during S/S is dependent on full understanding of the technology and the issues concerning field implementation. A well-designed and successful CQA program will ensure that the project meets its remedial targets, is timely and within budget.

The first step in designing a CQA program involves establishing clear data collection objectives for bench-scale testing to elucidate the key issues faced during full-scale implementation of S/S. This is particularly important as the bench-scale testing of potential binders for S/S often serves as the basis for contractor bidding and payment. Thus, the CQA program must encompass the entire S/S design process from pre-design data collection through full-scale application. Every step of the process must be scrutinised to ensure the work being carried out meets the agreed remedial performance targets.

Pre-bench-scale data collection objectives involve a careful assessment of the sub-surface soil conditions. The soil type encountered, its variability, moisture content and density data will all impact the amount (and type) of reagent(s) to be used during bench-scale testing and during the full field-scale operation. An inadequate assessment of field conditions can therefore have significant impacts on performance and cost if:

- Field soil densities are under or overestimated at bench-scale, then reagent application at full-scale may be too low or high. Low dosages may lead to S/S performance failure, whereas high dosages may lead to

unnecessarily higher project costs and potential leaching failures. As the field density of the treated S/S waste form is determined from the design proportions derived from bench-scale studies, this criterion is directly linked to contractor payment. Thus, great care is needed when obtaining representative field samples for subsequent evaluation
- Dry densities of the soils and reagents are not accurate since they are used as a basis for estimating the amount of binder addition during in-situ application. As this is also related to moisture content, it is important that variations in moisture are fully determined to ensure soil dry density is not under or overestimated
- The variation in untreated soil moisture content is not accurately understood since this will influence the effective water/binder ratios used during full-scale mixing
- The percentage and variability of clay in the soil is not accurately understood as this will directly affect the amount of reagent used and mixing effort required as well as the volume increase due to treatment. This can negatively impact projected cost or cause site-based space logistical issues

6.1.4 CQA objectives

The CQA/QC objectives for S/S can be grouped into two principal phases, design and construction.

Design

The design objectives ensure that the contractor performs S/S in accordance with the technical specifications and contract drawings. This covers the preparation and delivery of the reagents to achieve the agreed performance standards, extending from reagent delivery on-site to the examination of S/S-treated soil. A key aspect of the verification process involves documenting that the approach at pilot-scale is brought forward to full-scale and consistently implemented.

Construction

The construction objectives involve verifying the consistent application of the mix design at full-scale. A robust data set from the laboratory testing of samples will show both consistency and compliance with the design assumptions established during bench- and pilot-scale testing.

The CQA/CQC objectives for S/S sampling are interdependent and require a continuous assessment of S/S performance. This ultimately will provide a high degree of confidence that performance standards are met, particularly when the data generated from testing is only normally available 7 to 28 days (or longer) after the mixing/placement of S/S material in the field.

Monitoring construction objectives during the initial stages of the S/S operations will be critical pending development of a full-scale geotechnical data set indicating consistent and acceptable performance. It is desirable during bench and field pilot testing to obtain data on how critical parameters such as strength, permeability, and leaching, developed over time. Then by preparing a few additional performance molds during remediation, one can compare results at shorter time periods to assure that desired parameters are developing as expected so as to achieve proper results at the designated time period.

6.1.5 Roles and responsibilities in the CQA process

The successful implementation of CQA/CQC and the maintenance of the agreed design parameters is dependent upon key staff appointments:

Engineer or owners representative

The CQA program is the responsibility of the site engineer responsible for the S/S remedial design and/or the designated site-owners representative and may involve the following:

- Reviewing the contractor's CQC program and the quality procedures identified for each aspect of the S/S remedial process
- Coordinating the collection and testing of CQA samples, and their transfer to off-site laboratories
- Preparing treated S/S sample molds for onsite curing
- Reviewing the geotechnical and chemical performance data for completeness and ensuring the results obtained meet the required CQA performance criteria
- Reviewing the geotechnical and/or chemical laboratory analytical data
- Documenting the construction and CQC/CQA monitoring/testing process
- Preparing CQA sampling reports (for the contractor)
- Evaluating CQA/CQC testing performed and recording any relevant observations
- Reviewing the results of CQA/CQC laboratory testing
- Evaluating the testing results obtained that do not meet the agreed performance objectives

Contractor

The contractor is responsible for all aspects of CQC in strict accordance with the technical specifications, including the following:

- Coordinating activities with the site engineer and CQA team leader to meet the agreed schedule and requirements of CQA testing
- Implementing design changes through engineer-approved modifications (based on the assessment of the CQA program)
- Coordinating site surveys and material testing requirements
- Providing recommendations to and/or consulting with the lead project engineer

Geotechnical laboratories

A key part of the CQA program involves third-party laboratory testing by a laboratory that is independent from the contractors CQC program. Independent testing is used to verify that S/S is being conducted in accordance with the agreed technical specification and the S/S products comply with the performance objectives.

Geotechnical testing is conducted during full-scale operation, and two geotechnical laboratories, independent of the contractor's laboratory, are often selected to provide CQA. One laboratory can serve as the primary geotechnical laboratory for routine analyses, whereas the second undertakes additional CQA by analysing replicate-samples (e.g. 5 % of samples) being supplied to the primary laboratory. Generally not all analyses by the contractor's laboratory are repeated by the CQA laboratory. Often replicates for about 10 % of the samples analysed by the contractor's laboratory are selected for analysis by the CQA laboratory for the purpose of validating analyses by the contractor's laboratory.

6.1.6 General categories for the CQA process

Batch plant operations

CQA procedures for the batch plant operations should address the following:

Reagent delivery: this involves inspection of receipts for dry reagents delivered to the batching plant to verify that the correct reagents are used in the agreed mix designs.

Batch plant calibration: including CQA calibration data (from the contractor) to verify precision and accuracy of methods for calculating reagent densities.

Reagent densities: are required for grout-based binders, and are calculated from the water:reagent (or binder) ratio required for pumping. Reagent densities require independent examination as part of the CQA program, via sampling of the production grout from the batch plant and/or by verifying the mix weights/volumes prior to delivery to the S/S mixing equipment. Batch plants often have their mix tank placed directly on a load scale allowing for reagent and water addition based directly on weight. With this approach, measuring grout densities is not necessary. For *ex-situ* mixing, reagents are added on a weight basis to the untreated soil. With a pug-mill for example, the untreated soil is added via a weigh belt and reagents are added via weigh belts or through use of calibrated screw feed from silos.

Water to reagent/solids ratios (W/S): are directly related to reagent density and can be independently evaluated as part of the grout preparation process. During *in-situ* application, the W/S is kept as low as possible to reduce volume increase (the S/S 'swell'), but can be adjusted during application in response to changes in ground or other conditions encountered during mixing. For *ex-situ* mixing, water is added to achieve the desired moisture content for placement/compaction.

Grout pumping rates: can be compared with the results from the pilot-scale tests to verify that the rate of grout delivery is correct and of consistent quality.

S/S mixing operations
During in-situ S/S, the contractor is required to provide CQC documentation covering the field application of the binder system. The CQC documentation required may include the following:

- Grout injection pressure and rate of delivery
- Auger rotation speed and pressure
- The vertical speed of auger advancement
- Mixing tool type(s) and/or diameter(s)
- The number of vertical passes through the entire treated soil column
- The duration of mixing
- The depth of mixing
- The controls over both horizontal and vertical (auger) alignment and S/S column overlap

For *ex-situ* mixing with a pug-mill, CQC documentation may be required for:

- Logs of hourly (sometimes quarter hourly) and daily feed rate of incoming untreated soil
- Daily calibration logs for reagent feed rate
- Transit time for soil in the pug-mill
- Logs of water addition rates
- Horizontal and vertical survey to precisely locate the placement of each days treated soil

For other types of in-situ or ex-situ mixing, similar CQC data should be provided. The objective here is to document the reagent addition(s) and mixing operation(s), and to provide a precise record of where the treated material is located.

The parameters controlling S/S mixing will be periodically (and independently) verified for comparison with the contractors CQC data. As part of this evaluation, discrete freshly treated S/S sample material may be collected from a designated depth within a given S/S column or specified mixing location.

During ex-situ application, samples may be selected spontaneously and at pre-determined time intervals during the mixing process, and subjected to visual inspection for key qualitative parameters, including sample colour and homogeneity (presence of poorly-mixed media) and contamination. These samples can also be cured on-site, and inspected as for the freshly treated material.

The qualitative data obtained will provide supporting data for the overall effectiveness of the operation and for any proposed modifications to the agreed operating parameters, or the CQC program. A detailed discussion of ex-situ mixing methods is provided in Section 4 while a similar discussion on in-situ mixing methods is provided in Section 5.

Post S/S surveying and record documentation
The completed S/S columns or cells are typically the responsibility of the contractor for surveying. The results are provided for CQA review and evaluation to assure that the columns, or cells, were completed in accordance with the technical specification. The survey data obtained are used to verify the following:

- The number of columns or cells completed
- The locations of columns or completed cells
- The location/dimensions and overlap of columns or cells
- The total volume of soil treated by S/S
- The location of columns or cells with different mix designs (if applicable)

The post-S/S survey data may also be used for an analysis of the S/S operation including the following:

- The rate of S/S product production, including unforeseen change to production rates and potential impacts on the agreed time schedule, and/or conflicts with other elements of the construction
- A comparison with the CQC documentation and the identification of potential impacts on operational production efficiencies
- A comparison with engineering estimates and contractor applications for payment: to confirm if the S/S undertaken requires revision of project costs/ payments to the contractor

Assessment of geotechnical laboratory performance-related data
The geotechnical data established during pilot testing can be used to cross-reference the initial full-scale data obtained (e.g. for UCS and permeability). At full-scale the results obtained can be referenced to each mix design used, the soil type treated and any other major aspect of the S/S operation, and applies to both in-situ and ex-situ S/S application.

The data trends established can be used to monitor compliance with performance standards and to establish the 'average' data set for the S/S treatment. In addition to confirming maintenance of performance standards, this data can also be used to support the CQA defined construction and design objectives, including:

- A comparison of geotechnical testing data with defined operating parameters to verify that S/S meets the agreed technical specifications and performance standards required
- An evaluation of data trends to indicate any change in subsurface conditions that may require modification to the mix design
- The re-treatment and/or re-testing of S/S material as the forward projection (e.g. strength development) of performance data indicates a pending failure to meet agreed standards

The secondary laboratory generated geotechnical testing data can be used to corroborate the results from the primary laboratory. The comparison of results can highlight issues (e.g. internal QA/QC protocols or procedures), whereby the following specific actions can be undertaken:

- The QA/QC data from one, or both, laboratories indicates failing treatment properties, or failure to properly follow testing protocols. Additional testing can be performed (by the failing laboratory) on archived samples that are cured for a longer period of time (e.g. 28 days), or by the immediate repeat testing of replicates samples
- If no laboratory testing 'errors' are identified, then replicate samples should be tested by both laboratories at the next sampling interval. If this involves 28-day old samples, then both laboratories will perform the test, on the first available day that both laboratories are able to do so
- If a confirmatory test from the same laboratory fails as a result of this procedure, then additional testing data will be requested from the primary and secondary laboratories, and may be used to identify any systematic laboratory procedural errors
- If confirmatory testing meets the agreed specification, then the failing sampling event will be considered resulting from a random error in one of the sample testing results

6.1.7 Standards and decision processes for S/S performance

Performance standards
The development of appropriate site-specific S/S performance standards is complex and is covered in detail in Section 7.

Decision making framework
In the event of the need to change the specification for S/S, or for any other eventualities where change is required to the remedial action, a pre-defined decision making process should be agreed between the engineer and contractor. This is particularly important when:

- Retreatment is required as performance standards are not met
- Changes are required to the contractors CQC program, to monitor operational metrics more effectively
- Modification is required to the full-scale S/S operation to meet the performance standards

A framework/process for decision-making can be established in the form of conditional 'if/then' statements, designed for implementation at specific times during testing. This approach provides a structured decision-making process for both the engineer and the contractor, and an example is provided in Table 6.1.

Evaluation of the decision making framework

An evaluation of the decision-making process is required as part of the CQA process. This will evaluate whether the appropriate corrective actions meet with the agreed decision–making process when specified performance criteria are not being met. Three general categories for potential S/S performance-related issues arise:

Sampling and analysis: particularly errors during laboratory geotechnical testing resulting in false negative or false positive test results.

Design: substantively different sub-surface conditions from those encountered during pilot testing.

Operational: operating parameters that are not consistent with those demonstrated during pilot testing, or not effective for delivery or mixing of binder to meet agreed performance standards.

Each issue should be reviewed to assess whether the mix design and operational parameters identified will meet the agreed performance standards during full-scale implementation of S/S.

Particular attention is required to potential random and/or systemic errors, such as irregular curing of moldsed S/S samples, or improperly calibrated measuring equipment, respectively. Then the full-scale data set obtained can be compared with the base-line data (established during pilot testing), to enable a review of the changes to agreed procedures, including:

- An evaluation of field sample compliance with SOP's (for sample collection and preparation) and field inspection of archived S/S molds for compromises of sample integrity
- Submission of replicate S/S samples to the second geotechnical laboratory for a comparison with the results from the primary geotechnical laboratory
- Collection of untreated soil samples in areas where S/S fails to meet the agreed performance standard, to evaluate if subsurface conditions have deviated from the baseline assumptions
- Consideration of the effects of variations in operational parameters on geotechnical laboratory test results
- Evaluation of contractor CQC procedures to determine if the established inspection and monitoring procedures should be revised to address deficiencies

Outcome	Decision
Met at 7 days	Considered effective at 28 days
Not met at 7 days	Option to re-treat the failed column/section or re-test samples at 14 days
Met at 14 days	Considered effective at 28 days
Not met at 14 days	Option to re-treat the failed column/section or re-test samples at 28 days
Not met at 28 days	Contractor to remove or re-treat failed column/section or, at engineers option, allow more time for curing

Table 6.1: Example of a simple decision making framework agreed between the site engineer and contractor

Figures 6.1 to 6.3 are example decision-making flow charts to address sampling and analysis, design and operational issues.

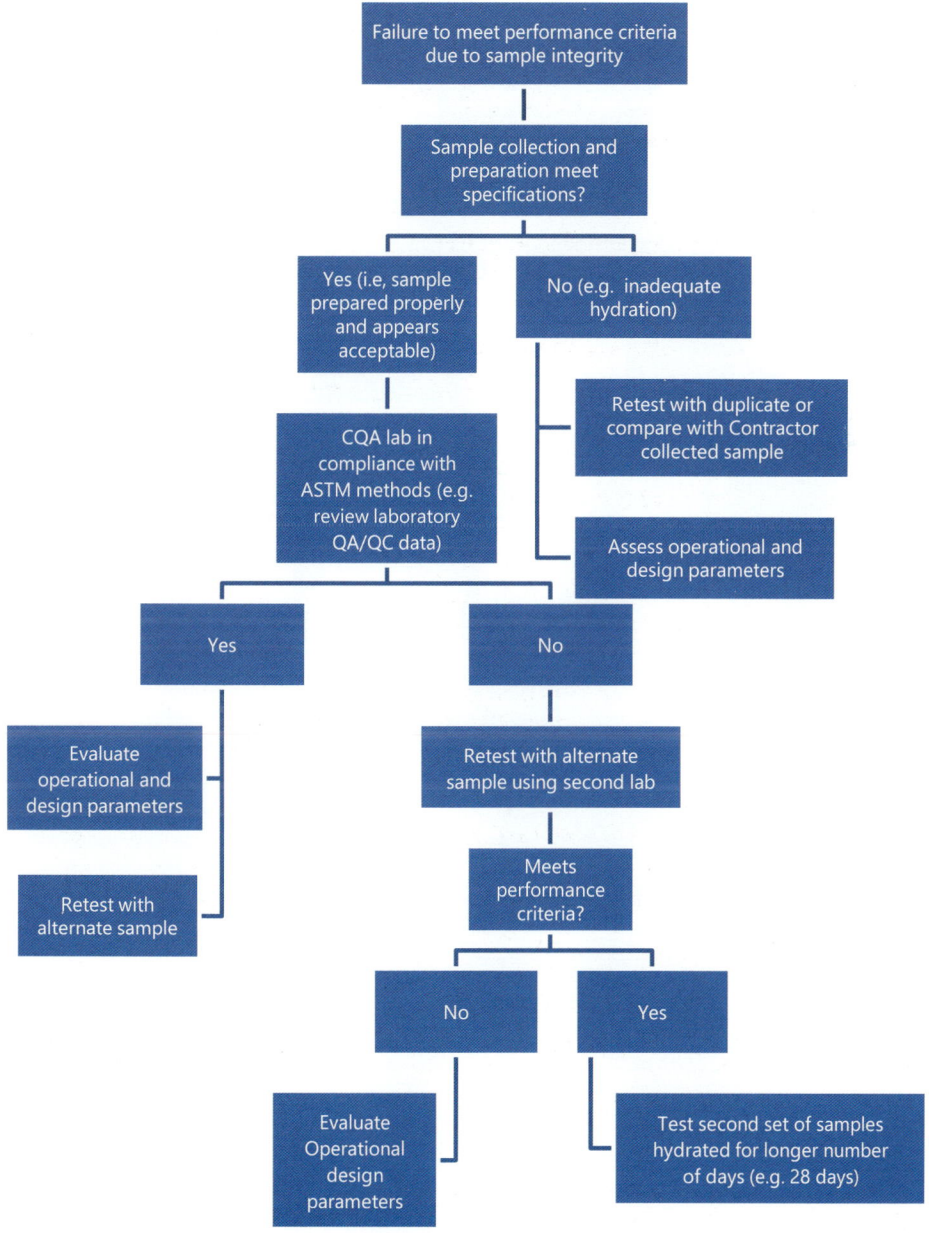

Figure 6.1: Flowchart for sampling and analysis-related decision making

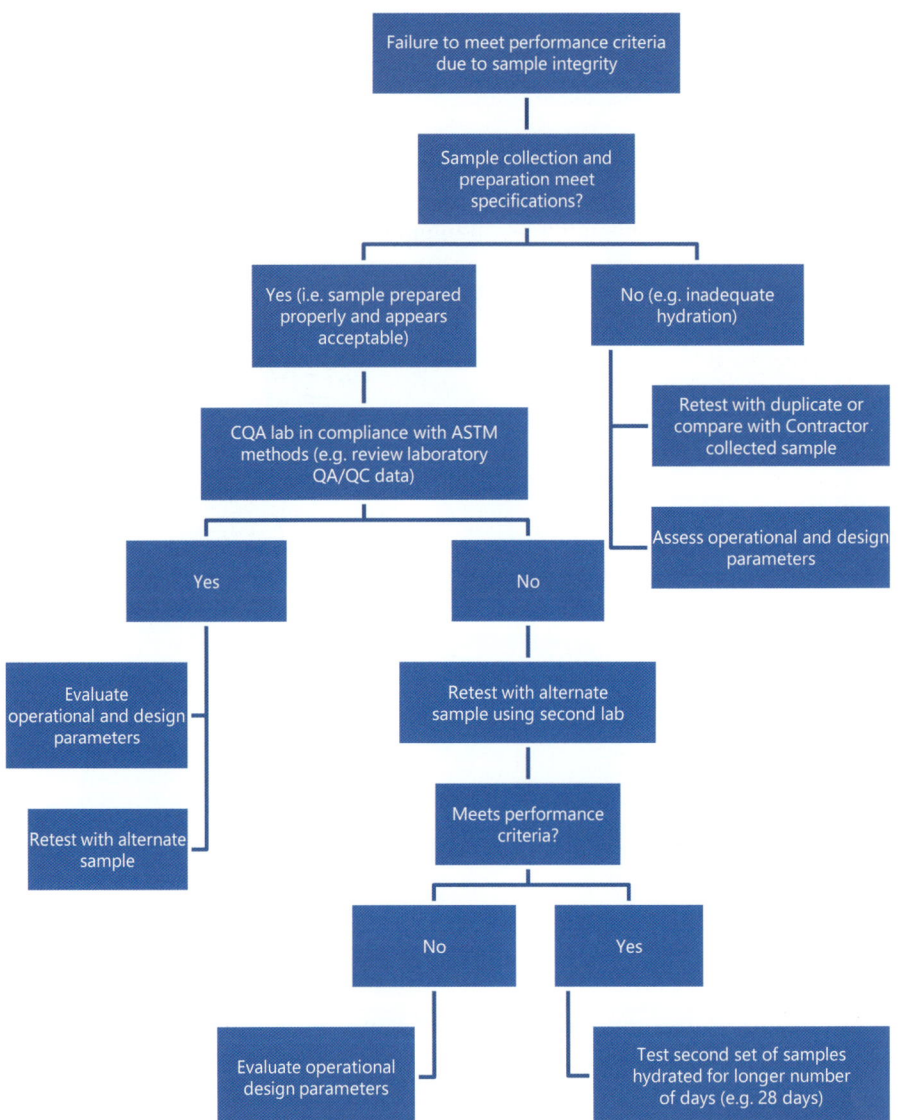

Figure 6.2: Flowchart for design-related decision making

QUALITY ASSURANCE AND QUALITY CONTROL | 131

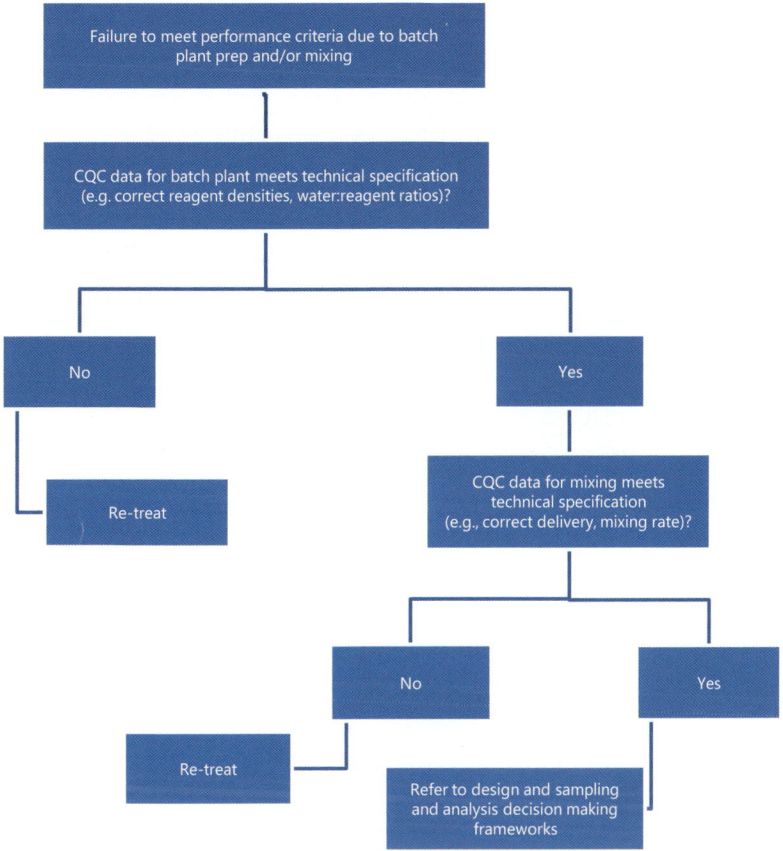

Figure 6.3: Flowchart for operational issues-related decision making

Corrective action

Potential corrective actions can be identified on the basis of an evaluation of decisions and may include the following modifications:

Design mix(es): change could include increasing and/or decreasing the amount of reagent and/or reagent ratios (e.g. ratio of Portland cement to slag or bentonite)

S/S operational procedures: these may include increasing the duration of mixing, the number of vertical auger passes through the column, and adjusting the water to binder ratio(s).

CQA/CQC procedures: can include modification to the frequency of inspections, the types of inspection carried out, on data reporting, and ways in which key S/S parameters are measured by the Contractor.

No Action: a 'no' action result can be considered if a failed S/S column or section will not interfere with meeting the overall project objectives, and appropriate corrective measures have been implemented for future S/S operations.

6.2 Sample collection - the performance sampling plan

The performance-sampling plan is critical to the effective execution of an S/S remedial action and should be carefully prepared and reviewed (and endorsed) by all stakeholders in the project. It should contain, at a minimum, the procedure for collecting samples, the frequency of the sampling events, the test methods that will be employed, the number of sample coupons to be prepared per sampling event, preparation of the specimens (or *coupons*) prior to testing including their curing and storage, and the agreed procedure for handling sample failures.

Quality Assurance is discussed in Section 6.1, whereas performance test methods are discussed in Section 7.3. Sample preparation and curing are discussed in Section 6.3, and handling performance sample failure is addressed in Section 6.4. The frequency of sampling events, the appropriate number of sample coupons to prepare, and the methods for collecting the bulk performance samples are discussed in this section.

Performance sample collection during S/S is not simple and requires careful thought to assure that an adequate number of samples are collected at sufficient frequency to meet the project quality assurance objectives. The precise method of collecting the samples depends on both the equipment used for treatment, and in the case of in-situ treatment, the depth of soil mixing. Failure to carefully consider these factors ahead of time can result in collection of too few performance samples to document treatment performance or the introduction of bias that jeopardises validation that the treatment met the S/S project objectives.

6.2.1 Frequency of sampling

The frequency of performance sampling depends on the overall size of the S/S project, the daily treatment rate, and the variability of soil or COC properties observed, as there is no single answer! However the following guidance is suggested:

- At least once each day of significant production
- Every 500-1000 cubic meters (or cubic yards)
- A large enough number of sampling events to be able to demonstrate statistical significance

Good practice will generally lead to 1 or 2 sampling events during a typical 10-hour work-day, since generally 500-1000 yd^3 (383-765 m^3) will have been treated if there was no significant downtime. During a typical treatment day when 500-1000 yd^3 (383-765 m^3) are mixed, the authors prefer 2 sampling events, the first scheduled about two hours into the treatment day and the second about two hours prior to stopping treatment for the day. This is to help assure that treatment operations were consistent. It is suggested that the exact collection times be varied to both suit convenience and to avoid the treatment vendor knowing in advance when they will occur or in which cell, column, or batch they will occur. It should be noted that anytime the oversight engineer suspects that the treatment may be questionable (soil property change, apparent non-homogenous mixing, issue with reagent addition rate, etc.) an additional sample should be collected.

The frequency of sampling is also influenced by the need to have had an adequate number of sampling events by the end of the project to be able to draw a statistically valid conclusion regarding the overall effectiveness of the treated material properties. This is not an issue for large projects, say over 100,000 yd³ (765,000 m³), but may become important for small projects.

All stakeholders should agree ahead of time regarding what would be a minimum number of samples to conclude that the project was successful. For small projects, the authors have suggested 20 sampling events as a default minimum. By way of reference, an unnamed creosote-contaminated site using a pug-mill for treatment recorded 131 performance-sampling events during S/S of 81,000 tons (74,000 tonnes) of soil; whereas a former MGP site contaminated with coal tar, employing in-situ augers for the treatment of 136,000 yd³ (104,040 m³), carried out 211 sampling events.

6.2.2 Number of samples per sampling event

The rule of thumb is to reasonably determine the minimum number of samples needed per sampling event, and then increase these by 50-100 %. The minimum number of samples should be adequate to provide individual samples for each performance test to be carried out, such as for strength, permeability, and leaching.

Multiple samples will also be needed for testing at different cure times as for the determination of strength by a standard method (typically at 7, 14, and 28 days of age). For the determination of permeability testing this commonly takes place at 14 and 28 days, to allow for maturation.

Leach testing is commonly carried out at 7, 14, and 28 days of age to document improving retention of contaminants within the S/S waste form matrix. It has often been the practice to discontinue testing of longer cured samples once the required specification has been met. However it should not be assumed that early achievement of design properties will be achieved and it is important to ensure that an adequate number of samples for the full extended testing programme are collected, adequately stored and available, including 'reserve' samples (which should always be collected as insurance against unforeseen events).

Additional samples are inexpensive to collect and prepare initially, but there is no option to 'go back' to obtain them at a later date. Thus it is recommended that an additional 50-100 % more samples (reserve samples) be collected than the minimum (to accommodate planned testing).

Occasionally a target property may not have been met after the full cure period, yet the data obtained indicates that the material is maturing slower than expected, but is likely to meet the required specification at a later date. If the property is met at the longer cure time, then the treated material in place is likely just fine. In such a case it may make sense to continue curing from e.g. 28 to 56 days to enable the target criteria to be met. This pragmatic approach to maturation requires additional samples from the reserve to accommodate this extended testing.

It is not unusual for some performance samples to produce results that are outliers. The cause may be apparent, such as when a sample develops a crack and records a high permeability, and sometimes the cause is unknown. It is then desirable to repeat the test on two replicate reserve samples to either confirm the result or to show that it was anomalous and not representative of the S/S material being treated.

An example of a performance-sampling plan might look like the following. However each case is different and should be agreed upon by all stakeholders in advance:

- Strength testing at 7,14, 28 days, 3 samples plus 3 reserve = 6 samples
- Permeability testing at 14, 28 days, 2 samples plus 2 reserve = 4 samples
- Leach testing at 7,14, 28 days, 3 samples plus 3 reserve = 6 samples
- With a total of 16 samples per sampling event, two events/day = 32 samples/day

Some economy regarding the number of samples and sample preparation time can be realised by utilising the same sample molds (container) size for multiple tests. Often the same size/shape sample can be used for strength, leaching, and sometimes permeability testing. This then allows for any individual reserve sample to be a reserve sample for any of the criteria tests. In such a case, 50 % reserve samples should be quite adequate.

A large number of samples can be produced very quickly during S/S and appropriate facilities are required on site for their curing and storage (see Sections 6.3 and 6.4) until they can be shipped to the testing laboratory. It is important that no sample be excessively handled or shipped until it has taken a hard set. This will prevent damage to samples when they are immature. The result of damage to the samples at this stage may mean that the sample fails to meet the design criteria even though the in-place treated material is fine (see Section 6.4 for a discussion on variation and failures). In most cases, setting will occur in 3 to 5 days, but in the authors' experience, there have been exceptions that required a much longer cure time to achieve a hard set, yet the samples met the treatment criteria. Once treatment criteria have been met, the samples may be disposed, though it may be wise to hold them until the end of the project if feasible.

6.2.3 Sample collection methods

There is no one best method for collecting performance samples during full-scale treatment by S/S. A number of successful collection methods exist and these share common key features:

- S/S material must be fresh from the treatment process before any field "setting" of material commences. Ideally, samples should be collected immediately after mixing, as from a pug-mill discharge or from the cell or column as soon as mixing is completed
- Samples must be collected in sufficient bulk for all the anticipated testing required, plus contingency reserve samples
- S/S material should be collected when the S/S treatment process is operating normally
- Fresh S/S material should be examined immediately to assure it appears homogenous
- Staff and equipment are in place to immediately prepare the samples/coupons/molds as required for testing

The quantity of fresh S/S material to be collected should be at least 150 % of what is needed to prepare the samples, and preferably 200 % to provide for rejected material and waste (Section 6.3.4). It should be noted that during sample preparation, the last 20 % of the material obtained may have undergone change in moisture content or particle size due to drying, draining, or during holding and sample preparation. Generally one or two 5 gallon buckets (19 L) are used to temporarily hold the bulk sample while coupons are prepared.

The method for collecting the fresh samples for performance testing depends on the equipment being used for S/S treatment. With a pug-mill, the sample is usually obtained at the point of discharge. Some pug-mills employ an open discharge box and treated material is dumped and removed by front loaders to the location of disposal or into trucks for onward transport. Pug-mill mixers may also discharge onto a stacker belt that lifts and discharges the treated material into a truck for transport. In either case, the bulk performance sample should be carefully collected immediately after discharge into buckets (approximately 70 % full) using a shovel. It is also acceptable to collect the fresh sample from a dump truck at the disposal area, providing the time between discharge and collection does not exceed 15 minutes.

If ex-situ mixing by an excavator in a mix pit is being carried out, then a sample can often collected by the excavator bucket, immediately following mixing. If the mixing is in-situ using an excavator or a rotating tiller (and the depth of mixing is shallow - say about 5 ft or 1.54 m), the excavator can be used to collect the sample from the treated cell.

For in-situ mixing below 5 ft, it is preferable to use a sampler designed for collecting a sample from a specific designated depth. This approach assures that the fresh S/S sample does originate from a designated depth, and is commonly used when in-situ treatment uses augers, injection tillers (such as the Lang), or for excavator mixing. For example, if in-situ auger mixing is used to treat material to a depth of 30 ft (9.2 m), then it is common practice to collect a performance sample from designated intervals, e.g. 5-10 ft, 15-20 ft, or 25-30 ft (1.5-3.1, 4.6-6.2 and 7.9-9.2 m, respectively).

For any specific sampling event, the sample should be collected at the completion of mixing to full depth. Typically one sample is collected with the designated depth specified just before sampling. Across subsequent sampling events, the collection depth will be varied randomly, though a specific depth may be selected more often if it presents a greater concern. Under special circumstances such as at project start-up or a full-scale field test, samples may be collected from multiple depths to assure uniform treatment to full-depth. Figure 6.4 shows a hydraulically operated piston tube sampler being used to sample a freshly mixed auger column. The excavator both inserts and retrieves the sampler, which in Figure 6.5 is fitted with a hydraulically controlled gate that is opened at the desired sampling depth.

A number of sample collecting equipment designs have been employed to gather samples from different depths. However the most successful designs all have common features, which are:

- The sampler is rugged and suitably engineered for sampling at depth in abrasive soil slurry
- A sampling chamber is fitted that is capable of holding a suitable quantity of sample for testing/evaluation
- The sampler can be pushed into, and retrieved from, the S/S material using an excavator
- The sampling chamber remains closed until target depth is reached
- At the target depth, a piston or gate can be opened hydraulically or pneumatically to allow the sample to be collected

Regardless of the method used to obtain the discrete sample, it should be closely inspected to assure that the mixing appears uniform. If clumps of untreated soil larger than a predetermined size (generally 2-4 in (5-10 cm) in diameter are observed, additional mixing of the column/waste form may be required prior to making another attempt to obtain a performance sample.

Figure 6.4: Collection of sample using piston tube inserted by excavator

Figure 6.5: A sampler employing a hydraulic gate

6.3 Sample preparation and curing

Once the required quantity and quality of bulk samples has been collected, it is necessary to prepare the designated number of individual replicate specimens for the sampling 'event'.

The size and shape of the specimen molds is often specified by the test method. For example, a commonly used method to assess strength is: ASTM D 1633 (ASTM), which requires the use of right cylindrical molds with a length equal to twice the molds diameter.

Although this provides choice as to actual size used (2 in or 5 cm, 4 in or 10 cm, 6 in or 15 cm diameter molds etc.), the authors advise that molds smaller than 3 in (7.5 cm) diameter may produce lower strength values, while molds larger than 6 in (15 cm) diameter are cumbersome to handle, and do not yield significantly better results. Thus, molds in the range of 3 to 6 in (75-150 mm) are often the best choice, whereas specimens for permeability or leach testing may, or may not, demand different sizes of specimen.

6.3.1 Sample preparation

The bulk sample obtained from the S/S waste form or from the mixer will require careful placement into the designated sample molds for curing. The objective in loading the sample molds is to try to replicate, as closely as practical, the conditions of the treated material curing in place in the field.

If the material curing in place in the field is moist but not sloppy (as from a pug-mill), it is most likely being placed in the field using compaction equipment such as dozers and perhaps rollers. Thus the performance sample is placed into the molds, in lifts and compacted using a rod so as to mimic, as closely as feasible, the field compaction effort, and avoid loose material and voids. If the material in the field is quite fluid and sloppy (as is often the case for in-situ treatment), then it is placed into the molds and a rod may need to be gently used to eliminate air bubbles (Figure 6.6). In both cases, it is necessary to screen the material placed into the molds to eliminate oversize material For example, a 2 in (5 cm) piece of debris in a 3 in or 7.5 cm diameter molds will result in a high permeability value from the lab test. However this same oversize debris in the field S/S waste form will have little effect on permeability of the monolithic waste form. A common practice is to screen the bulk samples through a 0.5 in (1.25 cm) screen to remove oversize debris (Figure 6.7). In this image, note that all molds are carefully labelled, including the top cap, to indicate the sampling event, location, time and project (Figure 6.8).

Capping cylindrical molds is essential to prevent loss of moisture during curing and to avoid having one end of the cured specimen becoming overly dry and brittle, leading to erroneous test results.

Once the specimens have been filled, they need to be cured to allow S/S reactions to take place. The intent is to mimic as far as is possible, the conditions experienced by the waste form in the field, as curing progresses slowly over days and weeks. If the specimens are allowed to sit out in the field, or if placed in a storeroom under atmospheric conditions they will dry out and not cure properly, and will not be representative of the field-cured material. To avoid this, the freshly prepared samples are placed in a containment environment with very high or saturated atmospheric moisture, but at ambient temperatures.

Figure 6.6: Preparing a 3 in x 6 in (75 x 150 mm) molds

Figure 6.7: A 0.5 in (12.5 mm) screen to remove oversize debris

Figure 6.8: Labelling of specimen molds and cap

Figure 6.9: Sample molds curing on-site in water bath

Figure 6.10: Checking set/strength by a pocket penetrometer

Figure 6.11: Detailed record keeping of properties of cured sample

Figure 6.12: A slump test on an S/S bulk sample

A specially controlled room can be used for storage for curing, but most often a standard picnic cooler containing water saturated towels, or (if the molds are sealed on the bottom), approximately an inch of water is placed in the bottom of the cooler (see Figure 6.9). Plastic caps should be placed over the top of the molds to protect them. The literature provides at least one standard method for curing (ASTM D1632-07).

Stored samples should be checked every day to determine if they have set. Stored samples should not be transported off-site for testing until they have hardened, as unset S/S material can be damaged producing unrepresentative results when tested. Reserve replicate samples can be used to track setting either by physical examination/handling or by the use of a pocket penetrometer (Figure 6.10). Once they have set, the molds can be carefully wrapped and sent to the laboratory for testing at the designated time.

It is very important to ensure that detailed records are kept of all sample collection and preparation activities (Figure 6.11). These records become critical if apparent failures are reported later in the performance sample-testing program.

Details should include: date, time, location and a detailed visual description of the sample, its apparent moisture content and the presence of debris, etc. A slump test (Figure 6.12) is sometimes carried out on a fresh bulk sample as a quantitative descriptor of sample rheology.

It is also good practice to record the condition of cured samples in the field or in the laboratory prior to testing, and to photo-document the samples prior to testing. This is a good assessment of the quality of the prepared cylinders, as poor preparation will induce air voids (Figure 6.13) and other imperfections. A suitably prepared cylinder is shown in Figure 6.14.

Figure 6.13: Poorly prepared field-specimens

Figure 6.14: A well-prepared specimen with minimal air voids

6.4 Variation and failures

The testing of cement-based S/S soils contains many similarities with the testing of civil/geotechnical engineering materials, including commonality of test methods (see Section 7.3). As is normal practice, replicate samples are tested to obtain a statistically valid result. It is normal for performance samples to exhibit variation, which can be caused by a number of factors, including:

- The natural variability of soils and the COCs
- The length of time the sample has been cured before testing
- A non-homogenous matrix, caused by inadequate mixing during treatment
- An irregular rate of reagent application
- Variations in sample moisture content (and the consequent degree of hydration of binder experienced)
- Variations in the way the sample was placed into the molds
- The inclusion of debris in the sample
- Cracking of the sample, due to curing or handling
- Variability in the application of the testing method
- Errors in the reporting of data
- Poor quality control in the testing laboratory
- Differences introduced by methods applied by different testing laboratories

With so many potential factors introducing variability in sample test results, it is important to differentiate between normal variability that does not affect the viability of the remedy and those variations that indicate treatment failure.

6.4.1 Normal variation

As discussed in Section 7 (Developing Performance Specifications) for material treated and disposed on-site, an overall performance goal is established. For example, this might include achieving an average overall permeability value $\leq 1 \times 10^{-6}$ cm/sec ($\leq 1 \times 10^{-8}$ m/sec) for the solidified mass of treated material. However this does not mean that every result obtained from a suite of samples of treated S/S material must individually meet this criterion, in order to meet the treatment goal.

With this in mind, the variability that can normally be expected without undue concern that something is going wrong is summarised from the authors' experience below:

- Strength ≈ 20 % low
- Permeability ≈ half order of magnitude more permeable
- Leaching ≈ 20-50 % target value

With respect to leaching in particular, the amount of normal variation is very sensitive to how low the target was set. For example, if the target number was 282 µg/L (as it was for one Pb contaminated site) then a 20 % variation means 56.4 µg/L, which is not unreasonable. However if the target number was 15 µg/L (as experienced at another Pb contaminated site) then 20 % equals a variation of only 3.0 µg/L. Variability this small is likely to be exceeded given all the possible factors affecting the test result, and would not necessarily indicate remedy failure.

Days Curing/sample	Pb in Leachate (µg/L)					
	1	2	3	4	5	6
14	1.2	1.1	18	7.9	15	15
21	0.6	0.6	NA	7.9	5.3	0.6
28	0.6	0.6	9.1	2.5	3.4	7.8

Table 6.2: Construction performance samples - United Metals site

Table 6.2 presents some construction performance leaching results (SPLP) from remedial actions at United Metals, a battery-recycling site, which was heavily contaminated with Pb. The performance target was Pb $\leq$ 15 µg/L, which is quite restrictive, and the same as the USEPA drinking water standard.

All performance samples met the target value, but there was scatter in the data obtained, with the increase in Pb leaching values at 28 days vs 21 days for sample number 6 explained by micro non-homogeneity within the sample(s) selected for testing. From a 3 in x 6 in (75 x 150 mm) molds, only 100 grams were selected and used for determining the SPLP lead content. Hence micro non-homogeneity, unimportant in the large monolith, impacted the laboratory result.

An example of what is considered normal variation in physical properties is illustrated in Table 6.3, which shows the results from replicate samples from the successfully remediated American Creosote Site in Tennessee. The data show not only variation between "replicate" samples, but also within the range of values from multiple sampling events across the entire course of the remediation. More information about the successful remediation of this site contaminated with creosote, pentachlorophenol and dioxins is available in Bates (2002). Acceptable ranges for variation from target values are often stated in remediation contracts. Development of acceptable, but pragmatic, ranges for variations should be developed during design with input from the key stakeholders in the project.

6.4.2 Recognising failure

If the performance specifications include a target value derived from "an average of all treated" values, it is especially important to recognise failure of the treatment as early as possible.

Performance specifications are often written to include a value to be achieved after 28 days of curing, or longer. However, it is desirable to conduct early testing so as to recognise impending failure. Being able to 'recognise' that failure is occurring from early testing allows one to intervene and apply corrective action long before the 28-day deadline passes, thus reducing the amount of off-spec material produced.

Since some normal variation in data is to be expected, with some data falling below the overall average target value, it may be difficult to determine that treatment is failing. One of the best methods is to compare the values being obtained during the early stages of remediation against those achieved during the bench or pilot-scale treatability study.

UCS at 7 Days (psi)		Permeability, cm/sec (m/sec)	
Sample	Replicate	Sample	Replicate
224	125	6.3×10^{-6} (6.3×10^{-8})	2.3×10^{-5} (2.3×10^{-7})
201	210	7.9×10^{-7} (7.9×10^{-9})	9.9×10^{-8} (9.9×10^{-10})
160	178	1.4×10^{-6} (1.4×10^{-8})	8.2×10^{-7} (8.2×10^{-9})
284	399	1.3×10^{-7} (1.3×10^{-9})	1.3×10^{-7} (1.3×10^{-9})
445	267	8.5×10^{-6} (8.5×10^{-8})	8.4×10^{-6} (8.4×10^{-8})
245	223	4.0×10^{-6} (4.0×10^{-8})	2.9×10^{-6} (2.9×10^{-8})
285	235	2.7×10^{-7} (2.7×10^{-9})	1.2×10^{-6} (1.2×10^{-8})
135	193	1.6×10^{-6} (1.6×10^{-8})	3.6×10^{-6} (3.6×10^{-8})
260	258	6.5×10^{-7} (6.5×10^{-9})	4.4×10^{-7} (4.4×10^{-9})
315	383	4.5×10^{-7} (4.5×10^{-9})	7.2×10^{-7} (7.2×10^{-9})

Table 6.3: Results from duplicate performance samples, American Creosote site

During the later stages of the treatability study (Section 8.3), replicate samples should have been produced to demonstrate the viability of the treatment formula by testing at, for example, 7, 14, and 28 days of age. This database can serve as the reference template to identify anomalous results. For example, the strength of performance samples at 7 days of age can be directly compared with the 7 day data recorded by the treatability samples. Once a significant database has been built using performance samples from the remediation, then this replaces reliance on the treatability database as the template for comparison. Figure 6.15 illustrates how the strength increases over time for typical S/S treated material.

6.4.3 Handling failure

For S/S treatment, handling failure is neither easy nor straightforward. The best approach is to place substantial emphasis on construction quality control and on early detection of any treatment problems.

With some remedial technologies, it is possible to re-treat material that is considered to have failed. However for S/S, re-treatment is often not viable, especially if leaching was one of the performance criteria, because S/S processes usually employ cement or other alkaline reagents and these binders will have irreversibly changed the properties of the soils, especially with respect to pH.

As the solubility of many metals (and some organics) are pH dependent, retreatment, with the addition of more alkaline reagents, may cause the soil to exhibit increasing leaching of, for example, amphoteric metals like Pb and As.

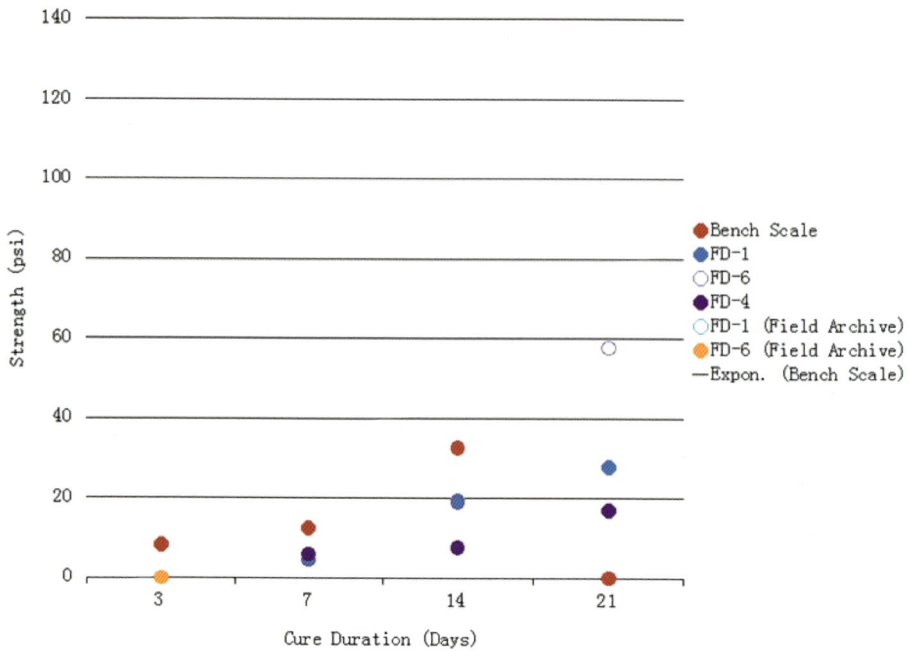

Figure 6.15: Increase in strength by S/S soil over time

If however there are no concerns with leaching criteria, and the failure is one of strength or permeability, it may be possible to re-treat the material. However, even in these cases, the re-treatment formula likely will need to be changed from the original formula and designed to treat the cause of failure. This may require a new bench-scale treatability study and a consequent delay in completing the remedial action.

If the testing results indicated that it did not, or is likely not ever to, meet specifications, then the failing test result must be examined to see if it is valid and not simply an inaccurate value.

As mentioned, factors inducing erroneous data include sample cracking, poor compaction, and the inclusion of debris. Thus, an investigation of the laboratory test procedures used, the quality of samples tested, record keeping, and calculations used, especially dilution factors applied during leachate analysis, is necessary. Then one may have the laboratory undertake repeat testing, using reserve replicate samples that have been kept for such eventualities. If the reserve samples meet the remedial targets and samples from other sampling events have been fine, then it is likely that the earlier 'poor' result was due to a poorly prepared or non-representative sample.

If the performance testing does reveal a treatment failure, then it is important to know exactly where the 'failed' material is located and how much was produced on the same day of production so as to judge the potential significance of this event. For example, if the permeability specification is that the average of all treated material should be $\leq 1 \times 10^{-6}$ cm/sec ($\leq 1 \times 10^{-8}$ m/sec) and that no single sample exceed 1×10^{-5} cm/sec (1×10^{-7} m/sec), then a sample reporting 2×10^{-5} cm/sec (2×10^{-7} m/sec) is outside the specifications and a failure. However, if this represented only half a day's production, out of 60 days of treatment, and is located in the interior of the monolith, this significance is not great. If, however, it is located on the edge of the monolith, it may be more significant. Thus position may weigh into the decision regarding whether to take an action, considering that it may be necessary to rip out a lot of good treated material to reach a piece that is slightly off-spec.

If it is determined that an action is warranted, there are several options. The off-spec material could be removed and sent off-site for disposal. The off-spec material could be removed and retreated, but note earlier discussion regarding potential difficulty in retreating. With either of these options, it may be necessary to remove and manage a significant amount of treated material that does meet specifications, to gain access to the target material. In this case, an alternative option is to leave the material where it is, but take a compensating action. A compensating action may take the form of providing additional isolation from the environment, for example if the material is above the water table a GCL might be added as part of a cap design that did not previously include it. This was done on one of the authors' sites.

If the treated but failed material is readily accessible then one should examine the S/S material before beginning expensive or difficult removal and/or retreatment options. Sometimes the bulk treated material may prove to be within specification, and it was the performance samples that were not representative.

With reference to the remediation of a Pb-contaminated site, performance samples from 2 sampling events indicated a serious failure in strength. The S/S-treated material was accessible and field inspection revealed that some of the material was out of specification, leading to removal and retreatment. However it was also found that a significant portion appeared well cemented and within specification. This material was tested in place using a field cone penetrometer to measure penetration resistance as a surrogate for compressive strength (Figure 6.16).

Figure 6.16: Using a field penetrometer to assess strength of field placed S/S material

Figure 6.17: Excavating top of in-situ column

Figure 6.18: Cores obtained by sonic drilling of S/S material

Although, unconfined compressive strength in the laboratory and penetration resistance/compressive strength in-place in the field, are not directly comparable, the in-place strength results were sufficiently impressive, that it was agreed that this material did not require removal and retreatment. Likewise if the performance sample failure had been of a leaching nature, samples from the actual field placed material could have been obtained and tested.

In another example, in-situ augers were used to treat soils contaminated with coal tars from operation of an MGP. Performance samples indicated that the top parts of a small number of columns (out of the hundreds treated) did not meet requirements for strength. Since the tops of these columns were accessible, it was decided to excavate the suspect columns until obviously well cemented competent material was encountered, then retreat and replace the poor quality material (Figure 6.17).

Handling failure is difficult and expensive in all cases, however the best approach is to avoid failures in the first place by implementing strict construction quality control measures and by using a binder formulation that provides a safety margin, above minimum treatment criteria. If performance samples do indicate a failure, the use of reserve samples to verify the result, augmented by field investigation of the affected material, is encouraged. If failure is verified, then be imaginative in looking for possible solutions. Retreatment is sometimes, but not always, a viable solution.

6.4.4 Coring of in-place S/S material

Coring in-place material has sometimes been proposed for collecting performance samples. It has also been tried in order to conduct evaluations of in-place treated material several years after the remediation. However, this approach can introduce significant bias in the results obtained, and is not recommend, except in special circumstances.

The physical aspect of cutting a core can introduce micro-fractures that increase and may invalidate permeability testing results. These fractures will also have a negative effect on strength. Furthermore, if during the coring water is used as a lubricant and coolant, then the core will be pre-leached before recovery. If coring is dry, then substantial heat is generated which adversely affects the core properties.

The authors have attempted the field extraction of cores several times using differing drilling equipment, but the result is almost always the retrieval of cracked, shattered, bits and pieces of material (Figure 6.18).

This is partly due to the fact that S/S soils generally are weakly cemented materials with typically 2 to 10 % of the strength of structural concrete. Treatment specifications for S/S usually require about 50-100 psi (0.3-0.7 MPa), compared to 2000 psi (14 MPa) or greater for concrete applications such as sidewalks and driveways.

PART SEVEN

Performance specifications for S/S

The development of performance specifications for S/S is a critical step in the planning, design, and implementation of this technology.

As a "technology class", S/S:

- Uses different binders, reagents, and additives (Section 8)
- Employs varying methods of implementation, utilising different equipment (Sections 4 and 5)
- Can treat waste/contaminants with contrasting behaviour (Section 3)
- Relies upon different chemical and physical mechanisms to achieve treatment including: encapsulation/containment, source control, solubility and/or leaching reduction, etc.

Thus, the setting of performance expectations for S/S materials is not a "one size fits all" approach but requires thoughtful site-specific planning. One should avoid simply adopting performance criteria, test methods, and material properties from other sites/projects, from case studies, or misappropriating "regulatory tests" (such as the Toxicity Characteristic Leaching Procedure, TCLP, (EPA method 1311)) without regard for their original purpose.

The inherent dangers in adopting others' approaches or defaulting to promulgated "regulatory tests" without due consideration of site-specific conditions, risk pathways, and objectives can lead to failure to treat, or over-treatment resulting in excessive cost or failure of the remedy, and the prospect of litigation.

To date the majority of available guidance emphasises the technical aspects of S/S technology and appropriateness for use at a given site and waste/contaminant types. More recently, the ITRC published guidance on developing performance specifications for S/S (ITRC, 2011) that includes useful flowcharts illustrating how material performance goals and specifications are developed and implemented to achieve successful S/S.

The basic concepts and logic flow can also be applied to both in-situ and ex-situ S/S with some modifications of the "considerations" column. This ITRC flowchart, adapted for use in this Manual of Practice, is shown in Figure 7.1. Key concepts are used as the basis for this section.

7.1 Setting overall goals for treatment

The key terms for setting goals for treatment were introduced in Section 1.2 and are summarized here as follows:

Remedial goals: overall objectives of the S/S remedy to address the identified risk pathways

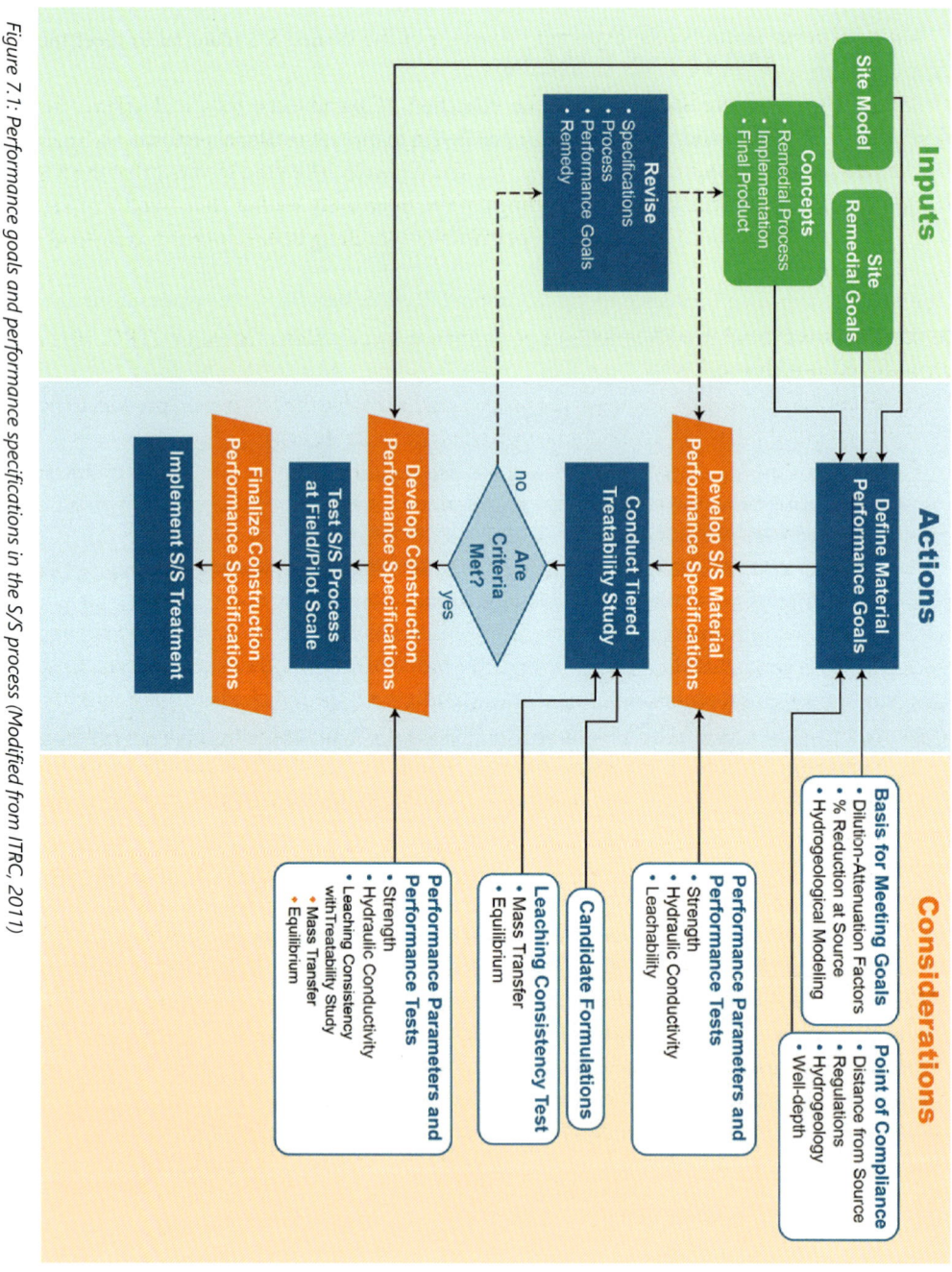

Figure 7.1: Performance goals and performance specifications in the S/S process (Modified from ITRC, 2011)

Material performance goals: expected behaviour of the treated S/S material to meet the remedial goals

Material performance specifications: the collection of parameters, tests and criteria for developing a mix design and the testing of the mix design to meet the material performance goals

Construction performance specifications: the data/criteria to be used to verify that the S/S treated material is consistent with the findings of the treatability testing phase, and that key performance characteristics (e.g. strength, permeability, leaching behaviour) are consistently met during treatment

In addition, the following components of performance specifications (based on ITRC, 2011) are useful to define here:

Performance parameters: the materials properties that enable the S/S-treated product to be 'fit' for its intended purpose (e.g. leaching, hydraulic conductivity etc.)

Performance tests: testing of the S/S-treated material that can result in one or more representative measurements (e.g. leaching evaluation by ANS/ANSI 16.1, hydraulic conductivity by ASTM D5084)

Performance criteria: design values used to demonstrate that acceptable performance has been achieved.

Performance criteria may be established for the remediated contaminated site based on regulatory criteria (e.g. groundwater concentrations derived from water quality criteria or other risk-based standards), or a material parameter/value considered suitable for meeting established remedial or regulatory goals, such as e.g. a maximum acceptable hydraulic conductivity value, or a maximum acceptable leachate concentration for COC's to achieve the remedial goals/targets.

The process of developing and implementing performance specifications for S/S begins with setting site remedial goals. These are developed from the Conceptual Site Model (CSM, see Section 7.2) and provide meaningful and appropriate performance goals to be achieved by S/S.

The remedial goals are intended to address the impacted media, risk pathways and exposure endpoints. The role of S/S is site-specific, ranging from a sole technology (single treatment option) to a component of an overall remedial strategy/treatment train. Examples of the way S/S has been utilised include:

- Ex-situ S/S combined with a composite surface cap and a perimeter slurry wall as a source control to limit mass flux of a mixed-contaminant site (i.e. VOCS, PAHs, pesticides, PCBs and metals), to meet groundwater quality standards at a down-gradient well point of compliance (Peak Oil Site described in Appendix C of ITRC [2011])
- In-situ S/S to encapsulate soils containing coal tar and non-aqueous phase liquids (NAPL) mixed with soil at the former Columbus Georgia manufactured gas plant (MGP site). The outer perimeter of this site was adjacent to a river, where a binder-rich mix was developed to function as a barrier/containment wall facilitating construction of a riverfront park (EPRI, 2003; USEPA, 2009)
- A number of former MGP sites where in-situ S/S encapsulated soils containing coal tar/NAPL to reduce the mass flux to groundwater (EPRI, 2009). Some of these sites used S/S as the sole remedy to manage impacted soil and groundwater, while others included additional components, such as covers

S/S can be used alone or in combination with other technologies (see Section 7.4) and the material performance parameters and tests will reflect the site-specific approach adopted.

By way of example, at sites where the source material is accessible for treatment, S/S can be the primary remedial remedy. Here, a source term defined from leaching tests coupled with a groundwater transport model can verify that the (groundwater-related) remedial goals can be achieved by S/S alone. At other sites where the contaminated material is not completely accessible, S/S may be used as a containment system, whilst recognising that some groundwater treatment may be necessary. In this scenario, a groundwater model and a detailed leaching evaluation may not be necessary, as the impacted groundwater will be managed by other technologies.

However, when using S/S in a treatment train, it is important to consider compatibility between different technologies. For example, if the raising of groundwater pH may result from S/S, the effect of this change on down-gradient treatment design, e.g. chemical oxidation/bioremediation etc. will need to be considered.

Having established site remedial goals and how the S/S technology will be used and implemented, the material performance goals can then be established. Examples of material performance goals include:

- A bearing capacity to support overlying soil cover, structures, or equipment
- Chemical treatment of contaminants to achieve lower soluble species and limit mass flux to groundwater (i.e. stabilisation)
- Containment via reduced permeability (i.e. solidification) limiting groundwater contact
- Containment to prevent migration of NAPL
- Reduced leaching rates (i.e. mass flux) to meet groundwater remedial goals at the designated point of compliance at some distance from the treated material

Most S/S projects will utilize one or more material performance goals similar to those examples presented above to develop a mix design to meet the site remedial goals. The use of other supplemental technologies in conjunction with S/S (e.g. barrier walls, caps), as described in Section 7.4, may also influence the degree to which source materials need to be treated, considering that secondary barriers may be in place as well.

An important concept to recognize in developing a mix design for S/S is that the material performance goals are "goals". That is, they are desired properties of the treated material. The material performance goals will be used to identify the material performance specifications which will include the performance parameters to be evaluated, the performance tests to be conducted, and a preliminary set of performance criteria.

As various mix designs are developed and evaluated and the performance tests (and other material property tests as desired) are performed, it may not always be possible to achieve the performance criteria (e.g. a desired leachate concentration of a specific contaminant) or the reagents or specialty additives required may be too expensive or in limited supply for the quantities/proportions needed based on the treatability testing. The cost and availability of reagents is often taken into account during the choice of mix designs selected for testing to assure that the formulas tested are economically viable for full-scale treatment. Failure of economically viable formulations to fully achieve the desired property does not have to mean that the treatability testing has failed, or that S/S will not work for the site, rather this may be a point in the treatability testing where the material performance goals and how S/S fits into the overall remedial strategy are re-evaluated.

At this point, the introduction of supplementary technologies (e.g. capping, vertical barrier walls, down-gradient groundwater treatment) might also be deemed necessary to meet site-specific goals. For example, a more cost-effective S/S mix design (that does not meet the preliminary leaching criteria) may be selected and used in conjunction with a barrier wall or an "outer ring" (i.e. at the interface between the S/S-treated material and groundwater) as a more cost effective solution to achieve the site remedial goals.

The material performance specifications developed in conjunction with the treatability testing will ensure that the S/S mix design achieves the material performance goals. However, the testing of cured specimens must recognise that under some circumstances extended curing times beyond 28 days are appropriate, especially where contaminated sediments and other contaminated materials require treatment.

The concept of testing materials at 28 days of cure time comes from the concrete industry where it has important construction considerations, but during S/S, this time limit does not fully account for development of desirable properties. Under some circumstances, therefore, the performance evaluation should take many months, e.g. when difficult materials are being treated or diffusion controlled leaching of COC's is being evaluated.

However, it is often desirable for a quicker 'turn-around' of results to meet deadlines or construction productivity targets, whilst also maintaining the material performance goals. Under these (or changed) circumstances, the tests to be used may differ from the material performance specifications developed during the treatability testing. By way of example, the evaluation of the strength development during the treatability study will allow for shorter specimen-cure times in the field. It is possible to "predict" (from the rate of strength-gain achieved) if the treated material is "on-track" to meet the target strength. An example strength-gain plot of bench vs pilot-test data is shown in Figure 7.2. Although the plot shows some variability in pilot-test data (compared to the bench testing), the data was used to make periodic adjustments to the quantity of reagents applied at the particular site in question.

For the evaluation of leaching, it is often not practical to perform leaching tests to monitor construction progress and compliance with the performance specifications. In some cases, therefore, strength and permeability are used to demonstrate consistency with the data obtained from treatability testing, with the leaching evaluation being undertaken later. In other cases, batch leach test or a single-point test at a relatively short curing time (e.g. 7 days) can be useful. However, it is critical that shorter-duration performance tests also be evaluated during bench-scale testing (Section 8.3) to determine what the construction performance criteria should be, as they may need to be different from the material performance criteria as, for example, contaminant flux rate decreases with increasing S/S curing time. ITRC provide example applications in this respect (ITRC, 2011). Other tests that may be useful during construction include slump (ASTM C143), moisture content and grout density. Performance parameters/test methods are discussed further in Section 7.3.

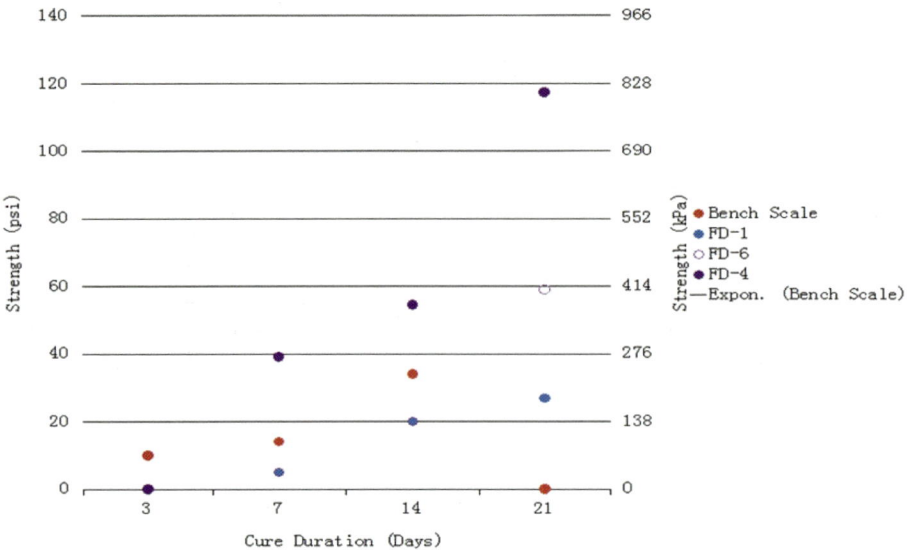

Figure 7.2: Strength-gain of bench vs. field demonstration (FD) data

7.2 The importance of site characterisation and the conceptual model

The development of a remedial strategy and choice of technologies requires a thorough understanding of COC release, the distribution and fate of these contaminants and their likely exposure pathways, and the risks posed to human health and the environment.

A site characterisation program based on specific data quality objectives (DQOs) should be developed in accordance with USEPA and/or state-specific regulatory programs (see: www.itrcweb.org for other appropriate guidance, including accelerated site characterisation: Publications ASC-1 to ASC-4; sampling, characterisation and monitoring: SCM-1 to SCM-3; and for DNAPL: DNAPL-1 and DNAPL-4).

A conceptual site model (CSM) is a vital tool in understanding the site conditions and the applicability of S/S. CSMs typically describe key site features including:

- Hydrogeology
- Flow regime(s) (granular media, karst, or fractured bedrock)
- Stratigraphy (simple or complex with preferential flow zones)
- Aquifer characteristics (hydraulic conductivity, depth of confining units, state groundwater classification)
- Contaminants
- Type, concentration, mass released, toxicity and mobility
- Contaminant distribution in soil, bedrock, groundwater, surface water, sediments, and soil vapour
- Presence of NAPLs
- Groundwater plume geometry, direction, and rate of movement
- Site topography, structures, utilities, the surrounding land use and potential receptors

A sample conceptualisation of a contaminated site and its relationship to identified receptors is illustrated in Figure 7.3.

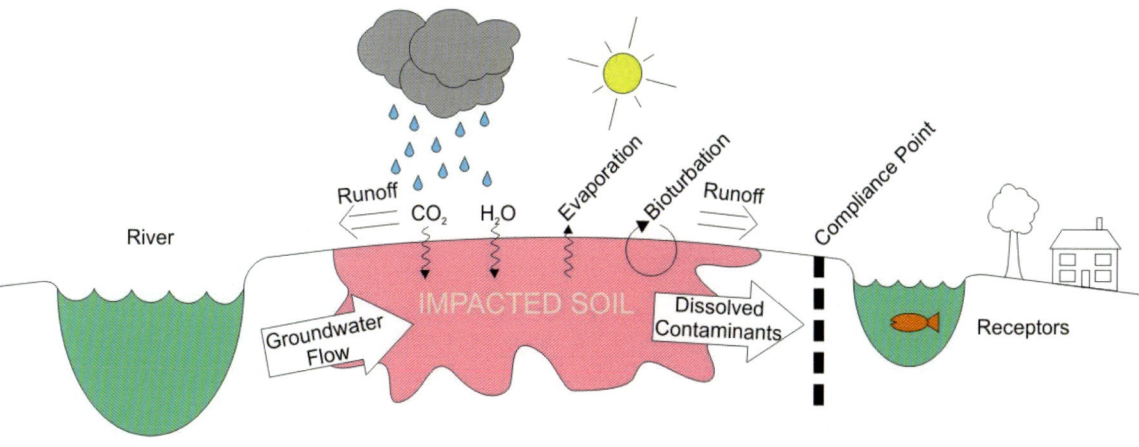

Figure 7.3: A simplified conceptual site model

The CSMs should be carefully prepared in accordance with regulatory (or other) guidance such as ASTM E1689 Standard *Guide for Developing Conceptual Site Models for Contaminated Sites* (ASTM, 2008).

The development of a CSM also helps determine if there are data gaps in the site characterisation, which could affect the site characterisation or the remedial technology evaluation, and what data gaps need to be addressed (e.g. additional site characterisation, pre-design investigation, treatability evaluation).

Once a detailed understanding of the site is obtained, the applicable or relevant and appropriate requirements (ARARs) from a regulatory perspective should be developed, the present and potential future site risk analysis can be performed, and the potential clean-up standards that may apply can be identified.

The risk analysis will identify potential receptors, what COCs and risk pathways are present, and those that may pose unacceptable risk to human health and the environment. The level of detail required is however, site-specific and dependent on the regulatory setting. For guidance one can consult the ITRC Web site (www.itrcweb.org) for appropriate guidance documents concerning contaminated site risk assessment (ITRC Publications RISK-1, RISK-2, and RISK-ALL).

Following the detailed analysis and risk assessment of the site to be treated, the remedial action objectives can be formulated. Then one can identify what technologies should be considered (and how they should be used) and the remedial goals for the site can be determined (see Section 7.1).

An important consideration at this stage is the assessment of site geology and hydrogeology, contaminant-related factors, and the limitations of remedial technology options to meet the remedial targets identified. Sites with DNAPLs and complex geology have been recognised by the USEPA as often impracticable to fully manage, and these should be evaluated as such (USEPA, 1993).

Alternate remedial strategies for complex sites where restoration is technically impracticable or cost prohibitive, may include actions such as exposure control (through institutional controls), source control by removal, treatment, or containment, and where practicable, groundwater restoration by in-situ or ex-situ treatment or natural attenuation.

S/S is a commonly used source control remedy with the treated materials either left in-place, consolidated into an on-site disposal cell, or disposed off-site at a licensed landfill facility. S/S can be used as a stand-alone remedy or in combination with other remedial actions (see Section 7.4).

In order to support the selection of S/S there are a number of factors that should be considered in the site investigation, CSM, and risk analysis phase of the project. Table 7.1 summarises the primary site assessment considerations required, addressing soil, groundwater and contaminant properties and treatment effectiveness. In addition, the method of S/S implementation will have limitations that may be due to soil type (e.g. clay content, rocks, cobbles, debris), presence of debris, obstructions or rubble (which may require removal through pre-processing), or aggressive soils or contaminants (e.g. acidic conditions).

The behaviour of S/S materials in the environment and site conditions which influence treatment by S/S are described in several useful documents including: ITRC (2011), EPRI (2009) and Environment Agency (2004a and 2004b). Waste/solidification reagent interactions and interferences are addressed in PCA (1997 and 1998), Spence and Shi (2005) and Environment Agency (2004b).

S/S Evaluation Factor	Project Phase	Typical Analyses/Observations	Significance
Soil Classification/ Physical Characteristics	Site Investigation, Treatability Study, Pre-Design Investigation	• Gradation • USCS Classification • Atterberg Limits • Moisture Content • Unit Weight • Debris content • Organic content • Porosity • Density • Permeability	• Soil physical properties and their variability at a site have a significant impact on the number of mix designs evaluated/required and the choice of equipment and reagents to perform solidification. • Some physical characteristics such as cobbles and boulders, significant debris content, and very dense soils may eliminate in-situ S/S from consideration.
Soil and Groundwater Geochemistry	Pre-Design Investigation, Treatability Study	• Salinity • pH • Sulfate • Other cement hydration interference parameters as needed • Contaminant metals • Solidification reagent compatibility with site contaminants (adverse reactions)	• Affects cement hydration reactions and long term durability. • Additives may be required to overcome interference mechanisms. • Ettringite (a form of calcium aluminum sulfate) formation due to high sulfate concentration causing excessive swell.

Table 7.1 Typical S/S site assessment considerations

S/S Evaluation Factor	Project Phase	Typical Analyses/Observations	Significance
Contaminant Characterisation and Distribution	Site Investigation, Treatability Study	• History of contamination and location of source areas • Classes of contaminants • Presence and distribution of NAPLs • Distribution of contaminants in geologic strata • Contaminant concentrations and mass released • Contaminant properties (phase, solubility, volatility) • On- or off-site impacts • Extent of soil and groundwater impacts • NAPL physical properties • Leaching behaviour	• Affects applicability of S/S and possible selection of additives for contaminant attenuation or mix-ability improvement. • ISS generally limited to 60-80' below ground surface depending on geology. Grout rheology may need to be modified to reduce mixing shear resistance. • Off-site impacts require access agreements, possible temporary loss of use, and/or utility reconstruction and relocation. • Waste acceptance criteria for off-site disposal of treated material. • Leaching behaviour of untreated soils provides a baseline against which to judge results of treatability studies and full-scale application.
Water Table Depth	Site Investigation	• Water Table depth and seasonal variability • Vadose zone and/or saturated zone contaminant and/or NAPL impacts • Presence of perched water table	• Amount of water available for cement hydration affects grout design. • Up-gradient groundwater mounding due to in-situ solidified soils may be a concern at shallow water table sites. • Penetration of perched water table may create new migration pathways for contaminants.
Hydrogeology	Site Investigation	• Geologic strata including geometry of geologic units • Hydraulic conductivity of impacted zones • Groundwater flow direction & gradients	• Affects contaminant distribution and accumulation zones. • Affects amount of solidification reagent to reduce hydraulic conductivity.
Land Use & Infrastructure	Site Investigation, Feasibility Study, Remedial Design	• Site and surrounding land use (e.g. residential, industrial, commercial, etc) • Site access • Ability to impose institutional controls • Site ownership • Archaeological/historic features • Ecological status • Site activity (actively used or abandoned?) • Presence of active or abandoned buildings and subsurface infrastructure, utilities, foundations, tanks, etc. • Proximity to water bodies • Groundwater use	• Affects ability to leave contaminant mass in the ground. • Active infrastructure and/or buildings can limit accessibility to impacted soils. • Buried remnant structures and utilities may require demolition and removal prior to S/S. • Groundwater pH and alkalinity changes due to ISS need to be considered.
Future Site Use	Site Investigation, Feasibility Study, Remedial Design	• Potential for future contact with solidified soils • Geotechnical properties of solidified soils	• Future intrusive work and future building construction affect mix design, thickness of clean soil buffer above solidified soils, and need for a geo-membrane barrier over solidified soils or a vapour intrusion barrier. • Creating clean corridors may facilitate future underground utility installation needed for site redevelopment and reuse.

Table 7.1 Typical S/S site assessment considerations (continued)

S/S Evaluation Factor	Project Phase	Typical Analyses/Observations	Significance
Receptors/Risk Pathways	Site Investigation, Risk Assessment, Feasibility Study, Remedial Design	• Routes of exposure • Compounds or contaminant classes driving the risk • Severity of the risk • Ability to implement engineering and/or land use controls • Remedial action objectives	• Affects S/S implementation, mix design, need for additional containment or separation of solidified soils, and vapour/emissions controls during remediation.
Aquifer/Waterbody Status	Site Investigation	• Regulatory classification of groundwater and surface waters	• Affects remedial goals, clean-up criteria, and determination of practicability to achieve clean-up criteria

Table 7.1 Typical S/S site assessment considerations (continued)

7.3 Performance parameters, tests/methods

7.3.1 Performance parameters

As described, the performance/properties of S/S treated materials are characteristic of the ability of the waste form to meet its intended purpose. These parameters are derived from the material performance goals established for a site, and can be grouped as follows:

- Environmental performance (i.e. release of constituents to the environment through leaching)
- Characteristics affecting leaching (e.g. permeability to reduce water contact, pH when chemical stabilisation is a desired mechanism)
- Characteristics affecting placement (e.g. strength, bearing capacity, durability, etc.)

The key performance parameters most often used to evaluate S/S materials are strength, hydraulic conductivity, and leaching. The importance of these is summarised below, and discussed in ITRC (2011), EPRI (2009), Spence and Shi (2005), USEPA (1989), and EA (2004b).

Strength: Strength is the ability of the treated material to withstand an applied load. For S/S materials, strength may be used to assess suitability for placement of ex-situ treated material in a defined disposal area (i.e. adequate bearing capacity), to assess the creation of a monolithic mass (i.e. unconfined compressive strength), or as a surrogate parameter to ensure that the chemical reaction of the binder (e.g. Portland cement), and water has occurred and has not been subject to significant interference by the waste constituents or other site-specific geochemical reactions (e.g. sulfate inhibition). Strength can also be used as a surrogate measure of durability, considering that, in general, it would be expected that higher strength materials would be more resistant to deterioration over time.

Hydraulic conductivity: Hydraulic conductivity provides a measure of how easily water can pass through a material. For S/S materials where containment and encapsulation of contaminants (i.e. solidification) is the desired treatment mechanism, reducing the hydraulic conductivity reduces the ability of water to come in contact with the contaminants, and therefore influences the rate of leaching. A reduction of two to three orders of magnitude in hydraulic conductivity through treatment of S/S materials is typically needed to change the mode of water contact from a "flow-through" scenario to a "flow-around" scenario (Atkinson, 1985).

Leachability: Leachability is the materials ability to release a contaminant from a solid phase into a contacting liquid. The leachability of S/S materials can be influenced by contact with an eluent and the chemical reactions involving the S/S binders, and contaminant solubility. Leaching into groundwater is a principal pathway for contaminant release into the environment, and is the key target parameter addressed by S/S.

7.3.2 Performance tests and methods

Numerous test methods exist for evaluating the performance of S/S materials. Some useful references on methods and their limitations include ITRC (2011), Spence and Shi (2005), and EA (2004b). Two categories of assessment testing that are considered in S/S (Cajun and Shi [2005]) are:

- Basic information or index tests, which measure fundamental material properties such as gradation, moisture content, dry density, plasticity, contaminant concentrations, etc., typically applied to untreated material characterisation
- S/S material performance tests which relate to the performance of the treated material

For S/S where solidification is desired, testing is performed on samples cured in cylinder molds, over enough time for waste-binder-water reactions to progress satisfactorily, typically at least 14 days and preferably 28 days, or even longer where necessary (see Sections 6.3, 6.4, for treated sample preparation and curing considerations). When chemical stabilisation is the only desired treatment, curing and testing of physical properties such as strength and hydraulic conductivity may not be necessary.

Strength: Unconfined Compressive Strength (UCS), (ASTM D1633; BS 1924:4; ASTM 2166-06), is the most common test method used in the US for evaluating the strength of molded soil-cement cylinders. This test is the measure of the ability of a monolithic specimen to resist an applied load without breaking, and is only applicable to "cemented" or otherwise cohesive specimens.

If the treated S/S material is a granular product (e.g. resulting from ex-situ metals stabilisation) determination of a UCS test may not be appropriate. A soil bearing capacity test, such as the Cone Penetration Test (ASTM D3441), may be a better test to determine a treated material's ability to support a load after placement. In either scenario, the S/S material should be able to support the loads that will be present after treatment, whether as overlying soil/cover, or future construction or structures.

Other tests that may be used include BS-1924-2 1994, ASTM D 2166-06(cohesive soils), and CBR according to ASTMD1883-99 with prior compaction via ASTM D0698-00 or D1557-00.

A value of 50 psi (0.3 MPa) UCS is often used as a minimum acceptable value in practice, but is really a site-specific determination. A minimum strength value should be selected based on the anticipated loads the treated material will encounter where it is created or placed and to demonstrate if it is a desired property that the solidification reactions occurred. In some situations, post S/S construction may require driving piles through treated material, in which case a maximum strength may be desirable. Also in some cases a much lower strength may be acceptable based upon site-specific considerations.

Hydraulic conductivity: ASTM Method D5084 is the most common US method for the determination of hydraulic conductivity using a flexible wall permeameter. The selection of constant-head or falling-head procedures will be based on the expected hydraulic conductivity range (see discussions in Cajun and Shi [2005] and EPRI [2009]). Typical hydraulic conductivities of solidified soils range from 1×10^{-4} cm/sec to 1×10^{-8} cm/sec (1×10^{-6} m/sec to 1×10^{-10} m/sec), comparable to that of clay materials. The USEPA recommends that the permeability of solidified soils should be at least two orders of magnitude below that of the surrounding soil (USEPA, 1989). For many sites, a hydraulic conductivity criteria of 1×10^{-5} cm/sec to 1×10^{-6} cm/sec (1×10^{-7} m/sec to 1×10^{-8} m/sec) is used. In Europe this is often site-specific: leachability and permeability are combined in a model, as the effects of both influence the mobility of the pollutant. Sometimes very low permeabilities ($< 10^{-7}$ cm/s or $<10^{-9}$ m/s) are specified to compensate for the leaching potential of a pollutant.

Leachability: Leachability tests are typically performed on solidified or stabilised materials to evaluate mass transfer from a solid to a liquid. Extraction-based or equilibrium tests evaluate the mass of material that can be leached under a specific set of test conditions. Mass transfer, or flux-based testing, evaluates the rate of leaching over time from a mass or surface area of treated materials. Detailed discussions of the various test methods are provided in Cajun and Shi (2005), ITRC (2011), and EA (2004b) among others.

Common US extraction tests include the Toxicity Characteristic Leaching Procedure (TCLP, EPA Method 1311) and the Synthetic Precipitation Leaching Procedure (SPLP, EPA Method 1312). TCLP has limited use with S/S materials, unless they are destined for disposal in a landfill and hazardous waste characterisation is necessary. SPLP has been used for S/S materials. However the test crushes the specimen and provides a "releasable" amount of the waste constituent which provides no information on the time rate of leaching. SPLP may be most useful when evaluating the effectiveness of chemical stabilisation treatment mechanisms as opposed to solidification treatment mechanisms.

In Europe various leaching protocols exist, based on granular or crushed samples, or based on monolithic samples (diffusion testing). Many countries adopt these procedures in their own legislation.

Simple and short time batch leaching tests, as the EN 12457 series, are often applied in S/S validation, although the tests are based on granular or crushed material, and in that respect do not simulate the in-situ situation. Moreover, these tests do not give information about the kinetics of the leaching process.

More complex tests, as the column test NEN 7343 (for granular materials) or the tank test NEN 7345 (monolithic materials) are longer time, expensive (up to 8 eluates are analysed), but yield information on the kinetics of the leaching, and can be used to predict the long term leaching behaviour.

Since S/S most often relies on the formation of a solidified monolithic material such that the majority of the groundwater or infiltration flows around the treated material, diffusional release of contaminants from the treated material over time is the controlling behaviour. Therefore, flux-based leaching tests are considered more appropriate as they evaluate the time-dependent release of contaminants. The American Nuclear Society (ANS) test method 16.1, Measurement of the Leachability of Solidified Low-Level Radioactive Wastes by a Short-Term Test Procedure (ANS 2003), is a commonly used flux-based test that was initially designed for cement-stabilised low-activity nuclear waste, but has been adopted for use with other cement-stabilised wastes.

Some emerging EPA leaching methods for S/S materials are being developed and validated currently. These methods characterise leaching behaviours as a function of pH, liquid-solid ratio, and mode of water contact. These emerging leaching methods were initially developed as part of Vanderbilt University's Leaching Environmental Assessment Framework (LEAF; www.vanderbilt.edu/leaching), and a discussion of these LEAF methods is provided in ITRC (2011). References for many of the European Methods can be found in Cajun and Shi (2005), EA (2004b), Kosson *et al.* (2002) and Perera (2004).

How to determine an acceptable level of leaching is a challenge many projects face and has been addressed in many different ways. Three common approaches are:

- Demonstrating a reduction in leaching through treatment
- Determining an acceptable attenuation between the treated material and a down-gradient point-of-compliance
- Use of hydrogeology and contaminant fate and transport modelling using an S/S material source term derived from leach testing and simulating time and distance-dependent groundwater concentrations

Relating leaching performance goals and various methods to assess compliance are provided in ITRC (2011) and the reader is referred to that document for more detail.

Durability testing is sometimes performed as a relative comparison among mix designs. The two most common durability tests, 'Wetting and Drying Testing of Solid Wastes' (ASTM D4843) and 'Freezing and Thawing Test of Solid Waste' (ASTM D4842), are applicable only to materials that will experience these cyclic conditions.

For most S/S applications where material is below the frost line or below the water table, these cyclic conditions do not apply. As described in USEPA (1989), no standards exist for whether stabilised material has "passed" durability testing, although 15 % weight loss is often considered an acceptable amount. According to USEPA (USEPA, 1989) if the durability test results show relatively low loss of materials and retention of physical integrity after testing, the mix design is likely adequate for long-term stability. If the test results show a large loss of material and loss of physical integrity, a different mix design should be considered to provide long-term stability. Further discussion of durability tests is provided in USEPA (1989). However, it is important to note that such durability tests apply only to strongly cemented materials and not to cohesive S/S-treated material. European methods for durability testing can be found in EA (2004b).

7.3.3 Phases of performance testing

Performance testing is conducted in three phases from initial mix design development through implementation verification as follows:

- Bench-scale treatability testing
- Pilot-scale field demonstration
- Full-scale implementation

Bench-scale testing determines one or more mix designs capable of treating the contaminated material and meeting the material performance goals. In this phase, the selected mix design is tested on replicate samples of waste material to ensure that material variability is accounted for. Both physical and chemical containment can be achieved by the S/S using a cementitious reagent, and further discussion of the bench-scale treatability study is provided in Section 9.3.

Pilot-scale demonstrations determine if the mix design applied at the field-scale (using the full-scale mixing equipment) will produce results consistent with those obtained in the bench-scale testing. This phase also allows for further assessment of the physical and chemical variability of the waste material and how the selected mix design performs, and is used to further develop and optimise the mix design (if needed) and the construction methods for full-scale implementation.

Tests other than performance tests may be used to determine if treated materials are "consistent" with those produced in the laboratory. These consistency tests are typically established in the bench and pilot phases for use during the construction phase, and can be performed on freshly mixed material or cured specimens.

The real-time testing of freshly mixed material is used to identify significant variations in material properties that can affect performance test results, such as grout density, slump, and homogeneous mixing. Short-term tests on specimens as they cure may include strength-gain rate using a pocket penetrometer or UCS tests at intervals such as 3, 7, 10, 14, 21, and 28 days to evaluate strength-gain rate, and/or visual observation of bulk field-cured samples (e.g. 5-gallon (19 L) bucket). Abbreviated leaching tests, or leaching tests performed with shorter cure durations (i.e. 7 or 14 days) may also be considered in this phase.

Full-scale implementation of S/S uses a combination of performance and consistency tests to monitor treatment progress in real-time and conformance with performance criteria.

As mentioned, leach testing may be limited to the bench-scale and pilot-scale phases with strength and permeability being used as surrogate measures for leachability during full-scale S/S. As long as the target strength and permeability are met, then the leaching performance should be met as well, although an exception to this may be for metals stabilisation, for which short-duration extraction tests, such as SPLP, may be needed to demonstrate chemical stabilisation has been achieved. Other indicative parameters (for chemical stabilisation) that can easily be measured in the field include the pH of the treated material.

7.3.4 Performance criteria in specifications

Performance criteria statements should be written to recognise that some limited variability in test results is often acceptable without compromising the overall success of the remedy (see Section 6.4). This approach relies on an assessment of failures in light of the overall aggregated performance of the treated material and the potential contaminant flux from the entire S/S mass to the environment, and not just from one sample point.

Incorporating tolerance intervals in the performance criteria specification can provide flexibility to field personnel, whilst recognising that variable field conditions are the rule, rather than the exception at S/S sites. An example of tolerance-intervals for strength, for use in performance criteria follows:

An average of all performance samples must not be less than 50 psi (0.3 MPa), no individual sample shall be less than 40 psi, and no more than 20 % of the performance samples shall be less than 50 psi (0.3 MPa).

The site owner/representative/designer and regulator should develop a consensus on what is acceptable based on the agreed remedial objectives whilst considering the location of failing test results, among other factors.

For example, a few failing test results that meet the lower tolerance limit that are randomly scattered throughout the S/S area may be less of a concern than a cluster of failures in one general area (possibly indicating differing soil/contaminant conditions, inadequate reagent batch preparation, or inadequate mixing) or along the outer edge (where groundwater or infiltration exposure may be greatest).

Retreatment can be expensive, is difficult to effect on previously cured materials, and may not result in significant improvement due to pH increases and/or solidifying broken cemented material. Thus the value of replicate specimens for retesting if a failing result is obtained cannot be under-estimated at this juncture (Sections 6.2 and 6.4).

7.4 Impact of other protective measures

It is increasingly common to combine S/S treatment with other protective measures as part of a package of technologies to remediate sites at lower cost while still achieving acceptable levels of protection. The objectives of these protective measures are often to isolate the S/S material from surface and groundwater.

Isolation from surface water can be achieved via low hydraulic conductivity caps (e.g. flexible membrane liners (FMLs), geo-synthetic clay liners (GCLs)), whereas for groundwater, vertical cut-off walls keyed into an underlying low permeability strata promote groundwater flow around the site (e.g. bentonite slurry walls) or vertical drains to capture groundwater flow coming into the site (i.e. vertical drains).

Most S/S remedial projects to date have used S/S to treat all of the contaminated media of concern. Incorporating multiple barrier systems with S/S can affect the required performance specifications for the S/S treated component. By incorporating such measures, groundwater flow through the site will be very low and any potential off-site migration will be due to slow diffusion limited transport. Methods to assess the impact of multi-barrier systems on contaminant migration are outlined by Lake (2013).

Examples exist on how both the extent of contaminated media and the S/S performance specifications can change for a multi-barrier system. The Whitehouse Waste Oil Pits site in Florida, USA (EPA, 2011), was remediated by using the following approach:

- Use of shallow in-situ S/S to treat the top two lifts of soil and oily acid sludge (approximate the upper 3 ft [0.9 m]) to create an S/S sub-cap of low hydraulic conductivity
- The installation of an RCRA cap (containing both a GCL (geo-synthetic clay liner) and an FML (flexible membrane liner))
- Construction of vertical confining walls, bentonite slurry walls, around the contaminated media, keyed through multiple underlying clay layers at depths of approximately 65 ft (19.8 m)

According to EPA (2011) "All substantial elements of the physical construction of the remedy were completed on May 4, 2006, and initial groundwater monitoring results indicate the remedy is effective in isolating groundwater contamination." Since the remedy incorporated a multi-barrier approach via a HDPE/GCL RCRA cap and vertical confining walls that were keyed into a confining clay (see Figure 7.4) the only leaching specification was that the S/S treated material did not leach contaminants of concern (COCs) in excess of the untreated material. A permeability of 1x10-6 cm/sec (1x10-8 m/sec), and a minimum strength of

Figure 7.4: Installing a slurry wall, Whitehouse, Florida

0.3 MPa (50 psi) for the S/S treated material were retained as specifications.

Another such example of utilising a multiple barrier approach is the Brunswick Wood Preserving Site in Georgia, USA. This disposal area contained several lagoons in which wood preserving wastes such as creosote and pentachlorophenol (PCP), with dioxins, were disposed.

The remedial approach called for excavation and ex-situ S/S treatment of the top 3-4 ft (1-1.2 m) of contaminated soil and sludges, and the return of these S/S treated materials as a sub-cap over the lagoons (shown in Figure 7.5). The formula used was 10 % Portland cement and 10 % fly ash, by weight, to the excavated soil/sludge.

Vertical confining walls keyed into the underlying confining clay at a depth of about 60-70 ft (18.3-21.3 m) were constructed using both trench/backfill, like those at Whitehouse and by using a Bauer panel cutter (Figure 7.6). The cap over the S/S treated material contained a GCL, but in this case no FML. The S/S treated sub-cap contained a hydraulic conductivity specification of 1×10^{-6} cm/sec (1×10^{-8} m/sec) and a strength specification of 100 psi (0.7 MPa).

Due to the 3-4 ft (1-1.2 m) of low hydraulic conductivity sub-cap and the vertical confining walls, the leaching specification for the S/S treated material was that the S/S treatment did not significantly increase the leaching potential for COCs in the S/S treated material.

Although the strength specification of 100 psi (0.7 MPa) may seem high, it was easily achieved by the formula designed to meet the permeability requirement, since the site soils were high in sand and silt content.

Figure 7.5: Constructing a S/S sub-cap, Brunswick

Figure 7.6: Constructing a vertical wall using a Bauer Panel Cutter, Brunswick

Figure 7.7: Oily water emulsion at Stauffer

A third and very challenging example can be found at the Stauffer Chemical site in Florida, USA. This site formulated insecticide and herbicide products into dusts, granules, pellets, and liquids that were packaged for commercial distribution. Compounds such as kerosene, xylene, clay, and diatomaceous earth were also handled on site.

The primary remedial approach was on-site excavation of contaminated soils and placement into an RCRA equivalent lined and capped containment cell located on-site.

During the remedial operation, a substantial amount of groundwater was collected and treated to facilitate the excavation. Upon preparing to decommission the temporary water treatment system, it was revealed that a substantial amount of pesticide and herbicide contaminated oily water emulsion had collected as a residual in some of the water treatment tanks (Figure 7.7). By employing treatability studies, it was determined that these contaminated emulsions could be solidified using an excavator mixing on an existing concrete pad.

The formula used was (by volume): 100 parts of oily water, 200 parts site sand, 25 parts proprietary oil absorbent, 5 parts cement, and 2 parts water absorbing polymer. In this case, solidification alone was used without stabilisation, to change the oily contaminated water emulsion into a soil-like material. It was not necessary (though it would have been possible) to turn the waste into a monolithic material.

However by converting the emulsions into a soil-like material (Figure 7.8), it was feasible to dispose this waste, as a solid, in the RCRA C-equivalent containment cell that was already being constructed on-site for other contaminated soils.

Figure 7.9 shows installation of the FML as part of the construction of the RCRA C-equivalent disposal cell. There were no specifications for hydraulic conductivity, strength, or leaching required for the solidified treated material, due to its being placed into a fully containing RCRA cell.

These three examples illustrate how the integration of multi-barrier systems with S/S can substantially change the performance specifications required from the S/S treated material. The level of change to the performance specifications will, however, vary depending on the given project and regulatory environment and the overall performance required for the site.

Figure 7.8: Solidifying oily water emulsion Stauffer

Figure 7.9: Installing an FML at Stauffer

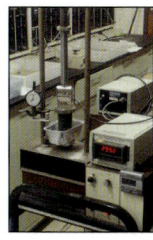

PART EIGHT

Developing effective S/S formulations

8.1 Common reagents

Common non-proprietary products used for S/S include Portland cement, quick and hydrated lime, cement kiln dust, limekiln dust, fly ash, bottom ash, magnesium oxide, phosphate (in various forms), sulfides, and carbon-based materials.

Combinations of these reagents are available under various proprietary names, but this section will discuss the basic generic reagents. Information on sourcing reagents is provided in Section 8.2.

8.1.1 Portland cement

Portland cement is made by heating a mixture of limestone with clay, shale, sand, iron ore, bauxite, fly ash and/or slag to about 1450°C in a kiln. The raw materials form a "clinker" which contains 70-80 % calcium silicates, with over 60 % of these silicates as tricalcium silicates, and trace amounts of tricalcium aluminate, tetracalcium aluminoferrate, magnesium oxide, calcium oxide, and alkaline sulfates (e.g. Na_2SO_4 and K_2SO_4).

The cooled clinker is mixed with calcium sulfate (gypsum) and pulverised. It is considered a hydraulic binder as it sets and hardens by chemical interaction with water. Portland cement retains its strength and stability after hardening, even under water.

Portland cement relies on both pH control and chemical reactions to chemically immobilise metals. pH control is important because many metals form low solubility hydroxides and stabilised materials which have a leachate pH in the range of 8 to 11 and typically have low levels of leachability. The efficacy of Portland cement as a binder for use in S/S is discussed in the companion volume of this work (Hills *et al.*, 2014).

Many of the metals can also isomorphically substitute into the calcium silicate hydrates (C-S-H) and other phases, which form when Portland cement reacts with water. The incorporation of the metals into these compounds decreases their availability. Furthermore, the cemented matrices can encapsulate free-phase organics or organics sorbed to waste particles, and there is evidence that in some circumstances solid–organic salts are precipitated and encapsulated in the hardened S/S waste form.

Portland cement is delivered as a dry powder, in bulk pneumatic trailer, bags, or 1-ton jumbo bags. Due to its reaction with water, Portland cement must be stored in a manner to prevent contact with water in liquid or vapour form. The interaction with water vapour means that Portland cement will degrade over time and its storage on-site is time restricted (less than 1 month). Portland cement that has become lumpy has lost a significant portion of reactivity and should not be used for S/S.

In bulk, the Portland cement powder is pneumatically transferred to silos, equipped with baghouses to contain dust during the transfer, for storage. The silo should be well sealed to prevent contact with liquid water. The surface area to volume ratio for Portland cement in the silo is low, minimising exposure to water vapour. Portland cement delivered in bags or jumbo bags should be stored on wooden boards above a plastic barrier and covered with plastic.

Portland cement can cause health effects by skin contact, eye contact, or inhalation. Risk of injury depends on duration and level of exposure and individual sensitivity. Portland cement contains traces of calcium oxide (that is corrosive to human tissue), crystalline silica (that is abrasive and can damage lungs), and hexavalent chromium (that can cause allergic reactions).

8.1.2 Quicklime and hydrated lime

Quicklime (CaO, or burnt lime) is produced by the calcination of limestone above 825°C. Due to aluminium, iron and/or silicate impurities in the limestone, the quicklime is often clinkered during the heating process, and thus grinding is often employed to produce a dry powder. Hydrated lime ($Ca(OH)_2$, or slaked lime) is produced by the reaction of quicklime with water (otherwise known as slaking). The slaked lime is then thermally dried at approximately 200°C and ground to a fine powder.

Quicklime and hydrated lime (Portlandite) impart pH control to chemically immobilise metals. The calcium hydroxide gel, formed by reaction with water, can also provide a medium for the sorption of organics of low water solubility.

Quicklime is delivered in bulk as both pebble and dry powder, by pneumatic trailer, 50 or 90 pound bags (23 or 45 kg), or 1-ton bags. Due to its exothermic reaction with water, quicklime must be stored dry. The interaction with liquid water can be both exothermic and violent, and the reaction with water vapour means that quicklime will degrades with time, and its storage on-site should be brief (less than 1 month). However, unlike Portland cement, quicklime that has reacted with water can be used for S/S purposes, though its efficacy is slightly reduced.

Hydrated lime is delivered as a dry powder, in a bulk pneumatic trailer, 50 or 90 pound (23 or 45 kg) bags, or 1-ton bags. Although hydrated lime contains water, it should be stored in a manner to prevent contact with liquid water. This is because water will degrade its handling properties. However, unlike quicklime or Portland cement, hydrated lime can be stored on-site for several months, and when in contact with water, can still be used for S/S purposes, though its handling properties as a powder will be affected.

In bulk, powdered quicklime or hydrated lime are pneumatically transferred to silos, equipped with baghouses to contain dust emissions during the transfer to storage. The silo should be well sealed to prevent contact with liquid water. Bulk pebble quicklime can be delivered in dump trailers and should be used immediately upon delivery. Quicklime (powder or pebbles) and hydrated lime delivered in bags should be stored on wooden boards above a plastic barrier and covered with plastic.

Quicklime can cause serious harm to health by dermal or eye contact and inhalation. Its very alkaline and energetic nature with water can pose a serious risk of injury. Similarly, hydrated lime is hazardous due to its alkalinity. Figure 8.1 shows hydrated lime being added to waste in a mix pit.

Figure 8.1: Hydrated lime added to waste in a mix pit

8.1.3 Cement kiln dust

Cement kiln dust (CKD) is the fine-grained powder removed from the exhaust gas of a cement kiln by its air pollution control devices. The CKD contains unreacted raw materials from the Portland cement production and is largely recycled back into the production process.

The CKD that is not recycled is typically disposed of to landfill; some is also used beneficially for S/S. CKD contains calcium silicates (though more dicalcium silicate than tricalcium silicate) and calcium carbonates, with trace amounts of tricalcium aluminate, tetracalcium aluminoferrate, magnesium oxide, calcium oxide, and alkaline sulfates.

As with Portland cement, CKD relies on both pH control and chemical reactions to immobilise metals. The formation of low solubility hydroxides along with isomorphic substitution into the calcium aluminosilicates forming on contact with water, decreases the leachability of certain metals. The cementitious matrices can also encapsulate free-phase organics or organics sorbed to waste particles.

CKD is delivered in bulk as a dry powder in bulk pneumatic trailer, or 1 ton bags. Like Portland cement, CKD must be stored dry to prevent degradation on contact with water in liquid or vapour form. Site storage should be brief (less than 1 month), and CKD that has become lumpy will have lost a significant portion of reactivity and should not be used for S/S purposes.

As for Portland cement, CKD is pneumatically transferred to silos, equipped with dust control and sealed to prevent contact with water. CKD delivered in bags should be stored on wooden boards above a plastic barrier and covered with plastic.

CKD can cause serious harm via skin and eye contact, or by inhalation. The risk of injury depends on the duration and level of exposure, but as CKD contains traces of calcium oxide (that is corrosive to human tissue), crystalline silica (that is abrasive and can damage lungs), and hexavalent chromium (that can cause allergic reactions), exposure should be minimised.

In the United States, CKD is categorised by EPA as a "special waste" and has been temporarily exempted from federal hazardous waste regulations under Subtitle C of the Resource Conservation and Recovery Act (RCRA). The US EPA is in the process of developing standards for the management of CKD and has published a set of proposed Subtitle D (i.e. non-hazardous, solid waste) regulations to govern CKD management. In Europe, CKD falls under 'REACH' regulation, and can be re-incorporated into cement products or be disposed appropriate to its hazardous/non-hazardous nature (see: www. Defra.gov.uk/environment/chemicals/reach/).

8.1.4 Lime kiln dust

Lime kiln dust (LKD) is the fine-grained powder removed from the exhaust gas of a lime kiln by its air pollution control devices. LKD is largely unreacted or partially reacted limestone; it is generally recycled back into the quicklime production process. The LKD not returned to the production process is sold for beneficial reuse, such as a reagent for S/S. LKD is a combination of calcium carbonate and calcium oxide, with traces of aluminium, iron, and silicate minerals.

LKD relies primarily on pH control to chemically immobilise metals. The calcium hydroxide gel formed by reaction with water can also facilitate sorption of organics with low water solubility.

LKD is delivered as a dry powder, in bulk pneumatic trailer or 1 ton bags. It is reactive with water, and should be dry-stored. LKD that has come in contact with water can be used for S/S purposes, though its handling properties as a powder will be affected.

In bulk, LKD is pneumatically transferred to silos, equipped with baghouses to contain dust during the transfer, for storage. The silo should be well sealed to prevent contact with water. LKD delivered in bags should be stored on wooden boards above a plastic barrier and covered with plastic. Similar to quicklime or hydrated lime, LKD's highly alkaline nature can cause serious harm to health.

8.1.5 Fly ash

Fly ash is produced in boilers burning pulverised coal and is removed from the boiler exhaust gases by electrostatic precipitators, baghouses, or scrubber systems. Fly ash is formed either by the agglomeration of mineral inclusion from the coal into hollow spheres (cenospheres) or by the condensation outside of the combustion zone of mineral inclusions vapourised in the combustion zone.

Fly ash is a heterogeneous mixture of silicon, iron, and aluminium oxides, carbon, and calcium oxide. Because it is collected from the gas phase, fly ash particles are typically smaller than 200 μm in diameter and the mean particle diameter for fly ash is often less than 50 μm.

Class C fly ash (produced from the burning of lignite or subbituminous coal) relies on both pH control and chemical reactions to chemically immobilise metals. pH control is provided by the CaO in the Class C fly ash. But many of the metals can also isomorphically substitute into the calcium aluminosilicates which form as the fly ash particles interact with calcium hydroxide, either produced from the reaction of Class C fly ash with water or by calcium hydroxide added in conjunction with the Class C fly ash.

Class F fly ash (produced from the burning of anthracite or bituminous coal) relies primarily on chemical reactions to chemically immobilise metals. Class F fly ash typically contains less than 20 % CaO and little to none of this is actually present as CaO ("free" lime). Metals can be isomorphically substituted into the calcium aluminosilicates which form as the fly ash particles interact with calcium hydroxide that is added in conjunction with the Class C fly ash.

High carbon fly ash (with greater than 10 % loss on ignition) can be used for the stabilisation of organics. The carbonaceous component of these high carbon fly ashes can sorb free-phase organics or organics sorbed to waste particles, reducing their leachability.

Fly ash is delivered as a dry powder, in bulk pneumatic trailer or dump trailers. Pneumatic trailers can be off-loaded to silos or pneumatically transferred to covered stockpiles. Dump trailer loads are typically dumped into stockpile areas and covered or dumped at the point of use and used immediately.

Due to its fine nature, fly ash is very dusty and may be reacted with water to reduce its dusting. In bulk, fly ash is often pneumatically transferred to silos, equipped with baghouses to contain dust during the transfer, for storage. The silo should be well sealed to prevent dust emissions during transfer.

The fine crystalline silica within the fly ash can cause damage to lungs if inhaled. A temporary fly ash stockpile is shown in Figure 8.2. The cover was temporarily removed so that a front loader could access the stockpile.

Figure 8.2: Fly ash stockpile on a treatment site with its cover removed

8.1.6 Bottom ash

Bottom ash is also produced in boilers burning pulverised coal. It consists of spherical or angular agglomerated mineral inclusion from the coal, the particles of which were too large to be carried in the flue gases.

Bottom ash is a heterogeneous mixture of silicon, iron, and aluminium oxides, carbon, calcium sulfate, and calcium oxide.

Bottom ash produced from burning coal in the presence of limestone relies on both pH control and chemical reactions, but primarily the latter, to chemically immobilise metals. pH control is provided by the CaO in the bottom ash, but many of the metals can also isomorphically substitute into the calcium aluminosilicates. These form as the bottom ash particles interact with calcium hydroxide, either produced from the reaction of bottom ash with water or by calcium hydroxide added in conjunction with the bottom ash. Bottom ashes typically contain less the 20 % CaO and little to none of this is actually present as CaO ("free" lime).

High carbon bottom ash (with greater than 10 % loss on ignition) can be used for the stabilisation of organics. The carbonaceous component of these high carbon bottom ashes can sorb free-phase organics or organics sorbed to waste particles, reducing their leachability.

Bottom ash is delivered as a dry solid in dump trailers. The trailers are typically dumped into stockpile areas and covered or dumped at the point of use and used immediately. Due to its sandy particle size, bottom ash has little dust associated with it.

The fine crystalline silica within the bottom ash can cause damage to lungs if inhaled.

8.1.7 Magnesium oxide

Magnesium oxide (MgO) is found naturally as periclase, but is more commonly produced by the thermal decomposition of magnesium carbonate or magnesium hydroxide. Due to the presence of aluminium, iron and/or silicate impurities, MgO is often pelletised and then ground to produce a dry powder.

The effective use of MgO relies on both pH control and chemical reactions to immobilise metals. The control of pH is important as it promotes the formation of low solubility metal hydroxides. Stabilised materials with pore solution pH's in the range of 8 to 11 typically have low levels of metal leaching. The use of MgO stabilises pore solution pH near 10. Furthermore, many metals can isomorphically substitute into magnesium aluminosilicate, which often forms when MgO is added to soil, with a consequent reduction in leaching.

Magnesium oxide is delivered as a dry powder, in bulk pneumatic trailer or in 1-ton bags. Magnesium oxide should be stored in a manner to prevent contact with water, to prevent degradation of its handling properties. In bulk, MgO is pneumatically transferred to silos, equipped with dust-control facilities. Storage silos should be well sealed to prevent contact with water, and when delivered in bags these should be stored on wooden boards above a plastic barrier and be completely covered with plastic sheeting.

Magnesium oxide can cause health effects by skin contact, eye contact, or inhalation, due to its highly alkaline nature.

8.1.8 Phosphates

Phosphates form low solubility salts with many metals, including aluminium, cadmium, iron, and lead. This fact makes phosphates a useful reagent for the chemical fixation of metals. There are a number of different phosphate materials that are typically used for stabilisation treatments. They are listed below, along with information on the material, and its delivery and handling.

Phosphoric acid

Phosphoric (or orthophosphosphoric) acid (H_3PO_4) is a mineral acid. It is typically available in aqueous solutions of 75-85 %. At these concentrations, it is a clear, odourless, syrupy liquid.

Phosphoric acid is delivered as an aqueous solution in chemical totes or by bulk chemical tanker. Bulk chemical tanker loads are typically transferred by pump to a holding tank or vessel for storage.

Because it is a concentrated acid, phosphoric acid is corrosive and can cause health effects by skin contact, eye contact, or inhalation.

Calcium phosphates

Calcium phosphates are a name given to a family of compounds containing calcium and orthophosphate (PO_4^{3-}). These include monocalcium phosphate, dicalcium phosphate, tricalcium phosphate, hydroxyapatite ($Ca_5(PO_4)_3(OH)$), and apatite ($Ca_{10}(PO_4)_6(OH, F, Cl, Br)_2$). Triple superphosphate, often called TSP, is a concentrated calcium monophosphate used as a common phosphate fertiliser and is typical of a calcium phosphate used for stabilisation treatment.

Calcium phosphates are typically delivered in 1-ton bags. Calcium phosphates should be stored in a manner to prevent contact with liquid water, as water will degrade its handling properties. These bags should be stored on wooden boards above a plastic barrier and covered with plastic or their contents transferred into silos for storage.

Calcium phosphates can cause health effects by skin contact, eye contact, or inhalation.

Sodium phosphates

Sodium phosphates are a name given to a family of compounds containing sodium and orthophosphate (PO_4^{3-}). These include monosodium phosphate, disodium phosphate, trisodium phosphate, and sodium aluminium phosphate ($Na_8Al_2(OH)_2(PO_4)_3$). Typically, trisodium phosphate has been utilised for stabilisation treatment.

Like calcium phosphates, sodium phosphates are typically delivered in 1 ton bags. Sodium phosphates should be stored in a manner to prevent contact with liquid water, as it will degrade its handling properties. These bags should be stored on wooden boards above a plastic barrier and covered with plastic or their contents transferred into silos for storage.

Sodium phosphates can cause health effects by skin contact, eye contact, or inhalation.

Bone phosphates

Bone phosphates are bone residues that have been treated with caustic and acid, then neutralised with lime and dried.

Bone phosphates are typically delivered in 50 pound (23 kg) bags or 1 ton bags and should be stored in a manner to prevent contact with liquid water, as water will degrade its handling properties. The bags should be stored on wooden boards above a plastic barrier and covered with plastic or their contents transferred into silos for storage.

8.1.9 Sulfides

Like phosphate, sulfides form low solubility salts with many metals, including cadmium, iron, lead, and mercury. This makes sulfides useful as a reagent for the chemical fixation of metals. There are a number of different sulfide materials that are typically used for stabilisation treatments. They are listed below, along with information on the material, and its delivery and handling.

Sodium sulfide

Sodium sulfide commonly refers to the hydrated version of Na_2S and is commercially available as an unspecified hydrate ($Na_2S \cdot xH_2O$) with the weight percentage of Na_2S specified. The commercially available grades are produced by the reduction of Na_2SO_4 with carbon and typically contain approximately 60 % Na_2S by weight.

Sodium sulfide is typically delivered in 50 pound bags (23 kg), or 1-ton (0.9 tonne) bags and should be stored in a manner to prevent contact with liquid water, as water will degrade its handling properties. The bags should be stored on wooden boards above a plastic barrier and covered with plastic or their contents transferred into silos for storage.

Sodium sulfide is highly alkaline and can cause health effects by skin contact, eye contact, or inhalation. Reaction of sodium sulfide with acids can liberate highly toxic hydrogen sulfide (H_2S) gas.

Calcium polysulfide

Calcium polysulfide typically refers to a family of calcium sulfides (CaSx) produced by reacting calcium hydroxide with sulfur, leading to its common name of lime sulfur. Commercially, calcium polysulfide is available as a solution with a specified weight of CaS_x, typically 29 % by weight.

Calcium polysulfide is delivered as an aqueous solution in chemical totes (an IBC container) or by bulk chemical tanker. Bulk chemical tanker loads are typically transferred by pump to a holding tank or vessel for storage.

Calcium polysulfide is highly alkaline and is corrosive. It can cause health effects by skin or eye contact. Calcium polysulfide has a distinct smell of rotten eggs. Reaction of calcium polysulfide with acids can liberate highly toxic hydrogen sulfide (H_2S) gas.

8.1.10 Carbon-based reagents

For metals, chemical fixation (stabilisation) often involves the precipitation or re-precipitation of soluble metal species as less soluble species, as hydroxides or sulfides. The immobilisation of the hazardous metals in less soluble species slows the potential release of the hazardous constituent into the environment and lessens the material's impact on the environment. For organic-contaminated wastes, reactions that alter the organic compound or physical processes such as adsorption and encapsulation are used to retard the movement of the hazardous constituents. Carbon-based reagents can provide organophilic surfaces for the adsorption and/or reaction of organic contaminants.

Organoclay

Organoclay is a name given to a family of naturally occurring clay material intercalated with organo-cations, typically quaternary alkylammonium ions. Intercalating these organo-cations to the ion exchange sites on the internal surfaces of the clay platelets produces an organophilic layer between clay platelets. Organic compounds can partition into and be adsorbed within this organophilic layer, allowing organoclays to sequester organics. This property makes organoclays useful reagents for the chemical fixation of organics. The specific capacity for partition and adsorption of organics depends both on the organic compound to be adsorbed and the organo-cation (s) employed by the specific organoclay.

Organoclay is delivered as a dry powder, in 1 ton bags. Organoclay should be stored in a manner to prevent contact with liquid water, as water will degrade its handling properties. Organoclay bags should be stored on wooden boards above a plastic barrier and covered with plastic.

Organoclays can cause limited health effects by inhalation or by skin or eye contact.

Activated charcoal and carbons

Activated charcoal and carbons are produced from natural materials such as coir (coconut fibre), lignite coal, nutshells, peat, and wood, by a combination of physical and chemical processes. These starting materials may be pyrolysed in the absence of oxygen (carbonisation), and then activated in an oxidising environment above 250°C. Often, the raw material is amended with mineral acids or salts to reduce the carbonisation temperature. The activation process produces a final material that is extremely porous and has a high surface area (>500 m^2/g) for adsorption. Since activated charcoals and carbons are organophilic and have a high surface area for adsorption, these materials are useful reagents for the chemical fixation of organics.

Activated charcoal or carbon is delivered as a dry powder, or granules, in bulk pneumatic trailer or 1 ton bags (Figure 8.3). In bulk, charcoal and activated carbon is pneumatically transferred to silos, equipped with baghouses to contain dust during the transfer, for storage. Activated charcoal or carbon delivered in bags is typically stored on wooden boards above a plastic barrier and covered with plastic, though these reagents are not adversely affected when contacted with water.

Activated charcoal or carbon can cause limited health effects by inhalation or by skin or eye contact.

Figure 8.3: Granular activated carbon

8.2 Reagent sourcing

Locating reagent sources is an important step in the design process for S/S treatment. Based on the performance goals for the S/S treatment, potentially applicable reagents need to be established.

8.2.1 Proprietary reagents

Proprietary reagents can only be purchased from their manufacturer or its distributer(s). However, non-proprietary reagents such as Portland cement, lime products, and fly ash can be sourced from multiple vendors. Literature and Internet searches can provide information on proprietary reagents that may be of interest.

Information on manufacturers of most chemicals (e.g. sulfides, phosphates, etc.) can also be garnered on-line from sites such as Alibaba.com® (http://www.alibaba.com/), ChemNet® (http://www.chemnet.com/), or ThomasNet® (http://www.thomasnet.com/).

8.2.2 Mineral and waste reagents

For most mineral and waste reagents, commercial associations or environmental agencies offer contact or listing information. These include:

Portland cement and cement kiln dust:

- The Portland Cement Association in the United States (http://www.cement.org/)
- the Cement Association of Canada (http://www.cement.ca/)
- the Mineral Products Association in Britain (http://www.mineralproducts.org/)
- the CEMBUREAU in Europe (http://www.cembureau.be/)

All provide information on the plant and sales office locations of their members. Since all Portland cement manufacturers create cement kiln dust, contacting the nearby cement plants can help identify locally available sources.

Quicklime, hydrated lime and lime kiln dust:

- The National Lime Association in the United States and Canada (http://www.lime.org/index/)
- the British Lime Association (http://www.britishlime.org/)
- the EuLA in Europe (http://www.eula.eu/)

Provide information on the plant and sales office locations of their members. As all lime manufacturers create lime kiln dust, contacting the nearby lime plants can help identify locally available sources.

Fly ash and bottom ash – these materials are produced in coal-fired power plants:

- For the United States and Canada, the respective national environmental agencies (US Environmental Protection Agency and Environment Canada) maintain lists of the coal-fired power plants operating within those countries
- The IEA Clean Coal Centre (http://www.iea-coal.org.uk/site/2010/home) maintains a list of coal-fired power plants in its member countries (Australia, Austria, Canada, EC, Germany, Italy, Japan, Poland, South Korea, South Africa, the United Kingdom and the United States)

Wikipedia also contains similar lists of the operating coal-fired power plants in most countries (http://en.wikipedia.org/wiki/List_of_coal_power_stations), though care should be taken to verify the reliability of the information posted on this site.

Bentonite – is produced worldwide in Argentina, Brazil, Canada, Cyprus, Germany, Greece, Hungary, India, Italy, Japan, Mexico, Poland, Romania, Spain, Turkey, and the United States:

- The Wyoming Mining Association (http://www.wyomingmining.org/minerals/bentonite/bentonite-members/)
- the European Bentonite Producers Association (http://www.ima-europe.eu/about-ima-europe/associations/euba) have membership lists, which provide points of contact
- Information on bentonite manufactures can also be garnered on-line from sites such as Alibaba.com or ThomasNet.com

8.2.3 Importance of location

Transportation costs are often on the order of 20 % of the total delivered reagent pricing for mineral manufactured products such as Portland cement and lime. For waste products, such as fly ash or cement or limekiln dusts, transportation can account for over 50 % of the delivered pricing. Therefore, the identification of local or nearby sources of applicable mineral or waste reagents is necessary to minimise reagent costs for the S/S treatment. It is important to emphasise that the specific sources of waste products contemplated for use in the S/S project be identified, and only these sources be used for waste product reagents used in the treatability testing (Sections 8.3 and 8.4).

8.2.4 Reagent samples for testing

Many vendors will supply a sample of their material for testing purposes. While this is not as important for commercial chemical or manufactured materials (e.g. Portland cement or lime products), this is very important for waste (e.g. fly ash, bed ash, cement or lime kiln dust) or off-specification (e.g. spent carbon), as the properties of these materials vary widely from source to source. Testing with the specific waste or off-specification material is necessary to ensure that these materials will perform as anticipated with the specific waste material to be treated.

8.3 Bench-scale treatability design and testing

Treatability testing for S/S is conducted at both the bench (laboratory)-scale and in the field at full-scale (pilot testing). This section describes bench-scale testing procedures, which are conducted before field- or pilot-scale testing.

Several excellent references for conducting S/S treatability tests are available in the literature (ITRC 2011, Environment Agency 2004, USEPA 1992, USACE 1995, EPRI 2009). The remainder of this section will focus on conduct of bench-scale treatability tests. The reader may want to refer to Section 8.4 for a discussion of field- pilot- or full-scale tests, Section 8.5 for a discussion on selecting the best sample for treatability testing, Section 6.3 for a discussion of sample preparation and curing, Section 7.1 regarding the setting of overall goals for treatment, Section 7.2 regarding the importance of the site conceptual model, and Section 7.3 for a discussion of test and analytical methods

A site specific bench-scale treatability test is needed to obtain essential information to evaluate the feasibility of S/S to treat the contaminated material, support selection of the remedy, establish treatment design parameters, develop the most cost effective formula and obtain information needed for scale-up to field tests.

Treatability testing for S/S may require 3-6 months, or more, to collect samples, select reagent formulas, perform the mixing, allow the treated samples to cure, analyse the results, and repeat the process at several tiers/levels of testing to derive the best and most cost effective treatment formula. Treatability testing is a systematic process to assure that the data generated is fit for the evaluation, identification of information for further testing, and development of full-scale specifications.

Analytical testing during bench-scale testing can be quite expensive if the practitioner is not experienced in selecting appropriate formulas to test, and in designing the sequence of testing activities. In addition, the collection of representative samples for bench-scale treatability testing is highly critical for the success of an S/S project, as selection and testing of the wrong materials from site will result in misleading results and the possibility of costly failure during full-scale treatment (see Section 8.5 regarding selection of the correct site material for treatability testing).

S/S bench-scale treatability studies usually include the following steps (after ITRC 2011):

1. Prepare a work plan
2. Collect test samples
3. Characterize the Initial Sample

- Homogenize raw materials
- Perform physical testing
- Perform chemical testing
4. Perform treatability testing
   - Identify appropriate reagents
   - Conduct testing by mixing reagents with contaminated material and prepare formulations for further testing
   - Optimize mix design
   - Selection of mix design verification phase
   - Prepare final mix design and test
5. Analyse, Assess and Validate Data
6. Prepare treatability study report

8.3.1 Bench-scale treatability study objectives

Prior to performing a bench-scale treatability test, the objectives of the testing should be identified. Typical objectives include:

- Demonstrate that the technology can meet remedial performance objectives
- Selection of appropriate reagent(s)
- Determine the most economical treatment formula (reagent(s) and dosage)
- Determine the impact of selected reagent(s) on contaminants and chemicals present in the waste material
- Develop treatment parameters and level of process control required
- Identification of contaminant/chemical emissions
- Identification of material handling issues and associated safety problems
- Assessment of physical and chemical uniformity of the treated material
- Estimate volume increase due to addition of reagent(s)
- Establish full-scale performance evaluation parameters and criteria
- Develop more accurate cost estimate
- Assure that the formula(s) developed can successfully treat all variations of soil and contaminant mixtures on the site

8.3.2 Importance of the conceptual implementation plan

A conceptual implementation plan should be prepared prior to conducting the treatability testing. This should address the proposed means of achieving the remedial objectives, including the areas for treatment, type of contamination to be treated, depth of treatment, type of equipment to be used for mixing reagents, etc.

The concept of how the remedial operation will be implemented is critical to designing the bench-scale treatability study. For example if the concept is to use in-situ augers to carry out S/S, then a candidate slurry-based reagent formula should be prepared, with a viscosity that is suitable for that purpose.

Thus, bench-scale testing will employ slurries (not solid reagents) to be mixed with the contaminated materials taken from site. However, if the concept is to mix using a pug-mill, then bench-scale testing should employ dry reagents, and supplemental water as necessary to achieve the desired final mix consistency.

The desired final mix consistency, especially moisture content, may depend on the transport and placement of the treated material. By way of example, if the treated material exits a pug-mill into on-site dump trucks (for transport to and compaction at the disposal area), then the material should not exit the pug-mill as a wet slurry (or it will not be capable of transportation and compacted as desired).

The treated material should however contain sufficient moisture to facilitate the chemical hardening reactions and facilitate compaction. Thus, the objective of the bench-scale treatability testing is to replicate as closely as possible the full-scale treatment, mixing, and placement of contaminated materials. Thus, development of a conceptual field implementation plan should precede the bench-scale treatability testing process.

8.3.3 Reagent selection considerations

The site remedial investigation data should be reviewed and potential reagents identified prior to treatability testing. The selection of potential reagents for testing depends on several factors, including the:

- Contaminants to be treated
- Concentration of contaminants detected
- Geotechnical properties of the waste material
- Agreed performance parameters
- Minimum acceptable performance criteria for the site

The identification of potential reagents often relies on a practitioners' experience. In the absence of previous experience, a survey of the technical literature is a good starting point, where a number of reagents may be identified. The 'narrowing' down of possible binders (reagents) will result in a lower treatability testing cost, over a shorter timescale. The successful selection of candidate reagents benefits greatly from experience. Selection of reagents relies upon knowledge of (after ITRC, 2011):

- Previous successful use of a reagent (a 'track record')
- Interference and chemical incompatibilities
- Organic and physical chemistry potential reactions
- Compatibility or reagents with the disposal site or reuse environment
- Availability of the reagent
- Cost of the reagent

A more detailed discussion on the above considerations is provided by the USEPA (EPA 1993), but it should be noted that the reagent used in the treatability testing should be exactly the same as planned for use during the full-scale remediation. For example a sample of F fly ash from plant "X" collected historically and 'archived' in a laboratory stockroom may not perform the same as a currently produced F fly ash from the same source, or from a different source!

8.3.4 Bench-scale treatability testing approaches

Bench-scale testing is most often conducted in the laboratory, though occasionally a limited test is conducted in the field. Relatively small amounts of site-contaminated material are treated using an assortment of candidate reagents and dose rates in order to determine the most cost effective formulation to achieve performance goals. Bench-scale testing can be conducted using a tiered approach (often the best), a shotgun approach, or a focused problem approach, as described below.

The **tiered approach** to bench-scale testing is a step wise approach, and is often the best for producing the most cost effective formula with the lowest treatability testing cost. It does takes longer, however, and usually requires at least 3 to 6 months or more for difficult to treat wastes. The results from each test tier are used to determine the subsequent steps and the next set of formulations to be evaluated. There are several different ways to define the tiers, but the following example is one that the authors have found usually works well. The four tier design process is illustrated in Figure 8.4.

Tier 1: Reagent screening involves the candidate reagents being tested individually and/or combined as initial formulations.

The objective is to assess the beneficial effect of the reagents and to establish upper and lower boundaries on the % of reagent needed for strength development. Generally, the focus is on achieving strength. Permeability may sometimes be evaluated on selected formulations that meet strength requirements.

Leaching is normally considered in Tier 2 or 3 testing. Tier 1 testing is specifically intended to highlight the most promising reagents, and determine the minimum and maximum dosages to be evaluated in step 2.

For example in Figure 8.4, Tier 1, if we assume the data from strength testing of formulas 1, 2, 4, and 6 have failed to achieve performance objectives, and formula 8 has achieved 5 times the strength objective, it is possible to extrapolate that 5-8 % w/w cement addition alone is not sufficient, and the use of Class 'F' fly ash did not make a significant difference, this latter reagent can be eliminated from further consideration.

However the inclusion of slag with cement did make a difference and formula 8 represented an excessive reagent dosage. Thus, the inclusion of permeability testing of formulas that met strength requirements might be beneficial, or (alternatively) one can delay all permeability testing to Tier 2.

Figure 8.5 shows UCS testing on cured cylindrical samples. Filter paper is sometimes added above/below the coupon/sample to determine if any free liquids were released whilst the specimen was subject to testing under pressure.

Tier 2: Formulation refining involves refining the reagents/formulations found to be effective in Tier 1 to ensure that all physical properties can be achieved and "leaning" (reducing binder content) of the formulations to reduce reagent costs. The most economical formulations that meet all the physical requirements can be leach tested and examined to determine whether additional reagent dosage, or possibly that new reagents, are need to achieve leaching targets.

In Figure 8.4, Tier 2, formulas 3 and 5 were carried forward from Tier 1, while formulas 9, 10, and 11 were modifications of 3 and 5. The original formula 7 without fly ash was formula 5, and since, in our example, fly ash was determined not to be beneficial, formula 7 was not carried into Tier 2. Formula 12 is the original formula 4 (also without the fly ash) but with the amount of slag being doubled.

In our example, we can assume all mixes meet the strength criterion but formulas 3 and 9 fail on 'permeability' (a typical permeability testing-rig is shown in Figure 8.6). We can conclude that slag needs to be used (with cement) in the mix design to meet target physical 'properties'.

186 | PART EIGHT

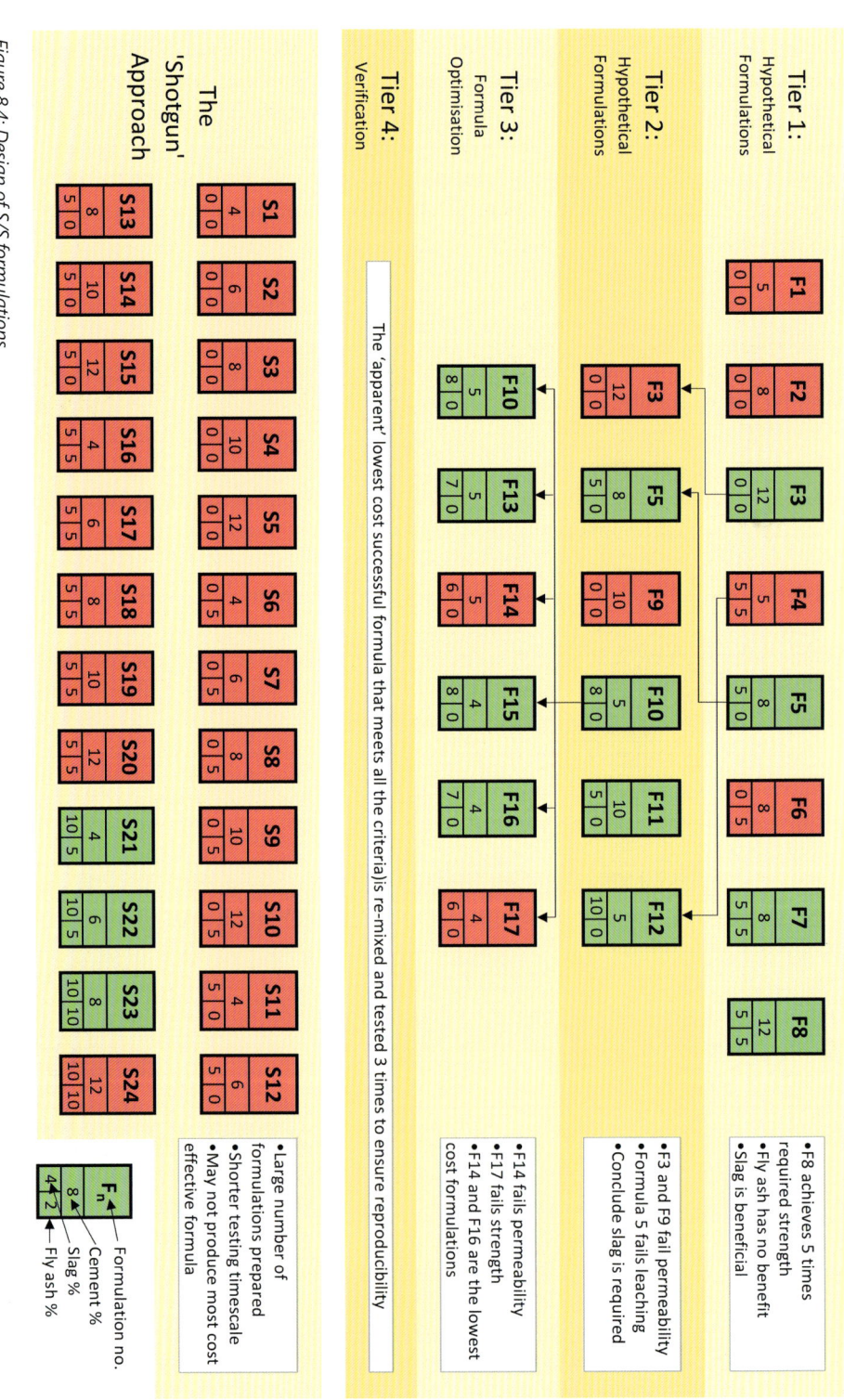

Figure 8.4: Design of S/S formulations

Those mixes that passed our physical criteria (formulas 5, 10, 11, and 12) are then leach tested, in which formulas 5 and 11 fail. Formulations 10 and 12 both meet leaching criteria.

On analysis, formulas 10 and 12 have lower cement contents (and consequently, lower pH's which can sometimes be important with immobilising amphoteric metals), but also contain higher slag contents. Although both formulas 10 and 12 meet all our criteria and could go forward to Tier 4 testing, it may be possible to reduce reagent dose rates, save treatment cost, and still meet our criteria. To explore this further, testing at Tier 3 can be used to determine the most cost effective formula.

Figure 8.5: Unconfined compressive strength testing

Figure 8.6: Permeability testing

Tier 3: Formula optimisation
It is tempting to accept formula 10 as the cheapest available. However, significant cost savings can be made if formula 10 can be adjusted to maintain performance but with a reduced cement and/or slag content. Tier 3 formulations are based upon mix design formula 10, but with reduced cement and/or slag content, hopefully achieving all criteria with considerable cost savings on treatment at full-scale.

As formula 10 has already proven successful it is not necessary to fully test it again in Tier 3. It is however shown in Figure 8.4, Tier 3, for reference as it has already passed all our criteria. The mix variations of formula 10 are first strength tested, where formula 17 fails. Formulations 13-16 are examined for permeability, where formula 14 fails. Formulas 13, 15, and 16 are then leach tested, and all pass the established criteria. The four mix designs passing Tier 3 are formulas 10, 13, 15, and 16. The lowest cost formulations are mix designs 13 and 16, and one of these is chosen and carried forward to Tier 4 testing.

Tier 4: Verification
At Tier 4 the 'apparent' lowest cost successful formula (i.e. meets all our treatment criteria) is re-mixed and tested 3 times to ensure reproducibility. Thus, repeated mixing, curing, and testing this formula using three replicate actions and sampling each mix is undertaken, not mixing once and then taking 3 sample sets. Should the results not meet all criteria, then it is necessary to return to Tier 3 to verify the second lowest-cost formulation that met the established performance criteria.

At some sites, where contaminated soil is variable, it is possible to determine that there exists more than one "reasonable worst case soil/contaminate type" to be treated (see Section 8.5). Under these circumstances, in Tier 4, the successful, selected formula is applied to these other soil/contaminant types to verify performance criteria are met. If not, then it may be necessary to develop a new formulation for these soils. The conclusion of Tier 4 testing is a successful, low-cost binder formulation for full-scale field pilot testing.

The 'Shotgun' approach.
The shotgun approach describes a mix designing process that is used when treatability testing is required at short notice. A large number of formulations are mixed, cured and tested, all at once. This approach takes less time, but generally costs are higher (more mixes are tested and analysed). It should be noted that this approach might not produce the most cost effective formula for field application, and one not suitable sometimes for 'difficult' to treat wastes. However, if time is not available for a tiered approach, the skill of the practitioner can be used to produce an effective binder formulation in about 6 weeks.

With 24 or more binder formulations to test against performance criteria, strength, permeability, and leaching tests could be conducted concurrently, or sequentially (i.e. with strength testing first, as it is the fastest and cheapest).

With a sequential approach, those formulations that passed the required strength would be subjected to the leaching evaluation, with those passing both (strength and leaching) submitted for permeability testing. If the tiered approach discussed previously was used, either 5 % cement with 6 % slag or 4 % cement with 7 % slag would meet all of our criteria. In the shotgun approach illustrated in Figure 8.4, formulas 21, 22, and 23 met all the criteria (formula 24 failed leaching due to high leachate pH). It is noteworthy that despite the shorter testing timescales and larger numbers examined, the resultant formulation is not optimised for cost effectiveness.

The focused-problem approach: applies to special conditions where a formula has been developed, used with success, but then generates failures, usually due to difficult field conditions and/or more challenging materials.

In this case one should first try to determine what changed in the field, then conduct a focused test to determine how the current formula can be modified to compensate. The first step is to quickly identify the changes in materials being treated. These could be physical changes or changes in contaminant concentrations. Often resolution is achieved by using more binder, but if this does not work, then a targeted treatability study can be undertaken, including adding another reagent (that experience has shown may be beneficial). For example, the addition of bentonite to reduce permeability, or a reagent to reduce contaminant solubility (see Section 8.1) is required. Occasionally a bench-scale treatability study may be conducted in the field during S/S. Figure 8.7 shows the development of a formula to solidify (not chemically stabilise) an oily pesticide-contaminated waste-water for disposal in an on-site RCRA C landfill.

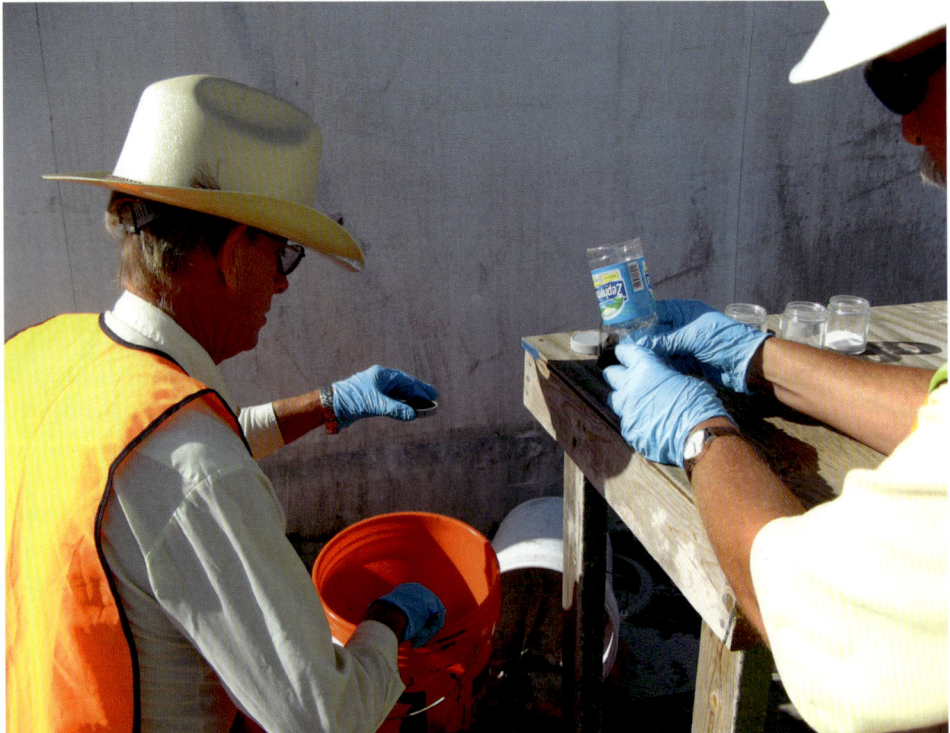

Figure 8.7: A limited bench-scale field treatability study

Regardless of the approach used, once an acceptable mix design has been formulated and verified through replicate testing, the mix design will need to be applied at field-scale in a pilot-demonstration phase (see Section 8.4).

In preparation for a field demonstration, it is useful to develop additional data in the laboratory to support the field efforts. For example, determining the slump of the lab-treated material will provide insight into mixing of reagent and contaminated material at full-scale. A very low slump e.g. <2 in (5 cm) may be difficult to mix in the field and require additional water. However, the slump to be achieved depends on the mixing equipment and method of disposal used at full-scale. If a wetter mix is needed in the field, this could have a detrimental effect on strength or permeability of cured materials. It may also be useful to evaluate strength and permeability as a function of time up to 28 days, to establish if shorter curing times will be possible, or if full curing is necessary to obtain the required testing results. In some cases a cure time longer, or shorter, than 28 days may be appropriate.

8.3.5 Initial sample characterisation

Once a sample is collected from the site, it must be carefully homogenised to ensure that as aliquots (of the sample) are selected for treatment by different candidate formulations, there is no significant difference in the untreated soil aliquots that would affect the treatment results. A common approach is to homogenise the bulk sample, and obtain three random sub-samples/aliquots for analysis. If the three untreated aliquots all produce similar results for key contaminants (totals or leaching), then the bulk sample can be considered homogenised.

An initial (baseline) characterisation of the samples can then be performed, including as a minimum, the physical and leaching characteristics, using the same leaching test method to be employed during treatment. Section 8.5 provides guidance on selecting samples for treatability testing. Prior to initial characterisation, oversized material should be removed by screening through a 0.5 in (1.25 cm) mesh sieve, as larger particles will cause failure of small test specimens used in treatability testing (even though oversize material of this nature would have little, if any, effect during full-scale treatment).

8.3.6 Laboratory procedures

Generally it is difficult to homogenise samples while being collected in the field. Field samples are collected and transported in two or three 5 gallon (19-20 litre) 'shipping' buckets, and at the laboratory these are combined in a large mixer or drum and blended until they appear to be homogenised.

A rotary drum-mixer or commercial paint stirrer is sometimes used for this purpose. However if the materials contain volatile contaminants, then to avoid loss of these compounds a zero head-space mixer may be required. Following homogenisation, sample characterisation is conducted (see Section 8.3.5).

Once a bulk sample has been collected and homogenised it is ready for bench-scale treatability testing. The reagent addition rate for each test is specified in the treatability test plan as a percentage (weight/weight) of the untreated soil, including its soil field moisture content.

Figure 8.8: Reagents are measured on a weight/weight basis to the untreated soil

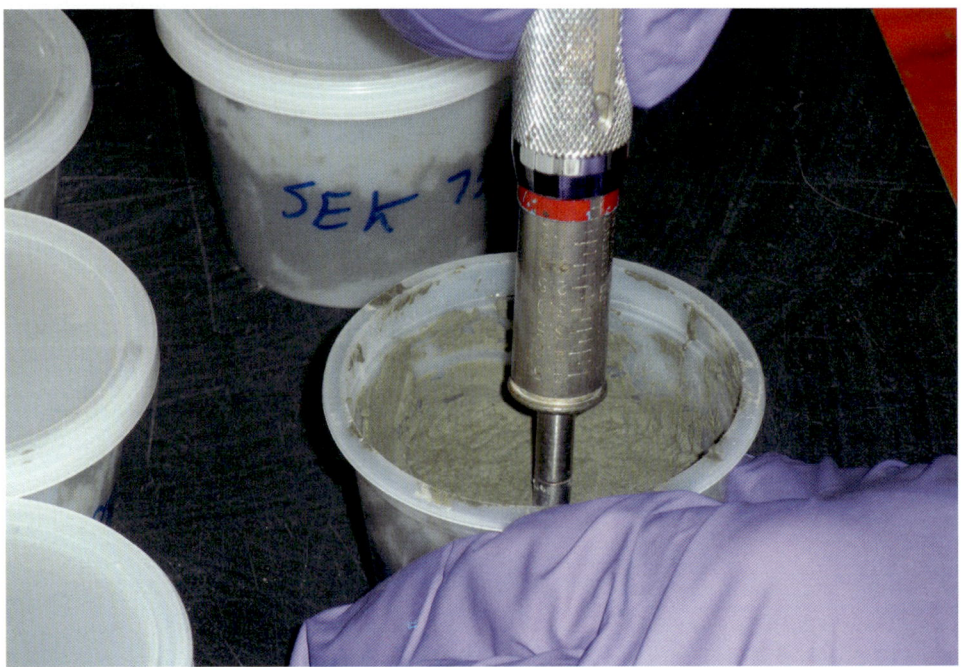

Figure 8.9: Pocket penetrometer testing for approximate strength

The materials are carefully measured as shown in Figure 8.8 ensuring there is enough mixed material to fill all the sample/coupon molds. The size/shape of these coupon/molds is often specified by the analytical method used. Water is also slowly added as a measured volume to the mixing bowl to facilitate the mixing process.

The reagents are sometimes blended with water to form slurry prior to mixing with the soil, if that is the approach envisioned at full-scale treatment. If the treatment approach involves in-situ injection of a binder-slurry followed by mixing, there will be limits set for viscosity of the slurry. Thus, the bench-scale test should also add the reagents as slurry at the viscosity compatible with the planned equipment to be used at full-scale.

If ex-situ mixing is planned, the reagents are added dry, but the amount of water added is limited by the handling properties required for the treated soil. For example, if the treated soil exiting the mixing will be transported by truck, followed by spreading/compaction by a dozer, then the water added during bench-scale testing must be capable of replicating the properties required for this scenario.

Samples produced for testing should be free of bubbles/air voids, and be allowed to cure in an undisturbed manner. In the field, treated material forms a large moist mass that cures over days and weeks. In the laboratory, the sample coupons are capped and placed in a humid environment at ambient temperature, with samples removed as and when necessary for testing, including one on a daily basis, to assess setting and initial strength development. Figure 8.9 shows this being done with a pocket penetrometer.

8.4 Pilot-scale tests

As described in Section 8.3, treatability testing for S/S is conducted at both the bench- (laboratory) scale and then in the field, either prior to, or as the initial phase of the full-scale implementation (i.e. the pilot testing). This section describes the role of the pilot testing phase in implementing an S/S remedy.

There is no "one-size-fits all" approach to pilot testing as each site and waste has its own unique characteristics and challenges. As such, the authors have not found any single comprehensive guide to pilot phase testing for S/S. Rather, there are a number of useful references which touch on the various aspects of pilot testing including: Environment Agency (2004a), USEPA (1986), USEPA (1989), Fleri and Whetstone (2006), ITRC (2011), Shi (2004), Perara, et al. (2004), and Butler, et al. (1996).

The remainder of this section will focus on the importance of, planning for, and implementing a field pilot test for S/S. The reader should refer to Section 8.3 for a discussion of bench-scale treatability studies for S/S, Sections 4 and 5 on S/S implementation equipment, and Section 6 on quality assurance and quality control in S/S applications as well the sampling plan, sampling methods, and curing of samples.

The properties of S/S-treated wastes may be different in the field, compared to the laboratory, due to variation in composition of a waste, the accuracy of the reagent dosage application, variation in site soil characteristics, and the type of S/S mixing and placing equipment and processes.

Pilot-scale testing should be performed using the equipment proposed for full-scale and the most successful mix design(s) determined during the bench-scale testing program. Pilot-scale testing will demonstrate how the mix(s) will perform at field-scale with relatively large volumes of waste, and be based on ability to meet performance criteria, ease of implementation and cost. The pilot-scale test provides the opportunity to:

- Refine the mix design, based on field-scale observations and testing
- Reduce construction risk
- Optimise S/S equipment processes and operational parameters
- Reduce variability in the treated product
- Determine the operational parameters under which the S/S equipment will be working
- Verify the thoroughness of mixing
- Verify the swell generated by S/S treatment
- Evaluate variability in site and waste characteristics
- Demonstrate the ability to meet performance criteria using the selected equipment/processes on a large scale under field conditions for comparison against the results of the bench tests

Although pilot testing can be costly and time consuming, it can be used to assess site safety considerations, reduce work stoppages, and increase product consistency and process reliability. Pilot-scale tests can also be used to train equipment operators (on the characteristics of the waste and the solidified product), and help a contractor optimise construction and process efficiency. An opportunity to develop or refine working practices and quality control procedures exists potentially resulting in cost savings during full-scale S/S, and reducing the risk of failure or the production of variable quality treated materials.

The pilot test phase is ideally implemented as a separate project phase between the bench-scale treatability study and full-scale implementation. This will provide the highest level of confidence in the selected mix designs/mixing process for field application and allow time to assess QC test results to reduce the overall construction risk. In practice, however, it is not uncommon for the pilot testing to be performed as an initial phase of full-scale implementation, after the S/S contractor has been awarded the contract. The decision to perform a separate pilot test or to perform the pilot as part of full-scale implementation is generally driven by project cost, timescales, project complexity or a combination of these factors.

The designer and owner, and in some cases the contractor (where the latter is tasked with developing the mix design/implementation method) will need to make this determination. In general, for sites or wastes where the potential for variability is low and previous experience indicates, or where pilot phase testing involves optimisation of the mix design only (i.e. the mix design has a proven track record with the wastes), pilot-scale testing is often part of the full-scale implementation program.

If, however, there is any doubt as to equipment suitability, waste variability, treated product consistency, site safety concerns related to reagent-waste reactions or S/S equipment, mix design scale-up to field conditions, cost escalation potential, or regulator acceptance of the S/S process, then a separate field pilot test is essential.

Another important consideration in determining how and when to implement the pilot test is how the owner desires to allocate project risk. The more information that is available to S/S contractors, the lower the risk of implementation issues arising, and consequently, more certainty over construction costs.

Additionally, the selection of performance-based contracting, design-build delivery approaches, or design-bid-build approaches, may influence the pilot testing approach. However, it is important to recognise that regardless of the contracting or risk-allocation approach, the pilot test often results in some modification of mixing procedures, reagent proportioning, and safety protocols. This in turn influences the cost of implementation, through optimised processing (lower cost) or through increased variability in site or waste characteristics (higher cost).

8.4.1 Pilot test objectives

The objectives of performing an S/S pilot test will vary slightly from project to project depending on the complexity of the site or wastes, the uniqueness of the reagents used, and the requirements of working within specific regulatory programs or regulatory jurisdictions. Typical objectives for the pilot test phase include one or more of the following:

- Enable visual observation of field-scale treatment processes
- Enable visual observation of treated materials
- Determine if debris, rocks, cobbles are present of sufficient size to impede in-situ mixing or require pre-excavation or pre-processing
- Evaluate treated material handling and placement characteristics and issues
- Assure that the formula(s) developed can successfully treat all variations of soil and contaminant mixtures on the site
- Demonstrate implementation of the construction quality control plan and adequacy of the selected tests and observations to control the treatment process
- Refine estimated production rates and overall S/S project schedule
- Evaluate the consistency of mix design performance between bench-scale testing and full-scale construction techniques
- Identify scale-up factors that will be useful in full-scale such as strength-gain rate and freshly treated material (uncured) consistency
- Assess the applicability and reliability of compliance criteria identified during the bench-scale testing and establish full-scale performance evaluation parameters and criteria
- Evaluate the consistency of treated product using full scale equipment
- Enable trouble-shooting and optimisation of S/S equipment, grout plants and grout pumping, calibration, process controls, material handling, and data collection systems
- Validate that the technology can meet remedial performance objectives under field conditions
- Refine reagent selection, proportioning, or sequencing of application
- Evaluation of contaminant/chemical emissions and controls
- Identification of material handling issues and safety problems in handling the waste
- Estimate volume increase due to addition of reagent(s)
- Develop a more accurate cost estimate

8.4.2 Confirming the conceptual implementation plan

As described in Section 8.3.2, a conceptual implementation plan should be developed prior to bench-scale testing to identify how the wastes will be treated, by what means will the wastes be mixed with reagents, and, for ex-situ treatment, how the treated materials will be placed/disposed.

The pilot phase provides the opportunity to validate and/or modify the conceptual implementation plan. For example, if the concept was to use in-situ augers to mix reagents, as a grout slurry, with the waste in-place, the field pilot may result in identifying difficulties with mixing thoroughness, penetration depth or grout viscosity for the batch plant and grout pumps.

The pilot phase also presents the opportunity to assess different auger diameters or configurations, vary mixing speed, penetration rate, or number of strokes of the auger in each column, or to modify grout properties to overcome field difficulties (see Section 5.1 on Auger Mixing). If the concept is to perform ex-situ mixing with a pug-mill or other suitable equipment, the material handling characteristics for hauling and placement may require modification of the moisture content to optimise these unit operations. In addition, one can vary and optimise the feed rate, which in turn controls the mixing time. Excavating a large volume of waste at the site for ex-situ treatment may also reveal the need for pre-processing such as screening or size reduction to avoid handling issues within the mixing equipment (Section 4).

The pilot test should also identify the optimum processing rate and any dust, emission, or safety issues. For example it may be necessary to add shrouds to lower dust emissions.

8.4.3 Implementing the pilot testing

The pilot-scale test phase should be guided by a work plan prepared for the site, identifying the objectives, selection of the pilot test location(s) and treatment volume, types of equipment to be utilised and evaluated, reagent mix(s) to be evaluated, quality control and quality assurance testing, process parameters to be monitored, and QA/QC recordkeeping.

It is also essential to have a safety plan in place for the pilot test. As mentioned previously, the pilot test should be performed with the same full-scale equipment planned for the remediation. Figure 8.10 shows equipment that might be used for an in-situ auger pilot test, and includes an 8 ft diameter auger (2.5 m), and support equipment. Not shown, are the slurry batch plant, pumps and hoses that would also be necessary. Figure 8.11 shows a 10 ft (3.0 m) in-situ auger conducting a pilot test. Note that although it was a pilot test, all equipment and procedures are the same as would be used for the actual remediation. Also, that a grout slurry blanket was laid down first to reduce release of volatile emissions. Figure 8.12 shows a pilot test for in-situ bucket mixing. Here also the equipment and procedures are the same as would be used for the actual remediation.

Figure 8.13 shows a full-scale ex-situ pug-mill system including reagent silos and a stacker for loading treated soils into trucks for transport to the disposal area. Pilot tests should not be conducted with scaled-down equipment or equipment differing from that actually planned for use in the remediation.

In general, it is good practice to collect samples for testing at a higher frequency (including more cure times) in the pilot phase than is anticipated for the full-scale in order to further evaluate variability and to have replicates available for potential additional testing to assess results that are significantly different from what was expected. It is important to view the pilot phase not as a pass/fail on S/S implementation, but rather an opportunity to further assess site conditions and treated waste behaviour and characteristics, and to make modifications to the mix design or application methods to address identified scale-up issues.

Quality control test results from pilot-scale may vary from results obtained in the bench-scale testing. The pilot-scale test is the opportunity to perform additional tests with longer curing times than the performance criteria, to understand if apparently failing results (if encountered) will improve with additional curing time, or other process modifications.

The selection of the amount of waste to be treated during a pilot demonstration and the location to treat the waste or soil are determined on a project-specific basis. Typically, at least 500-1,000 yd^3 (380-765 m^3) of material should be treated.

The volume selected should be sufficient to evaluate materials variability, cover at least one full-day of operation, achieve the maximum depth of S/S anticipated (for in-situ mixing), and allow for mixing of multiple batches/cells/columns. For ex-situ mixing, the pilot test should produce a sufficient quantity of material to evaluate preferred methods for field placement of the freshly mixed material.

The pilot-scale trials also present an opportunity to acquire treatment data at the field-scale to validate impacts due to minor alterations to the chosen mix design. For example, if the conceptual implementation plan involves excavator bucket-mixing in cells using a 10 % binder, then treating several cells and varying the mix over 9 %, 10 % and 11 % will provide insight into the performance of the S/S material, if (during full-scale implementation) deviation from the desired 10 % binder-content occurs. It may also indicate whether the mix developed in the bench-scale testing, over or under performs at field-scale.

Pilot-scale tests provide the opportunity to assess the adequacy of the quality control program and to ensure that the correct field-specimen preparation (see Figure 8.14) procedures are being followed. Field tests, including evaluation of consistency (via the slump tests - see Figure 8.15), and grout testing (via e.g. the mud balance test Figure 8.16) can be evaluated to ensure the results are within the desired range. Figure 8.17 illustrates a visual field QA/QC check to assure adequate homogenisation and mixing. The requirement here is that no clumps of unmixed soil exist that exceed fist size (about 4 in/10 cm). The excavator removes several buckets of material from various locations and depths.

Reagent metering and weighing systems (see Figure 8.18) are also critical to ensure that the right dosage levels are being applied. The reader is referred to Section 6 for a more detailed discussion of QA/QC monitoring and data collection.

The pilot test should also include post treatment evaluation of the treated material in-place after curing which affords the opportunity to view mixing homogeneity, overlap between treatment columns or cells, identify if segregation of materials and reagents occurs, as well as verify that material has cured adequately under field conditions.

Figure 8.19 illustrates a test pit excavated through solidified contaminated soils. Test pits are also useful, in shallow groundwater conditions, to demonstrate the absence of free water (visual) in solidified materials, providing confirmation that groundwater flow is essentially excluded from within the solidified soils.

Figure 5.11 (Section 5.1) illustrates exposure of solidified in-situ soil columns to evaluate cured conditions.

Figure 8.10: 8 ft (2.5 m) in-situ auger and carrier assembly

Figure 8.11: Pilot test using a 10 ft (3 m) diameter in-situ auger

Figure 8.12: Pilot test for in-situ bucket mixing

Figure 8.13: Ex-situ pug-mill with support equipment

Figure 8.14: Field quality control test specimen preparation

Figure 8.15: Field slump test performance

Figure 8.16: Grout density test by mud balance

Figure 8.17: Field check to assure homogenous mixing

Figure 8.18: Reagent silo electronic scale and calibration weight

Figure 8.19: Trial pit in an S/S soil monolith

Test pits also afford the opportunity to verify the strength and durability of the treated mass through observation of relative excavation effort, and in cases where future excavation ability is desired, to assess whether that goal has been achieved.

8.4.4 S/S safety considerations

The pilot-scale testing phase will enable the designer and the contractor to assess safety considerations that may not have been obvious or observable during bench-scale testing. As described by the USEPA (1986), safety considerations upon scaling-up to field-scale treatment can involve fuming, heat development, and volatilisation of organic compounds.

When vapour emissions are moderate, collection using a shroud, followed by treatment with activated carbon for example, may be sufficient. Figure 5.2 includes a picture of an in-situ auger rig with a shroud in place.

In more severe cases, pre-treatment, low-heat reagents, or alternate means of reagent sequencing or blending may be required, as the addition of large volumes of reactive alkaline reagents can result in excessive or explosive release of VOC's and the potential for excessive heat generation from exothermic hydration reactions.

An example of this is provided in Figure 8.20, when a pilot test (conducted as the initial phase of remediation) injected large quantities of alkaline reagents into the sulfuric acid-laced waste oil sludge, resulting in a violent exothermic reaction. The steam generated rose about 30 ft (10 m) into the air, and dangerous concentrations of sulfur dioxide gas were released. The issue was resolved by pre-treating the acid sludge with ground agricultural limestone (calcium carbonate) to raise the pH above 4.5 before injection of the alkaline reagents was carried out.

Figure 8.20: In-situ auger test producing steam, sulfur dioxide

Figure 8.21: Collecting a bulk treatability sample from auger flights

8.5 Selecting samples for treatability testing

Selecting a representative sample for treatability testing is an important part of the development of the mix design for use in S/S. The result of the treatability testing is the mix design (or formulation), which will be used for full-scale treatment. Therefore, the samples used for the treatability study will directly influence the mix design, and through it, production rate, production sequencing, and cost of the full-scale treatment.

Following the site investigation, the concentrations of the COC's and their variability (both horizontally and vertically) will be known and the area(s) requiring treatment by S/S delineated. The contamination levels are also often used to determine where to obtain samples for S/S treatability testing. The following sections summarise the 3 most common schemes for determining where to obtain treatability samples.

Figure 8.21 shows a bulk-sample being collected from the flights of an auger. Great care and attention are needed to collect a representative sample at this particular location.

8.5.1 Option 1 : maximum contamination sample

Under this option, the samples to be evaluated are collected from the area(s) of the site where the contamination levels are considered to be at a maximum. The objective of the sampling is to obtain a 'worse case' sample which represents the most difficult material to treat on-site.

Mix designs that achieve performance criteria using these samples will likely be applicable to any material encountered that requires treatment.

Advantages: The mix design developed during the treatability study should be applicable to any material encountered on the site requiring treatment. Therefore, the risk of failure (and/or retreatment) during full-scale S/S will be minimal.

Disadvantages: Typically, higher levels of COC require greater binder usage. A mix design developed for the maximum contaminant levels will often require more reagents thus be more costly to apply. Since the full-scale S/S treatment utilises stockpiles, cells or columns typically involving hundreds of cubic yards (cubic metres) of material, the act of treatment normally imparts some homogenisation to the material and thus, moderation of the contamination levels. Therefore, it is unlikely that the maximum contaminant levels will be encountered during full-scale treatment. Thus, higher reagent addition levels than are actually needed will be applied.

8.5.2 Option 2: high contamination sample

This approach involves sample collection at the area(s) of the site with approximately 75 % of the maximum contamination levels. The objective of the sampling is to obtain a sample representing the ***actual most difficult-to-treat*** material that will be encountered on-site, taking into account that excavation and stockpiling for ex-situ treatment or in-situ treatment of cells or columns will somewhat homogenise the materials contained within them. Mix designs that will meet the agreed performance criteria for this sample will likely be applicable to all of the material encountered during the remedial operation.

Advantages: The mix design developed will likely be applicable to all of the material encountered on-site. Therefore, the risk of failure (and/or retreatment) during full-scale S/S will be low (typically less than 2 %). The mix design developed for these high contaminant levels will often have lower reagent addition levels compared to Option 1, therefore the mix design will have a lower cost to implement.

Disadvantages: The risk of failure and retreatment under this option is typically in the range of 2 % to 10 %. Therefore, some treated S/S cells in the most contaminated area(s) of the site may require retreatment. Although retreatment of an S/S stockpile, cell, or column is only slightly more costly than the original treatment, the additional costs involved need to be recognised, but may not be significant.

8.5.3 Option 3: average contamination sample

Under this option, samples are collected from the area(s) of the site with 'average' contaminant levels. The objective of the sampling is to obtain a sample representing the typical material that will be encountered on-site, taking into account that some homogenisation will occur from stockpiling or the treatment process. Mix designs that will meet the design performance criteria using this sample will most likely be applicable to the bulk of the material encountered on-site.

Advantages: The mix design developed will most likely be applicable to much of the material on-site. The risk of failure (and/or retreatment) during full-scale S/S will be low to moderate. This approach will generally use the least binder/reagent of the three approaches, hence the lowest reagent cost to implement.

Disadvantages: In practice, the risk of failure and retreatment under this option is typically within the range of 10 % to 25 %, and in practice the most contaminated area(s) of the site will require retreatment by S/S and the additional costs may be significant.

8.5.4 Rationale for using the high contamination sample

The first option, Option 1 for the selection of treatability samples will produce a very robust mix design, which can be implemented during full-scale S/S with high confidence. However this mix design will likely be costly. Option 3 represents the 'other end' of the spectrum, and although the mix design is less robust (and less costly), there could be up to 25 % of the site requiring retreatment.

The second option (highly contaminated sample) typically provides the most economical mix design to implement by reducing reagent addition levels while reducing the retreatment rate to acceptable levels. Typically, the savings in reagent usage for Option 2 more than off-set retreatment cost providing the most economical choice for the selection of samples for treatability testing.

8.5.5 Sites with multiple materials requiring treatment

Many sites have more than one type of material to be treated (e.g. soil, sludge, paste, etc.), and within these there are many possible permutations (e.g. soil and DNAPL-saturated soil and sludge materials from different processes that were co-disposed). Typically each material 'type' should be sampled for the treatability study. As discussed in Section 8.3 (bench-scale treatability tests), good practice involves the collection of multiple (two or three) 5-gallon buckets of each material 'type' to assure an adequate quantity for all tests that may be desired. All the materials requiring remediation are included.

Conducting treatability testing on all the material types encountered at a site provides information on whether all material types can be treated with one common mix design or that separate mix designs will be required. The samples obtained for each material type should have high (approximately 75 % of maximum) contaminant levels (Option 2 above).

PART NINE

Post-treatment capping and monitoring

The general strategy for post treatment management of S/S material should be considered early in the design process and include collaboration with critical stakeholders and regulatory agencies.

The long-term performance and management objectives should be established with respect to future land use, post remediation operation and maintenance, required engineering and institutional controls, and the need for long term monitoring etc. Accordingly, an appropriate post-remediation management strategy requires an integrated approach where the long-term performance and management objectives are evaluated at various stages of the design process and during full-scale construction. These stages include:

- Pre-design data collection
- Assessment of reagent mix designs during bench-scale testing
- Verification of full-scale performance objectives during pilot testing
- Performance verification during construction
- Final "as built" configuration, and
- The requirements for post construction monitoring

The following discuss the factors most important for the selection and design of a cap for S/S treated material, and for post construction monitoring. The considerations for selecting a cap design and for post construction monitoring are the same for either in-situ or ex-situ S/S treated sites.

9.1 Overview of post treatment management of S/S material

A major design objective is to develop a capping layer that will divert surface water away from the S/S monolith to prevent ponding. Key issues to be addressed during the design process include:

- The reconciliation of the estimated quantity of S/S material following completion with actual quantities arising during treatment
- The modifications necessary to finalise geometry of the cap arising from the field conditions experienced
- Ensuring the cap meets future land use objectives established during the design phase

During the pre-design stage, careful consideration must be given to variables that will impact on the final site conditions and long-term operation and maintenance requirements. For an S/S project, this can be difficult, as it may require interpolation of the final site conditions in the absence of key data. Nonetheless, by carefully assessing potential factors of importance a management strategy for post-remediation conditions is possible and will help define data requirements from bench- and pilot-scale testing.

The following factors require consideration, and are discussed below:

- Climatologic and geographic considerations
- Site geometry
- Integration with existing site development
- Future site use
- Affected property owners
- Regulatory requirements

9.1.1 Climatologic and geographic considerations

Key design parameters can be obtained by considering surface water sources, and whether the site is located in a cold weather region, where freeze/thaw effects might impact on the long-term stability of the S/S monolith.

Surface water management requirements involve an assessment of a number of parameters such as the incidence and severity of rainfall, seasonal fluctuations in precipitation and the likelihood of wetting and drying cycles impacting on the monolith surface. The impact of cold weather requires an evaluation of the thickness of the cap, the final geometry and elevation of the S/S monolith relative to final site grading, and how any excess S/S material (swell from in-situ treatment) is managed and placed.

9.1.2 Site geometry

The geometry of the site will influence where and how S/S-swell material is placed, and how placement is sequenced within the full-scale S/S operations. Ultimately, the final S/S monolith elevations will need to take into consideration the pertinent design parameters such as surface water drainage and management that will be directly influenced by the site logistics.

9.1.3 Integration with existing site development

The post remedial plan needs to consider all existing structures, the site services including underground and above ground utilities and roadways, as these components are integral to the remediated site and influence the final cap configuration.

Site design plans will need to include specific requirements for completing the S/S monolith near existing structures/foundations and the maintenance of adequate cover for landscaping and surface water management. Clean utility corridors may need to be established across the S/S monolith to accommodate existing and/or new underground utilities, but do pose engineering challenges (for their integration into the final cap configuration and management of any surface waters) which must not be underestimated. If not properly designed, they may serve as containment vessels for standing water that could adversely impact on the long-term performance of the S/S monolith. For example, even if the monolith performs as designed with respect to leachability of target constituents, the presence of reagents such as Portland cement could lead to elevated pH levels in ponded contact water. In turn, this could cause the leaching of key contaminants, leading to surface water discharge standards not being met.

9.1.4 Future site use

Under some circumstances, the long-term development objectives for a site remediated by S/S may not be known at the time of treatment. However, where they are known early design consideration for the final cap configuration may be critical to achieving final post S/S management objectives. In addition to managing the amount of S/S material to be treated, a number of engineering parameters may require evaluation during bench-scale testing to meet the development objectives, including:

- Mix designs enabling lower UCS's for subsequent excavation of S/S material for foundations, infrastructure and surface re-grading
- Mix designs with higher UCS's for enhanced bearing capacity and shear strength
- Mix designs specific to soil types for management of bearing capacity and potential settlement induced by future load bearing structures

A number of future use requirements may also need to be considered that could range from mildly intrusive to major modifications to the cap and the S/S monolith, including:

- Landscaping, including the planting of trees and/or shrubbery
- Re-grading/modification of cover thickness and/or removal or repositioning of portions of the monolith (to accommodate new grading and drainage)
- Removal of monolithic material to accommodate new foundations and/or infrastructure

Flexibility in the design of the cap needs to be considered to allow the property owner(s) to accommodate future changes including:

- Reducing the elevation of the monolith to enable shallow building foundations
- Improved conditions for landscape plantings (facilitated by earthen materials in the cap)
- Shallow gradients to permit future access and usability

Additional assessments will also be required to establish agreed engineering and institutional controls with which the property owner must comply. These are normally incorporated as restrictive covenants and may include the following:

- Restrictions on modifications of the cover system and/or the monolith
- Planning and notification requirements for proposed significant modifications/new development
- Minimum technical specifications and performance specification compliance upon cover system reconstruction
- Management of regulated materials, such as disturbed S/S material and treatment requirements for excavation dewatering (if applicable)

9.1.5 Regulatory requirements

Requirements for the design, construction and maintenance of post remediation cap systems for S/S monoliths are regulated by local, regional (State) and national (Federal) regulatory agencies responsible for approving remedial design plans and specifications under various clean-up programs.

For example, projects under the USEPA's Superfund program require submission of a cap design for approval by the USEPA Project Manager that would comply with the established remedial performance objectives. The assessment of applicable design requirements could include comparison with current Subtitle D (municipal solid waste) and C (hazardous waste) specifications. Performance objectives meeting Subtitle D specifications include the following:

- Use of barrier materials that will have a permeability of no greater than 1×10^{-5} cm/sec (1×10^{-7} m/sec)
- Incorporate a barrier-layer of earthen materials of >12 in (>30 cm) in thickness
- Have an erosion control/topsoil layer constructed of earthen materials with a thickness of >6 in (15 cm), capable of supporting native vegetative growth and for frost protection
- Use of a geo-membrane (if the liner construction includes a geo-membrane)

However, under more rigorous 'Subtitle C' specifications, requirements may include:

- A cover-soil layer in addition to an erosion control/topsoil layer
- A drainage layer e.g. a free draining sand and/or gravel material, or a geo-composite consisting of geo-net-bonded with a geotextile (on one or both sides)
- A barrier layer of >24 in (90 cm) with a maximum hydraulic conductivity of 1×10^{-7} cm/sec (1×10^{-9} m/sec)

Typical example profiles for Subtitle 'D' and 'C' landfill cover are illustrated below:

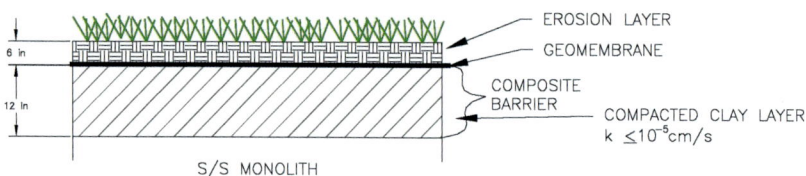

Figure 9.1: Typical profile for a Subtitle 'D' cap

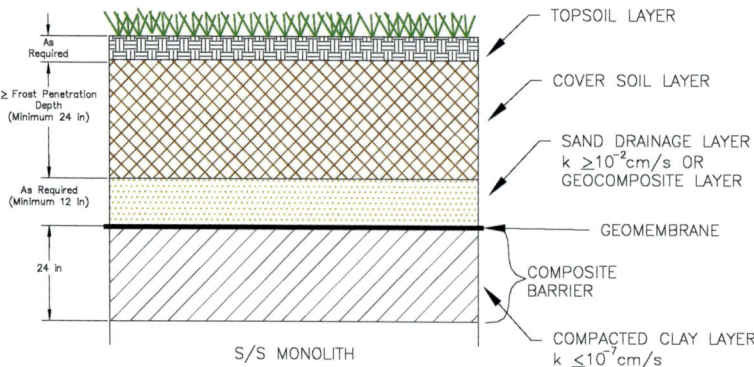

Figure 9.2: A typical profile for a Subtitle 'C' cap

The primary objectives of capping an S/S monolith are to prevent surface water ponding and maintain a direct contact barrier. Accordingly, there are a number of different types of caps that have been approved by regulatory agencies. These are dependent upon site-specific conditions, site location and the agreed site-specific requirements of the remedial action. Approved caps range from a simple single layer of soil to more complex multiple layer systems, combining earthen and geo-synthetic materials.

9.2 Post-treatment management of S/S

9.2.1 Key engineering parameters S/S post-treatment management

A major consideration for site management after S/S is the assessment of the total amount of material that will be generated during the treatment. This is influenced by the volume increase from addition of the reagent and water, and in the case of in-situ S/S treatment, the volume of swell that will be produced.

The final volumes of S/S material will directly affect the site grading and monolith surface configuration, and future surface water management requirements. An assessment of the anticipated volumetric changes should be estimated early in the design process (i.e. during bench-scale testing), from the total absolute volume(s) of the materials involved in the treatment (soil + reagent + water). This figure can then be compared to the volume experienced during treatment.

The volume increase due to ex-situ mixing is most affected by:

- In-place vs excavated volumes, where expansion occurs during excavation as voids are introduced thus reducing the density of the material being treated. (This volume expansion is further impacted by the addition of reagent and mix water, and will need to be reconciled with the space available for replacement.)
- The in-situ (or field) moisture content, which can be used to partly make-up mix water needed for the reagent mixing/activation needs
- The required reagent quantities that must be added

The amount of swell that will be generated during in-situ S/S is most affected by a several key parameters that include the following:

- A variation in the moisture content of soils to be treated will directly impact on the full-scale water to reagent ratios applied, and the mixing performance experienced upon reagent injection. Thus the amount of native water present along with water injected with the reagent slurry will affect the amount of swell that will be generated
- The percentage of fine drained soil (clay and/or silt) will directly impact on swell, with higher percentages of fines resulting in a higher swell. The finer grained materials often require higher reagent and water additions to achieve uniform mixing. In addition, highly heterogeneous soil conditions will make the estimation of swell volumes difficult, and swell volumes exceeding 30 % have been recorded at a number of S/S sites
- The reagent density is an important parameter, which relates directly to the quantity of reagent needed for treatment. This is based on the density of the soil to be treated and the estimated volumes and the

specified mix design. As different reagent formulations/densities may be required for different soil types, the characterisation of sub-surface conditions is extremely important
- The minimum water to reagent ratio is typically in the range of 1:1 and is directly related to reagent density and viscosity. Maximum viscosity limits are needed to prevent problems, such as the clogging of reagent injectors during mixing. Backhoe-mixing for example can tolerate more viscous slurries, and thus produce lower swell volumes upon treatment

9.2.2 Engineering approach for preparing S/S material prior to capping

Preparation of the S/S monolith prior to capping will require interim and post-construction strategies for managing the S/S treated material, including any swell that has arisen from treatment, to achieve the final grading requirements for surface water management and to conform with site-specific logistical and structural constraints.

Typically, during ex-situ treatment, the treated material is placed and compacted in lifts, with the final shaping (of the monolith) accomplished concurrently with placement. Dozers, and sometimes rollers, as shown in Figure 9.3, are used for this purpose.

During typical in-situ S/S utilising large-diameter augers, excess material from binder injection is managed as each column is formed, using a backhoe or excavator. Swell material is graded or removed prior to starting the next S/S column.

As discussed previously, if this excess material needs to be moved and/or stockpiled for final grading, additional reagent may be required to recondition the disturbed, partly set swell in order to meet performance targets. This is a key consideration as the volume of swell material may be large, if fine grained soils are involved.

Figure 9.3: A smooth roller being used to compact ex-situ treated material

Two basic approaches for managing swell material for final placement/grading prior to capping are as follows:

Direct wet placement and grading is suitable for sites that are not constrained by structural or topographical features, such as buildings or waterways. This form of management is preferred by contractors, as the swell material does not require multiple handling, enabling the S/S operation to be completed as one continuous process. Although less flexible for modifying the contours of the cover system in response to S/S treated volume changes, a key advantage of this method is that the material is placed under the same conditions and within the timeframe of the parent S/S material.

Staging and Reconditioning of swell material involves sequential handling by stockpiling and reallocation to other areas of the site. Disturbance of the S/S material results in a loss of treatment integrity and produces properties that deviate from those of the parent material. Consequently, the swell material will need to be re-conditioned (with additional reagent) to meet performance characteristics consistent with the undisturbed treated material. The multiple handling of material further increases the overall volume of S/S material that will need to be managed, but does provide greater flexibility for modifying final cap geometry.

The grading of the S/S monolith can be completed using excavators or dozers with no forced compaction requirement. Figures 9.4 and 9.5 show examples of final grading/contouring applying a direct placement approach using an excavator and a dozer, respectively.

It is usually desired that the characteristics of the swell material are consistent with the undisturbed treated material, and meet the same performance requirements. A smooth homogeneous appearance of the swell material is to be expected following mixing, providing a high degree of confidence that the final surface of the monolith will perform similarly to the underlying undisturbed treated material.

Figure 9.6 is an example of placement and final grading of swell material by reconditioning using grout slurry with a low percentage of binder/reagent in 'lifts'. A dozer is used to construct an embankment and access ramp to an industrial area for use by semi-tractor trailers.

Reconditioning involves re-mixing stockpiled swell material in a shallow pit with the reagent grout. The dozer then pushes bucket loads of the swell material through the reagent grout (located in the pit behind the dozer) and immediately places this freshly reconditioned material onto the embankment.

Assessment of the volume of S/S material involves the quantity of additional reagent needed to return the dried and disturbed swell to a consistency and appearance similar to the freshly treated material from the full-scale S/S operation. This approach can be evaluated at bench-scale level by drying and then reconditioning S/S material with different reagent additions, followed by performance testing for properties such as UCS and permeability and then comparing with long term performance objectives.

Figure 9.4: Direct placement of S/S swell by an excavator for final grading

Figure 9.5: Final shaping/contouring of in-situ S/S using a dozer

Figure 9.6: Reconditioning and placement of in-situ S/S swell

9.3 Capping technologies

9.3.1 Overview of design considerations and general types of caps applicable for S/S monoliths

As mentioned, two of the primary objectives of a capping strategy are to prevent surface waters ponding on top of the monolith leading to degradation and/or leaching, and to provide a long-term direct contact barrier.

As also previously discussed, the S/S treated material will be subject to regulatory approvals, depending on the nature of the remedial program involved. Accordingly, the cap design will be site-specific and dependent on the intended function of the site (especially site-specific performance goals). Site-specific design considerations for capping the monolith include:

- Compatibility with required grading and drainage contours/final monolith geometry
- The required thickness of the cap ensuring protection of the monolith from environmental loads e.g. due to climate, such as freeze/thaw or wet/dry exposure, and to meet the regulatory requirements from a direct contact barrier
- The availability of acceptable earthen materials for constructing the cap (e.g. a low permeability clay)
- Erosion control requirements for the management of surface water runoff
- Support for native vegetative growth and landscaping restoration

There are several types of caps/capping systems that can be used on an S/S monolith, including:

- Earthen caps comprised of clay and/or other low permeability geo-materials
- Geomembrane systems involving e.g. high (or low) density polyethylene (HDPE, LDPE) or polyvinyl chloride (PVC) with fabrication options such as smooth or textured surfaces
- Geo-synthetic clay liner (GCL) systems that involve a bentonite clay layer bonded by geotextile fabric or geo-membrane
- Evapotranspiration (ET) covers using fine- and coarse-grained layers of earthen materials with different grain sizes to create a capillary break to prevent the build-up of saturated conditions over the monolith
- Asphalt and concrete pavements that can be used as the primary cap or in combination with other capping materials/technologies

These capping technologies may be used alone or in combined (composite) approaches at a specific site. They may, but do not always, include drainage and cover, and surface or topsoil-layers above the barrier layer. A general description for each of these specific layers follows.

The drainage layer is located directly above the barrier layer and may consist of granular free-draining sand and/or gravel material, or a geo-composite consisting of a geo-net bonded with a geotextile on one or both sides. Construction using a free-draining soil may include the placement of a non-woven geotextile filter fabric between the barrier layer and drainage layer and between the drainage layer and upper materials, to prevent potential piping and/or migration of fines that could lead to clogging of the drainage layer. The typical recommended minimum design thickness for the drainage layer is 12 in (30 cm), as thinner layers are more difficult to construct uniformly and quality assure, and may facilitate damage to the barrier layer during placement.

The cover layer is located directly over the drainage layer, and is designed to meet site-specific conditions and the location of the site. Considerations may include protection from vegetative root penetration and/or animal intrusion into either the drainage or barrier layers, and protection from freeze/thaw and/or desiccation and erosion. In some applications, the cover layer may be designated as an erosion layer, with a design thickness highly dependent on site-specific requirements. Different soil-type materials may be used, depending on site location and availability. Under some circumstances, the cover layer may also be a vegetative layer, if it is integrated with the surface layer and/or contains sufficient organics to support vegetative growth.

The surface layer or topsoil sustains adequate vegetative growth to minimise erosion, and protects the cover and provides an enhanced overall visual appearance. Topsoil is the most commonly used material but in some locations where it may be difficult to support growth (such as arid/desert-like climatic regions) other materials such as sand, gravel or cobbles may be more suitably used. Different types of geo-synthetic applications are also available which include the use of synthetic turf layers that eliminate the need for maintaining vegetation, reduce long-term concerns for cap stability and provide the appearance of a manicured landscape.

Evapotranspiration (ET) caps present an alternate approach to single and composite cap systems and differ substantially in the engineering mechanisms for managing surface water infiltration and precipitation. In contrast to relying on barrier and drainage layers, which direct surface water into a drainage layer to promote dissipation of infiltration off the cap, the ET relies primarily on storage within a soil layer. Removal of the "stored" infiltration would rely primarily on evapotranspiration and transpiration mechanisms. Hence, these types of caps are more suited to arid regions, although some have been installed at sites in Florida, Maryland, Pennsylvania, Georgia, Illinois, Michigan and Wisconsin (USEPA, 2003).

A further examination of each of these technologies and their applicability to S/S post-construction conditions follows.

9.3.2 Clay and/or other low permeability earthen materials

The use of clay and/or other low-permeability earthen materials as cover over S/S-treated material has met regulatory approval at a number of remediated sites. This approach offers flexibility for specifying locally available material at lower cost. Generally two earthen materials are used in two layers:

- An initial low permeability layer of a thickness depending on site-specific requirements, but typically 1-2 ft (30-60 cm) to meet the direct contact pathway (promotes surface water runoff and negates standing or pond water)
- An upper vegetative layer consisting of 6-12 in (15-30 cm) of topsoil/suitable material to support vegetative growth and aid landscaping

Depending on site characteristics, the cap may not include a drainage layer. A typical cap comprising a single-component clay, or low permeability earthen materials is shown in Figure 9.7.

A variety of different materials may be used to construct these types of caps. Acceptable materials can include those classified as: SW, SP, SM, SC, CL, CH or a combination of these group symbols specified under the Unified Soil Classification System (USCS).

Unspecified soil types are often referred to as unclassified fill, and the recommended minimum thicknesses should be 2 ft (60 cm) to provide sufficient thickness for a rooting zone. Such caps are generally placed in lifts and compacted.

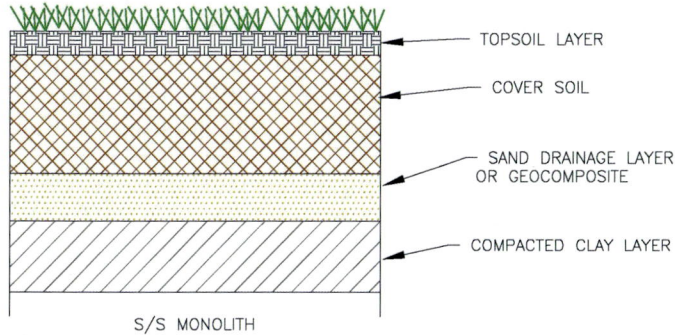

Figure 9.7: Example single component cap with a compacted clay layer

Figure 9.8: Cover layer comprising 2 ft of compacted soil, Schuylkill Metals, FL

Construction may or may not include a drainage layer and/or placement of a cover soil depending on the nature of the cap. An example of a successful capping layer comprising 2 ft of soil material is the Schuylkill Metals Site in Florida (Figure 9.8), where the S/S monolith and cap are located in the foreground. The lake in the background is not on the monolith.

Performance of an earthen cap will depend highly on the types of material and specified methods for construction. Low hydraulic conductivities may be possible with compaction, near optimum moisture contents and maximum dry densities based on standard or modified proctor test results. Some general design considerations include:

- Compaction should be evaluated against additional performance criteria, such as infiltration and suitability for long-term sustainable vegetative growth
- The capacity of the cap to effectively "store" surface water infiltration whilst promoting surface water drainage is a consideration (as in the design for ET caps) but is dependent on locally available materials/site location. In this respect, a lower level of compaction may provide a superior performance
- The cap thickness should be designed with storage capacity, long-term stability and potential for erosion as key design criteria
- Minimum index properties for the use of clay in low permeability caps include those indicated in Table 9.1 (Benson *et al.* (1994)

Property	Minimum Index
Liquid limit	20
Plasticity index	7
Percentage fines (i.e. particles passing the No. 200 sieve)	30
Percentage of clay	15
Activity	0.3

Table 9.1: Minimum index properties for the use of clay

However, there may be considerable variation in these properties depending on the availability of local materials, site-specific design requirements and flexibility in regulatory acceptance. Other considerations for selection of an appropriate material include the following:

- Greater cap thicknesses may be required where potential for freeze/thaw conditions exist, such as in the northern United States where frost can penetrate 6 ft below the ground surface, even if bench-scale studies indicate the monolith exhibits freeze/thaw resistance
- Lower permeabilities may be required where repeated wet/dry cycling are a concern, and/or site-specific conditions will not permit sufficient surface grades to promote effective drainage off the monolith
- Use of alternative materials with higher percentages of sand and/or gravel to provide subgrade conditions for potential future development, as plastic soils may not have the bearing strength for foundations/roadways

The engineering of soils to achieve low hydraulic conductivities may involve compaction on the 'wet' side of the optimum moisture content. Conventional specifications may include a minimum dry unit weight and a range of water contents >90 % of the maximum dry unit weight, and within 0-4 % (wet of optimum, modified Proctor), respectively. Figure 9.9 shows the construction of a 12 in (30 cm) thick, low permeability cap using locally available clay.

Figure 9.10 shows the final grading operations to promote positive drainage towards a waterway.

9.3.3 Geo-membranes

Geo-membranes or flexible membrane liners (FMLs) provide a relatively impermeable layer with readily available materials.

Geo-membranes are delivered in rolls that can be easily joined/seamed together, and include high density polyethylene (HDPE), low density polyethylene (LDPE), and poly vinyl chloride (PVC) although there are many other options.

These membranes are more flexible than compacted clay and can accommodate irregularities in an S/S monolith surface better than a clay cap. Water infiltration through a geo-membrane will occur through punctures, so care must be taken during the various stages of installation. Figure 9.11 illustrates the installation of a typical geo-membrane.

Figure 9.9: Construction of a low permeability clay cap

Figure 9.10: Final grading to direct surface drainage

Figure 9.11: Installation of a geo-membrane showing the welded seams between sheets

A consideration for using geo-membranes is the stability of the final cover slopes, as the smooth membrane surfaces have a low interface friction that can cause overlying soil layers to slide. Thus, textured membranes can be used on higher slope angles to reduce the possibility of sliding. A typical single component geo-membrane cap is shown in Figure 9.12.

The preparation of the subgrade is an important consideration during installation. Water infiltration through a punctured geo-membrane typically occurs before upper capping layers are in place. If the surface of the S/S monolith is used as the subgrade, it is recommended that particle sizes no larger than 3/8 in (1 cm) are allowed at the monolith surface to prevent puncturing of the liner. If subgrade conditions exceed this then an engineered fill of <3/8 in (1 cm) can be placed over the monolith. Alternatively, a thick (e.g. 12 ounce/380 g) non-woven geotextile could be used below the membrane.

9.3.4 Geo-synthetic clay liner (GCLs)

GCLs are readily available from manufacturers and can be easily installed. Performance of GCLs compares favourably to the applications using compacted clay with the advantage that it requires considerable less space for installation. A typical configuration for a single component cap using a GCL is illustrated in Figure 9.13.

An example of a GCL and soil cover over the S/S monolith is illustrated in Figure 9.14. This Figure also shows the use of a concrete flume to direct surface runoff across the cap.

A key design consideration is the very low internal shear strength of bentonite leading to instability on capped slopes. This can be compensated using fabrics, such as needle-punched geotextiles, which typically yield the highest shear strength followed by stitch bonded.

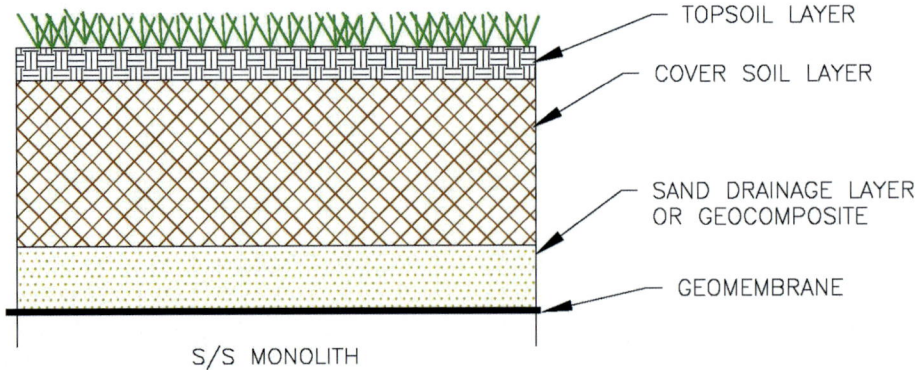

Figure 9.12: A single component cap with a geo-membrane

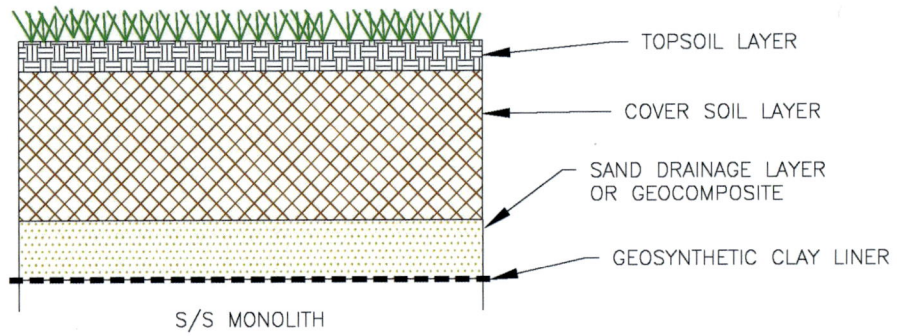

Figure 9.13: A typical single component cap with a geo-synthetic clay layer

Figure 9.14: GCL and soil cover with drainage channel, Peak Oil Site

Figure 9.15: Cover with a GCL, 2 ft of soil, and a gravel surface

However, the use of needle-punched backing will require a non-woven geotextile on one side Thus, the use of GCL's needs careful design consideration involving cap geometry, anticipated slope(s), interface friction angles and available cap materials used for layers above the GCL.

Cap design should include an acceptable safety factor to prevent sliding of the cover material over the liner system. GCLs using either needle-punched or stitch bonding can be used in this respect, as unreinforced GCLs are not recommended for slopes >10:1 (horizontal to vertical). In contrast, reinforced GCLs (needle-punched) have been successfully installed on slopes >3:1. A typical acceptable factor of safety for veneer stability is 1.5.

An example of a cover using a GCL and 24 in (60 cm) of soil, incorporating an added gravel surface for wear-resistance is illustrated in Figure 9.15, the American Creosote Site, Jackson, Tennessee. The gravel surface was added to facilitate site re-use as an equipment storage yard.

9.3.5 Clay with geo-membrane

When a geo-membrane is used over compacted clay, the clay provides a smooth base that minimises the potential for punctures during installation. Additionally, the geo-membrane can be quickly installed over the clay, protecting the clay from desiccation and cracking. The interface between the smooth clay and the geo-membrane can have low friction and may be susceptible to sliding. A textured membrane can be used to increase friction between the two layers. A typical composite cap using clay with a geo-membrane is shown in Figure 9.16.

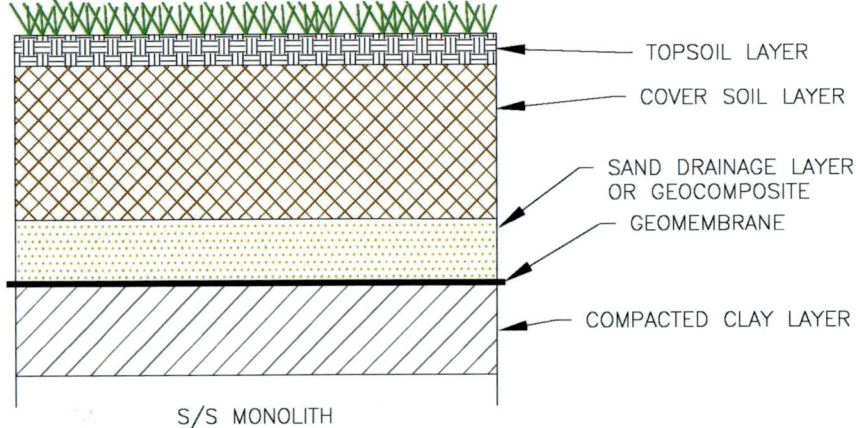

Figure 9.16: Profile of a composite cap with a geo-membrane over a compacted clay layer

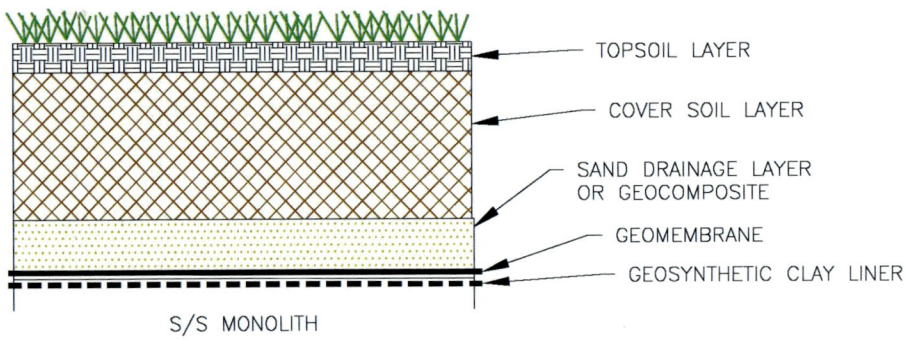

Figure 9.17: A composite cap with a geo-membrane over a GCL

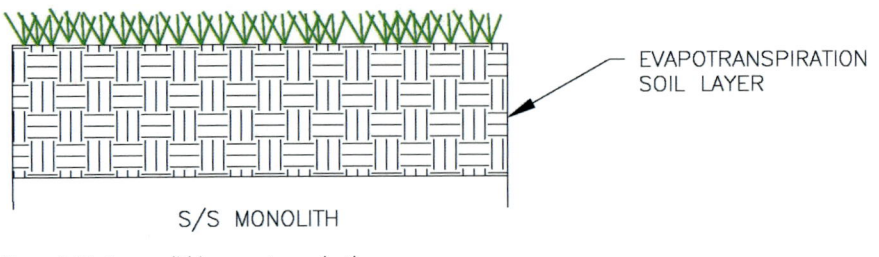

Figure 9.18: A monolithic evapotranspiration cap

9.3.6 Geo-composite clay liner with geo-membrane

Geo-membranes and GCLs can also be used together to provide a hydraulic barrier that is more flexible and resistant to differential settlement than compacted clay. A typical configuration for a composite cap using a GCL with a geo-membrane is illustrated in Figure 9.17.

9.3.7 Evapotranspiration (ET) caps

ET caps rely on a soil's water storage ability instead of low permeability layers to prevent the infiltration of water and ponding onto the underlying S/S mass.

There is an increasing use of these "alternative covers" but currently limited performance data to support their design guidance (USEPA, 2003). ET caps use soil layers to store water until the water is removed via evapotranspiration. Functionality is dependent upon the 'balance' between surface runoff, infiltration, soil storage, and evapotranspiration. ET caps are separated into two categories: monolithic, constructed in a single soil layer, and capillary break caps, constructed with two soil layers of differing grain size.

ET caps are generally only suitable for use in arid and semi-arid regions (e.g. the western United States) but have been occasionally used in areas with more humid conditions. ET caps may be lower cost than conventional caps if local soils are available to minimise transportation costs.

Monolithic ET caps rely on the water storage properties of a single soil layer, constructed to a thickness to promote storage at the peak rainfall times of the year. The soil layer used needs to have a storage capacity greater than the peak infiltration volume, to prevent infiltration into the underlying S/S material. A typical configuration for a monolithic ET cap is illustrated below in Figure 9.18.

Capillary break ET caps rely on the unsaturated hydraulic properties of soil to create a capillary break. This is facilitated by placing a fine-grained soil layer directly over a coarse-grained soil layer, with the former serving as a monolithic barrier for storing water, while the coarse layer acts as the capillary break. Essentially the water prefers to remain in the small pore spaces of the fine-grained material.

This system will prevent percolation as long as the coarse layer remains unsaturated, and therefore it is important to know the water balance of the area to ensure sufficient thickness of the fine grained storage layer. For capillary break ET caps, the fine-grained layer can range from 1.5-5 ft (45-150 cm) and the coarse-grained layer can range from 0.5-2 ft (15-60 cm) (USEPA 2003). A typical configuration for a capillary break ET cap is given in Figure 9.19.

An example of the placement of the coarse grained layer for an ET cap constructed over an S/S monolith at a project site in the south eastern region of the United States is illustrated in Figure 9.20.

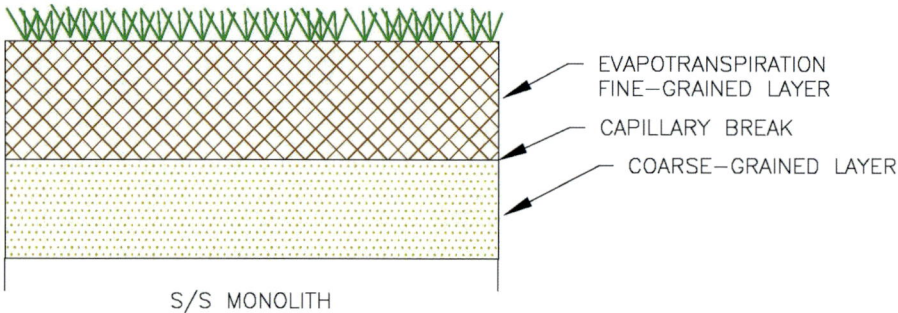

Figure 9.19: A capillary break evapotranspiration cap

Figure 9.20: A coarse-grained soil-based ET cap over an S/S monolith

9.3.8 Asphalt and concrete pavement applications

The integration of an asphalt and/or concrete pavement into a capping strategy can be important for S/S projects located in urban areas.

Urban applications include parking areas and/or streets that may need to be restored following completion of remedial construction. Design parameters that may need to be considered include:

- Final grades/slopes and elevations for the S/S monolith to meet required pavement structure subgrade and final grade requirements
- Utility corridors to accommodate surface water drainage for reconstructed streets or parking areas
- Integration of pavement structures with other components of the cap
- Transition of new pavement structures with existing structures at the limits of the S/S monolith

Figure 9.21: Reconstruction of an asphalt roadway over an S/S monolith in Sanford, FL

Figure 9.21 is an example of an asphalt roadway that was reconstructed over an S/S monolith. The landscaped areas on either side of the roadway consist of a capillary break ET cap. Prior to construction, detailed communications with the municipality were necessary to fully identify the reconstruction and restoration requirements, and detailed S/S designs were prepared for street alignment, profiles, grades, and the location of storm water collection points and conveyance. The final slope grading and construction requirements were also evaluated for the ET cap to meet final grades for the roadways.

9.3.9 Comparison of capping technologies

Table 9.2 provides a general comparison of capping technologies (excluding asphalt and concrete pavement) with respect to key criteria, consisting of the following:

Constructability: the ease or difficulty of construction which will be directly influenced by the types of materials selected and type of cap (i.e. single vs component or ET cap)

Permeability: of different capping scenarios is dependent upon site-specific drainage and local materials properties and availability

Freeze/Thaw Durability: is the susceptibility of capping materials to degradation on exposure to repeated cycles of freezing and thawing, and is highly influenced by geographic location, the type of materials used and final thickness of the cap.

Desiccation: is the susceptibility of the capping materials to the loss of moisture necessary to maintain the desired low hydraulic conductivity. Repeated cycles of wetting and drying can alter the properties of clay or GCL's. Design considerations include the available capping material, the sites geographic location and the engineering controls necessary to prevent moisture loss.

Type of Cap	Constructability	Permeability	Freeze/Thaw Durability	Desiccation
Single Component Caps				
Silty or clayey soils (SM, SC), Unclassified fills	Placement and compaction in lifts to meet specified compaction requirements	Varies but may achieve less than 1×10^{-5} cm/sec (1×10^{-7} m/sec)	Moderate durability	Moderate resistance
Compacted clay (CL, CH)	Higher level of placement and compaction QA/QC required than for earthen caps	Varies – typically between 1×10^{-5} (1×10^{-7} m/sec) to less than 1×10^{-7} cm/sec (1×10^{-9} m/sec)	Low durability if not provided sufficient cover for frost penetration	Low resistance if not provided sufficient cover for moisture loss
Geo-membrane	HDPE or PVC liners may require placement of appropriate fill to prepare subgrade prior to placement	Less than 1×10^{-7} cm/sec (1×10^{-9} m/sec)	High durability	High resistance
Geosynthetic clay liner (GCL)	GCLs could be placed directly over prepared ISS monolith surface	Less than 1×10^{-7} cm/sec (1×10^{-9} m/sec)	High durability	Low resistance if not provided sufficient cover for moisture loss
Composite Caps				
Clay with Geo-membrane	Clay layer would be placed directly over ISS subgrade followed by geomembrane	Less than 1×10^{-7} cm/sec (1×10^{-9} m/sec)	High durability	High resistance
Geosynthetic clay liner (GCL) with Geo-membrane	Use GCL bonded to geo-membrane and place directly over ISS subgrade	Less than 1×10^{-7} cm/sec (1×10^{-9} m/sec)	High durability	High resistance
Evapotranspiration (ET) Caps				
Monolithic	Place fine grain layer in lifts with limited compaction (e.g. rubber tyred or tracked equipment)	Not applicable because performance relies on storage of surface water infiltration	Low durability – generally not applicable for colder/humid climates	Low to moderate resistance depending on soil type
Capillary Break	Place fine and coarse grained layers in lifts with limited compaction (e.g. rubber tyred or tracked equipment)	Not applicable because performance relies on infiltration and storage of surface water with capillary break for drainage of excess water	Low durability – generally not applicable for colder/humid climates	Low to moderate resistance depending on soil type

Table 9.2: A comparison of capping technologies

9.4 Capping strategies based on slope, drainage, and climate

9.4.1 Slope and drainage considerations

Selection of the most appropriate capping strategy to meet anticipated slope and drainage conditions will be highly dependent on a number of site-specific factors and the intended function for the cap that include the following:

- Design limits for surface water infiltration
- Availability and type of local materials for cap construction
- Limitations for allowable grades to promote effective drainage
- Logistical constraints based on surrounding topography and existing land use

Potentially applicable capping strategies based on site-specific logistical and surface water drainage conditions are summarised in the Table 9.3.

Gently graded slope conditions will require significant reliance on barrier or ET layers to prevent surface water infiltration through the cap, and pose the most significant concern for ponding or standing water over the monolith.

Under most barrier design applications, it is not recommended to have cap design geometry with less than 3 % slope. In addition, dependent on geographic location, inadequate drainage may lead to extended saturated conditions leading to long-term operation and maintenance (O&M) challenges, for maintaining acceptable cap vegetation and landscaping.

As indicated in Table 9.3 moderately graded slopes may provide optimal conditions for a range of capping technologies to meet performance objectives. In contrast to gently graded conditions, the presence of steeply graded slopes may impact on soil veneer stability and/or erosion of cover materials, precluding the use of materials such as GCLs with low interface friction angles.

Construction using earthen materials may be applicable for steeply graded slopes if the material can be adequately compacted to minimise erosion. Erosion can be a concern for ET covers on steeper slopes due to the low compaction used to enhance water storage capacity.

9.4.2 Capping strategies based on climatologic conditions

The prevailing climatic conditions at a site are a key factor in determining the most appropriate capping technology and the types of materials that can be used (Table 9.4). Depending on the location, seasonal conditions will influence a number of design factors including:

- Desiccation of low permeability clay layers compromising cap integrity and increasing erosion
- Loss of cap integrity due to repeated cycles of freeze/thaw
- Slope instability and erosion due to cap materials becoming saturated
- Ability to establish adequate vegetative growth

The two critical design parameters most affected by geographic location are freeze/thaw durability and desiccation. Therefore, careful consideration should be given to the type of barrier selected in single component cap applications where extreme fluctuations in temperature and/or precipitation occur. Composite component cap applications using geomembranes enhance long-term performance.

Type of Cap	Applicability of Capping Technologies Based on Slope and Drainage Conditions		
	Condition 1 – Gently Graded Slopes (less than 5%) with poor or restricted drainage	Condition 2 – Moderately Graded Slopes (5 to 10%) with positive drainage	Condition 3 – Steeply Graded Slopes with rapid drainage
Single Component Caps			
Silty or clayey soils (SM, SC, ML, CL-ML). Unclassified fill materials	Less Applicable[1]	Applicable[2]	Applicable
Compacted clay (CL, CH)	Applicable	Applicable	Less Applicable
Geo-membrane	Applicable	Applicable	Less Applicable
Geosynthetic clay liner (GCL)	Applicable	Applicable	Potentially Applicable[3]
Composite Caps			
Clay with Geo-membrane	Applicable	Applicable	Less Applicable
Geosynthetic clay liner (GCL) with Geo-membrane	Applicable	Applicable	Potentially Applicable
Evapotranspiration (ET) Caps			
Monolithic	Potentially Applicable	Applicable	Applicable
Capillary Break	Potentially Applicable	Applicable	Applicable

[1] Less Applicable = significant engineering, construction, or maintenance issues that must be resolved for effective long-term implementation of the technology; [2] Applicable = can be applied effectively using standard engineering and construction practices. [3] Potentially Applicable = advanced engineering or construction practices may be needed to effectively implement the technology.

Table 9.3: Capping technologies vs slope and drainage

Freeze/thaw durability and desiccation are less likely for monolithic or capillary break ET caps because their performance is based upon water storage potential and the evapotranspiration of surface water infiltration rather than serving as a simple barrier layer.

9.5 Post-construction monitoring

Once construction of the S/S remedy, including the cap and any ancillary features, is complete, then an extended period of monitoring is often required.

The objectives/requirements of the monitoring program are often very site-specific, and dependent on the requirements of the site owner and regulatory agencies. It is best to define the objectives and general requirements for post-construction monitoring (including the frequency of monitoring events, longevity of monitoring, and general reporting requirements) during the initial design phase, before construction of the S/S remedy begins.

Details of the monitoring system, such as exact placement and design of monitoring wells, are best determined at the close of remedy construction as the precise quantities of soil treated and its final placement, including the cap, often vary in detail from the original pre-construction design. Considerations for post-construction monitoring of S/S sites are further discussed below. However, this section is not intended as a definitive reference for developing post-construction monitoring programs

9.5.1 Regulatory impetus and guidance for post-construction monitoring

Key factors for the development of a post-construction monitoring program are requirements of the responsible regulatory agencies. These may vary greatly depending on jurisdiction and upon site-specific factors.

An example is the USEPA guidance document on post-construction completion activities at Superfund sites (USEPA, 2005), which lists activities as:

Long-Term Response Action (LTRA) generally applies to the first 10-years of fund financed ground and surface water restoration.

Operation and Maintenance (O&M) includes the activities required to maintain the effectiveness and integrity of the remedy. Also includes continued operation of ground and surface water restoration remedies after LTRA.

Five-Year Reviews are required by statute to assure protectiveness for any remedial action that leaves hazardous substances on a site above levels that allow for unlimited use and unrestricted exposures. Five-year reviews are also conducted as a matter of policy in other situations.

Institutional Controls (IC) using non-engineered instruments, such as administrative and/or legal controls, that typically minimise the potential for human exposure to contamination and/or protect the integrity of the remedy by limiting land or resource use.

Table 9.4: Capping technologies and climate

Type of Cap	Applicability of Capping Technologies Based on Climate Conditions[1]			
	Humid (Dfa)	Humid and Moist Subhumid (Cfa)	Dry, Subhumid (Dfb)	Semi Arid (Bsk)
Single Component Caps				
Silty or clayey soils (SM, SC). Unclassified fills	Applicable[3]	Applicable	Potentially Applicable[4]	Applicable
Compacted clay (CL, CH)	Applicable	Applicable	Potentially Applicable	Less Applicable[2]
Geo-membrane	Applicable	Applicable	Applicable	Applicable
Geosynthetic clay liner (GCL)	Applicable	Applicable	Potentially Applicable	Less Applicable
Composite Caps				
Clay with Geo-membrane	Applicable	Applicable	Potentially Applicable	Potentially Applicable
Geosynthetic clay liner (GCL) with Geo-membrane	Applicable	Applicable	Potentially Applicable	Potentially Applicable
Evapotranspiration (ET) Caps				
Monolithic	Less Applicable	Less Applicable	Potentially Applicable	Applicable
Capillary Break	Less Applicable	Less Applicable	Potentially Applicable	Applicable

[1] The climatic zones are based on Thornthwaite (1948): humid, sub-humid, semiarid; with supplemental Koppen-Geiger classifications by Rubel and Kottek (2010) where Bsk=arid, steppe, cold; Cfa=warm temperate, fully humid, hot summer; Dfa=snow, fully humid, warm summer; and Dfb=snow, fully humid, warm summer. [2] Less Applicable = significant engineering, construction, or maintenance issues that must be resolved for effective long-term implementation of the technology. [3] Applicable = can be applied effectively using standard engineering and construction practices. [4] Potentially Applicable = advanced engineering or construction practices may be needed to effectively implement the technology.

Remedy Optimisation involves performing reviews to improve the performance and/or reduce the annual operating cost of remedies without compromising protectiveness.

NPL Deletion is the removal of sites or portions of sites from the NPL (National Priority List) because no further response action is appropriate.

Re-use involves working with the parties seeking to redevelop Superfund sites to ensure that their activities do not adversely affect the implemented remedy.

Except for NPL deletion, these activities may apply to any S/S remediation under any jurisdiction, since the S/S process leaves contaminants, though immobilised, in place within the S/S treated matrix.

Therefore, activities to assure the continued protectiveness and maintenance of the remedy are important, and include regular reviews such as five-year reviews, mentioned above.

For Superfund sites where contaminants are left in place after construction completion, a review is required every 5 years to assure that the remedy is still protective and is well maintained. Guidance for conducting such 5-year reviews has been prepared by the USEPA (USEPA, 2001). Although designed specifically (for the USEPA Superfund program) the guidance, which includes check-lists is a valuable resource for post-construction inspection, monitoring, and reporting, at any site where S/S has been employed.

A number of completed 5-year review reports are available on the USEPA website (USEPA, http://cumulis.epa.gov/fiveyear/index.cfm.

Additional guidance is available in the Environment Agency guidance document on the use of S/S (Environment Agency, 2004a), especially Section 5 and Appendix 4. The EA document (page 56) suggests that "Specific objectives for long-term monitoring for a re-use scenario" may be:

- To demonstrate whether S/S remains effective
- To provide a basis for implementing mitigation measures
- To identify detrimental changes in the re-use scenario (e.g. water table rise)
- To provide a basis for ceasing monitoring

The EA document also discusses what topics should be included in a monitoring report and the potential decision basis for discontinuing monitoring.

The ITRC (Interstate Technology & Regulatory Council) has also recently published guidance on Development of Performance Specifications for S/S, including in Section 7, post-construction monitoring, referred to as "stewardship" in their document.

The ITRC document states (ITRC, 2011, page 52): "Long-term stewardship of a completed S/S remedy may include monitoring of environmental media in contact with and potentially affected by the remedy, monitoring of institutional controls, monitoring and maintenance of engineering controls, financial assurances, and periodic review(s) by the controlling environmental agency.".

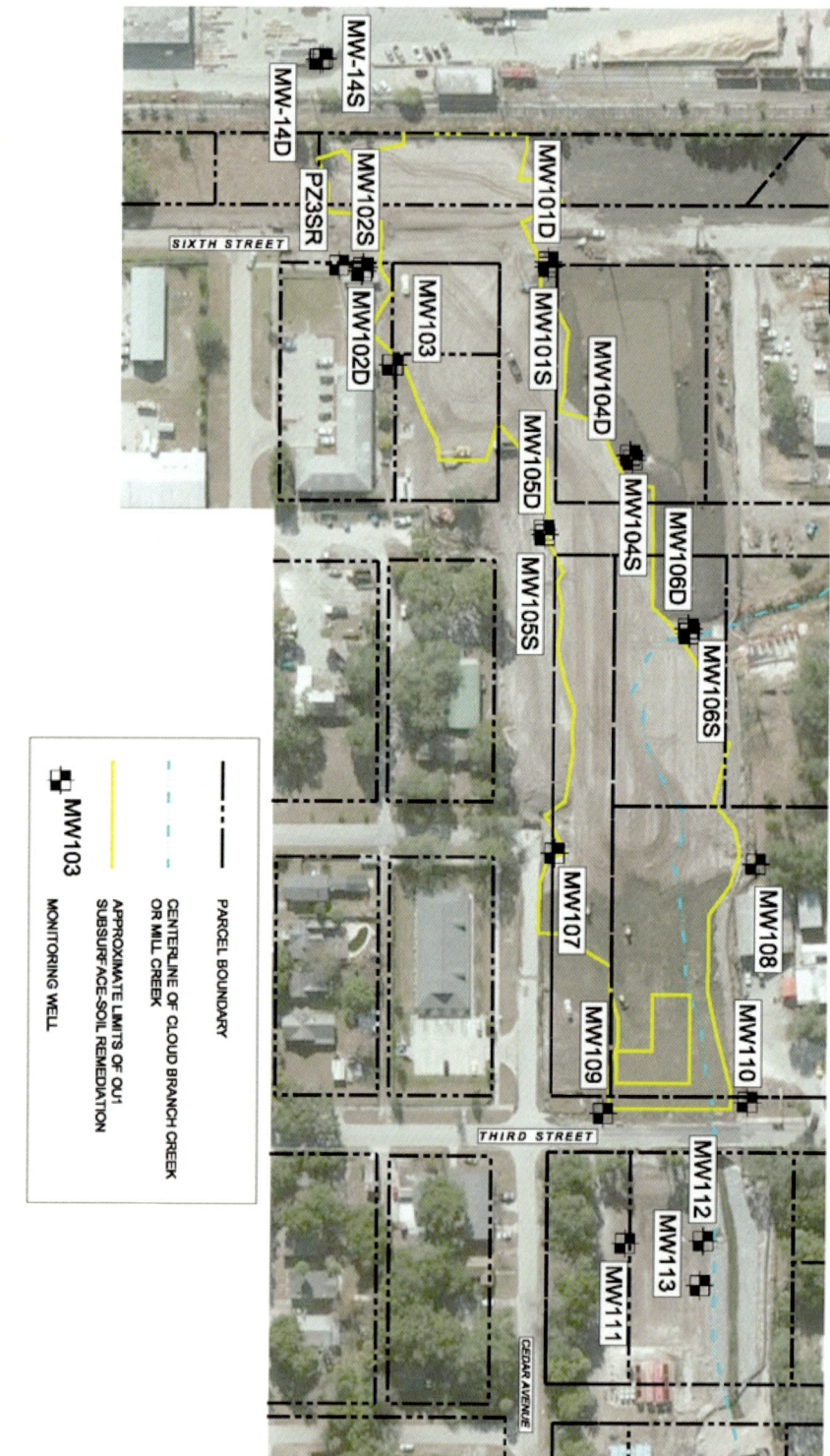

Figure 9.22: Groundwater monitoring wells around an S/S monolith at the Sanford gasification plant site

9.5.2 Specific features common in monitoring programs

Monitoring of remediated S/S sites is done for the following purposes:

To assure that the S/S treated material continues to meet its original design performance property of reducing the release rate of contaminants to groundwater or surface water bodies through low permeability and low leachability. As discussed in Section 6.4.4, it is impractical and generally not necessary to obtain representative samples of cured materials in the field long after treatment is completed. Rather, monitoring the effects of treatment through groundwater quality monitoring adjacent to and/or down gradient of the treatment zone is fairly standard practice.

Demonstration that release of contaminants to groundwater is adequately controlled over the long-term is usually made by locating groundwater wells immediately adjacent to the placed S/S material, and sampling for a selected list of COC's, along with field measurements for pH, Eh, and other selected indicator parameters.

'Immediately adjacent' is a relative term as it is influenced by the rate of groundwater movement and physical access including avoidance of compromising the cap. For example a distance of 10 -100 ft (3-30 m) from the edge of the monolith can be considered as normal. Often a subset of the COC's present in the S/S treated material is selected with preference for those most likely to be detected (most soluble) should the S/S material fail to maintain its control over release.

Although sampling/monitoring is often conducted quarterly (or seasonally) for the first couple of years, the frequency may be reduced thereafter to semi-annually or annually if no issues are detected. It should be noted that immediately after S/S treatment, especially in-situ S/S treatment, there will be a slightly elevated pH in the groundwater contacting the treated material. This is normal, should dissipate in a few months, and does not indicate a failure of the treated material. An example of a monitoring well network around an S/S monolith is indicated in Figure 9.22.

To document that groundwater quality is improving after remediation:
Often where S/S is implemented as the source control technology to treat contaminated soils, groundwater has also been impacted through the release of COC's from the contaminated soils. Thus, once the source has been successfully managed by S/S, it is reasonable to expect the groundwater quality to improve with time.

For some sites, this has led to selection of MNA (monitored natural attenuation) as either the primary, or secondary, method for remediating impacted groundwater. Therefore, monitoring wells are installed and a monitoring plan developed to assess groundwater improvement over time and eventual achievement of the agreed groundwater quality goals.

To document that engineering controls are functioning properly and maintenance is being conducted:
Engineering controls can include fencing, the cap itself, surface water diversion/runoff controls, and vertical impermeable walls/barriers.

Caps, surface water diversion controls and other surface engineering are inspected during the periodic site visits, for groundwater sample collection events, and after unusual precipitation events. Subsurface engineering controls like vertical impermeable walls are monitored indirectly through groundwater monitoring in the same manner as the S/S treated material. If the site has a vegetative cap, this is inspected and any necessary repair carried out. It is important that invasive trees and shrubs be removed from caps as their roots can not only cause damage, but may over time invade the S/S material.

To document the continued application of institutional controls:
Institutional controls include deed restrictions and other prohibition of excavation within the S/S treated material or construction of wells through the treated material without prior approval from the regulatory agencies. They may also restrict how the surface can be used. Controls are applied by regulatory agencies and often enforced at a local level. Any inappropriate site use is noted during periodic site inspections and a check can be made to assure that institutional controls are being recorded on property transfer deeds.

To document the exit-strategy:
This is highly variable and dependent on the nature of regulatory oversight and site-specific characteristics.

Under the USEPA Superfund program, S/S treated soils, which still contain the COC's, despite being treated to control release, fall into the same category as containment cells, and 5-year reviews are required in perpetuity.

Sites remediated under State programs in the USA follow the requirements of individual states, which are highly variable. The EA guidance document on S/S, as previously mentioned, provides information on the potential decision basis for discontinuing monitoring. In any case, the site owner, regulatory agency(s), and other stakeholders will need to reach a consensus on what aspects of monitoring can be discontinued and on what basis. The authors believe that if the S/S treatment is still performing well after 15-20 years, it will likely continue to do so indefinitely. Support for this is provided in the PASSiFy project report (PASSiFy 2010) evaluating the long-term performance of S/S treated material at a number of sites in the USA, France and the UK.

REFERENCES

1. ASTM International, http://www.astm.org/DIGITAL_LIBRARY/index.html
2. ASTM Standard D422(2007)e2. Standard Test Method for Particle-Size Analysis of Soils. ASTM International, West Conshohocken, PA, 2007, www.astm.org
3. ASTM Standard D698-07. Standard Test Methods for Laboratory Compaction Characteristics of Soil Using Standard Effort (12400 ft-lbf/ft^3 (600 kN-m/m^3)). ASTM International, West Conshohocken, PA, 2007, www.astm.org
4. ASTM Standard D1557-00. Standard Test Methods for Laboratory Compaction Characteristics of Soil Using Modified Effort (56000 ft-lbf/ft^3 (2700 kN-m/m^3)). ASTM International, West Conshohocken, PA, 2000, www.astm.org
5. ASTM Standard D1633-00 (2007). Standard Test Methods for Compressive Strength of Molded Soil-Cement Cylinders. ASTM International, West Conshohocken, PA, 2007, www.astm.org
6. ASTM E1689-95(2014). Standard Guide For Developing Conceptual Site Models for Contaminated Sites, ASTM International, West Conshohocken, PA, 2014 www.astm.org
7. ASTM Standard E2081-00(2015). Standard Guide for Risk-Based Corrective Action. ASTM International, West Conshohocken, PA, 2015,
8. ASTM Standard D2487-11. Practice for Classification of Soils for Engineering Purposes (Unified Soil Classification System). ASTM International, West Conshohocken, PA, 2011, www.astm.org
9. ASTM Standard D2488-09a. Standard Practice for Description and Identification of Soils (Visual-Manual Procedure). ASTM International, West Conshohocken, PA, 2011, www.astm.org
10. ASTM Standard D4318-10e1. Standard Test Methods for Liquid Limit, Plastic Limit, and Plasticity Index of Soils. ASTM International, West Conshohocken, PA, 2011, www.astm.org
11. ASTM Standard D5084-10. Standard Test Methods for Measurement of Hydraulic Conductivity of Saturated Porous Materials Using a Flexible Wall Permeameter. ASTM International, West Conshohocken, PA, 2011, www.astm.org
12. Barbisan, U., Guardini, M., (2007). Reinforced concrete: a short history. Tecnologos, Cavriana, Mantova, Italy
13. Bates, E.R., Akindele, F. and Springle, D., (2002). American Creosote Site Case Study: Solidification/Stabilisation of Dioxins, PCP, and Creosote for $64 per Cubic Yard. *Environmental Progress* 21, 2, pp 79-84
14. Benson, C.H., Zhai, H., and Wang, X., (1994). Estimating Hydraulic Conductivity of Compacted Clay Liners. *Journal of Geotechnical Engineering*, 120, 2, pp 366-387
15. Boyd, S.A., and Mortland, M.M., (1987). Use of Modified Clays for Adsorption and Catalytic Destruction of Contaminants, in *13th Ann. Research Symposium at Cincinnati*. U.S. Environmental Protection Agency, Cincinnati, OH. (July 1987)
16. Bozkurt, S., Moreno, I. and Neretnieks, I., (2000). Long Term Processes in Waste Deposits. The Science of the Total Environment 250, 1-3, pp 101-121
17. British Cement Association, (2004). The Essential Guide to Stabilisation/Solidification for the Remediation of Brownfield Land using Cement and Lime. BCA Report 46.112,
18. Butler, S. M., Barth, E. F., and Barich, J. J., (1996). *Field Quality Control Strategies for Assessing Solidification/Stabilisation*, Stabilisation and Solidification of Hazardous, Radioactive, and Mixed Wastes: 3rd Volume, ASTM STP 1240, T. (Michael Gilliam and Carlton C. Wiles, Eds.), American Society for Testing and Materials
19. Caterpillar Compaction Manual, (1989). Caterpillar Tractor Company, Peoria, Illinois
20. Chateau, Laurent, personal communication, (November 2012)
21. Christenson, H. and Wakamiya, W., (1987). In HazTech News, (June 18)
22. CFEM, Canadian Foundation Engineering Manual, Canadian Geotechnical Society, (2006)
23. Conner, J.R., (1990). Chemical Fixation and Solidification of Hazardous Waste. Van Nostrand Reinhold
24. Control of Pollution Act (1974): http://www.legislation.gov.uk/ukpga/1974/40
25. Control of Pollution (Amendment) Act (1989): http://www.legislation.gov.uk/ukpga/1989/14. Accessed 18.8.12
26. Cote, P., (1987). *Assessment of Solidification Technologies for the Immobilisation of Organic Compounds*, Draft by Wastewater Technology Centre, Environment Canada.
27. Delatte, N.J., (2001). Lessons from Roman cement and concrete. *Journal of Professional Issues in Engineering Education and Practice* 127, 3, pp 109-115
28. Dragun J., (1988). *The Soil Chemistry of Hazardous Materials*. Hazardous Materials Control Research Institute, Silver Springs, MD
29. Dragun, J., and Heller, C.S., (1985). Physicochemical and structural relationships of organic chemicals undergoing soil and clay catalyzed free radical oxidation. Soil Science 139, 2, pp,. 100-111
30. Ells, C., (2010). In: Proceedings of the International Solidification/Stabilisation Technology Forum. (Eds. Craig B. Lake and Colin D. Hills). June 14th -17th, Cape Breton University, Sydney, NS, Canada.

31. Emerick, K., (1995). The Survey and Recording of Historic Monuments. *Quarterly Journal of Engineering Geology* 28, pp,. 201-205
32. ENDS, (1988). Report 158, P8
33. ENDS, (1983). Report 120, pp 10-13
34. Environment Agency, (2004a), Guidance on the Use of Stabilisation/Solidification for the Treatment of Contaminated Soil, Codes and Standards for Stabilisation and Solidification Technologies (CASSST) Project, Science Report SC980003/SR1, September 2004, www.environment-agency.gov.uk
35. Environment Agency, (2004b), Review of Literature on the use of Stabilisation/solidification for the Treatment of Contaminated Soil, Solid Waste and Sludges. Codes and Standards for Stabilisation and Solidification Technologies (CASSST) Project, Science Report SC980003/SR2, September 2004, www.environment-agency.gov.uk
36. Environment Agency, (2004c). Model Procedures for the Management of Land Contamination, CLR 11
37. EPRI, An Integrated Approach to Evaluating In-Situ Solidification/Stabilisation of Coal Tar Impacted Soils, Electric Power Research Institute, EPRI, Palo Alto, CA, March 2009, Report 1018612
38. European Landfill Directive (1999). Directive 1999/31/EC. http://eur-lex.europa.eu/LexUriServ/LexUriServ.do?uri=CELEX:31999L0031:EN:NOT
39. Fernández Pereira, C., Luna, Y., Querol, X. D. Antenucci, D. and Vale, J., (2009). Waste stabilisation/solidification of an electric arc furnace dust using fly ash-based geo-polymers. *Fuel* 88, 7, pp 1185-1193
40. Fleri, M. A., and Whetstone, G. T., (2007). In-situ Stabilisation/Solidification: Project Lifecycle, *Journal of Hazardous Materials*, 141, pp 441-456
41. Garrido-Ramirez, E.G., Theng, B.K.G. and Mora, M.L., (2010). Clays and oxide as catalysts and nanocatalysts in Fenton-like reactions – A Review. *Applied Clay Science* 47, 3-4, pp 182-192
42. Gunning, P., (2011). Accelerated carbonation of hazardous wastes (PhD Thesis) oai:gala.gre.ac.uk:7135. http://gala.gre.ac.uk/7135/1/Peter_John_Gunning_Accelerated_carbonation_2011.pdf
43. Goddard, E.N., (1979). Rock Colour Chart, Geological Society of America, Boulder Colorado
44. Guo, X. and Shi, H., (2012). Self-Solidification/Stabilisation (S/S) of Heavy Metal Wastes (HMWs) of the Class C Fly Ash (CFA)-Based Geopolymers. Journal of Materials in Civil Engineering doi:10.1061/(ASCE)MT.1943-5533.0000595
45. Hall, C., (1976). On the History of Portland Cement after 150 Years. *Journal of Chemical Education* 53, 4, pp 222-223
46. Hills, C.D., Antemir, A., Leonard, S.A. and Carey, P.J., (2010). Applications of Stabilisation/Solidification for the Treatment of *Organically Contaminated Soil and Waste*. In: Proceedings of the International Solidification/Stabilisation Technology Forum. (Eds. Craig B. Lake and Colin D. Hills). June 14th-17th, Cape Breton University, Sydney, NS, Canada.
47. Hobbs, D.W and Taylor, M.G., (2000). Nature of the Thaumasite Sulfate Attack Mechanism in Field Concrete. *Cement and Concrete Research* 30, pp 529-533
48. Integrated Pollution Prevention and Control: http://www.defra.gov.uk. Accessed 18.8.12
49. ITRC, (2011). Development of Performance Specifications for Solidification/Stabilisation, Interstate Technology Regulatory Council, ITRC, 50 F Street, NW, Suite 350,Washington, DC 20001, http://www.itrcweb.org/gd.asp
50. Jana, D., (2007). Evidence from Detailed Petrographic Examinations of Casing Stones from the Great Pyramid of Khufu, a Natural Limestone from Tura, and a Man-made (Geopolymeric) Limestone. *Proceedings of the 29th Conference on Cement Microscopy*. Quebec City, Canada, May 20 -24, 2007
51. Johnson, R., (1987). History and development of modified bitumen. *Proceedings of the 8th Conference on Roofing Technology*, pp 81-84
52. Kanare, H., Milevski, I., Khalaily, H., Getzov, N., Nasvik, J., (2009). How Old is Concrete? *Concrete Construction*.
53. Kiefer, D.M., (2012). Its all about Alkali, in Chemistry Chronicles: http://pubs.acs.org/subscribe/archive/tcaw/11/i01/html/01chemchron.html. Accessed 18.8.12
54. Klich, I., (1997). Permanence of Metal Containment in Solidified and Stabilised Wastes. Unpublished Ph.D. Thesis, University of Texas A and M
55. Kostecki, P.T., Calabrese E.J., and Bonazountas M., (Eds.), (1992). Hydrocarbon Contaminated Soils (Volume 2). *Proceedings 5th Conf on Hydrocarbon Contaminated Soils*, Amherst, Mass. Sept. 24 -27, 1990. CRC Press
56. Kyles, J.H., et al., (1987). Solidification/Stabilisation of Hazardous Waste: A Comparison of Conventional and Novel Techniques, in *Toxic and Hazardous Wastes: Proceeding of the Nineteenth Mid-Atlantic Waste Conference*. Bucknell University, Lewisburg, PA.
57. Lake, C.B., (2013). Assessing Geo-Environmental Performance of Cement-Based Containment Systems, 2011 Canadian Geotechnical Society Colloquium Address, to be submitted to the Canadian Geotechnical Journal
58. Landfill of Waste: http://ec.europa.eu/environment/waste/landfill_index.htm. Accessed 18.8.12
59. Lange, L.C., Hills, C.D. and Poole, A.B., (1997). Effect of Carbonation on Properties of Blended and Non-blended Cement Solidified Waste Forms. *Journal of Hazardous Materials* 52, pp 193-212
60. Lange, L.C., Hills, C.D. and Poole, A.B., (1996). Preliminary Investigation into the Effects of Carbonation of Cement-Solidified Hazardous Wastes. *Environmental Science and Techology* 30, pp 25-30

61. Lear, P.R., and Conner, J., (1992). Immobilisation of Low Level Organic Compounds in Contaminated Soil, in P. Kostecki et al. (Eds.). *Hydrocarbon Contaminated Soils, Volume II*, Lewis Publishers, Chelsea, MI.
62. Leonard, S.A. and Stegemann, J.A., (2010). Stabilisation/Solidification of acid tars. *Journal of Environmental Science and Health* A, 45, 8, pp 978-91. doi: 10.1080/10934521003772394
63. MacKenzie, K. J. D., Smith, M.E., Wong, A., Hanna, J.V., Barry, B. J., Barsoum, M. W., (2011). Were the casing stones of Senefru's Bent Pyramid in Dahshour cast or carved? *Materials Letters* 65,2, pp 350-352
64. MacLaren, D.C., White, M.A., (2003). Cement: Its Chemistry and Properties. *Journal of Chemical Education* 80, 6, pp 623-635
65. Mays, L.W., (2010). A Brief History of Water Technology During Antiquity: Before the Romans. In: Mays, L.W. (Eds.) *Ancient Water Technologies*, Springer Science
66. Miller, R., (2009). Diamonds and Cement: A Review of Concrete Over the Ages. International Polished Concrete Institute. Available at http://www.ipcionline.org/. Accessed 14.08.12 http://www.ipcionline.org/index.cfm?fuseaction=journal.showArticle&mainArticle=EE5D6012-D9BE-42B5-B04F-027A3C7068DD&categoryID=CS
67. McConnell, J.D.C., (1955). The Hydration of Larnite (B-Ca2SiO4) and Bredigite (A1-Ca2SiO4) and the Properties of the Resulting Gelatinous Plombierite. *Mineral Magazine* 30, pp 672-680
68. Montgomery, D.M., Sollars, C.J., Perry, R., Tarling, S.E., Barnes, P. and Henderson, E., (1991). Treatment of organic-contaminated industrial wastes using cement-based stabilisation/solidification – 1. Microstructural analysis of cement-organic interactions. *Waste Management and Research* 9, 2, pp 103-111.
69. Moorey, P.R.S., (1994). Ancient Mesopotamian Materials and Industries: The Archaeological Evidence. Oxford University Press
70. Mottershead, D.N., (2000). Weathering of Coastal Defensive Structures in South-West England: A 500 Year Stone Durability Trial. *Earth Surface Processes and Landforms* 25, pp 143-1159
71. Ortego, J.D., (1989). New Chemical Technology for the Containment of Hazardous Substances, personal communication, (June, 1989)
72. PASSIFy, (2010). Performance Assessment of Stabilised/Solidified Waste Forms, Final Report. CL:AIRE Project RP16. http://www.claire.co.uk/index.php?option=com_cobalt&view=record&cat_id=23:stabilisation-solidification&id=298:performance-assessment-of-stabilisedsolidified-waste-forms-passify&Itemid=61
73. Penseart, S., De Groeve, S., Staveley, C., Menge, P and De Puydt, S., (2008). Immobilisation, Stabilisation, Solidification: A New Approach for the Treatment of Contaminated Soils. Case Studies: London Olympics and Total Ertveldt. 15th Innovation Forum Geotechnique
74. Perara, A.S.R., Al-Tabbaa, A., and Johnson, D., (2004). *State of Practice Report, UK Stabilisation/Solidification Treatment and Remediation – Part VI: Quality Assurance and Quality Control*. www-starnet.eng.cam.ac.uk
75. Pojasek, R., (1978). Stabilisation, Solidification of Hazardous Wastes. *Environmental Science and Technology* 12, 4, pp 382-336
76. Posasek, R. B., (1980). (Ed.) *Toxic and Hazardous Waste Disposal*, Vol. 4. Ann Arbor Science Publishers, Inc., Ann Arbor, MI.
77. PCA, (1997). Guide to Improving the Effectiveness of Cement-Based Stabilisation/Solidification, PCA Publication Number EB211
78. Pollution Prevention and Control Act, (1999): http://www.legislation.gov.uk/ukpga/1999/24. Accessed 18.8.12
79. Reid, J.M. and Clarke, G.T., (2001). The Processing of Contaminated Land in Highway Works. TRL Report 489
80. Rogers, R.W., (1900). A History of Babylonia and Assyria: Volume 2. Assyrian International News Agency
81. Rubel, F. and Kottek, M., (2001). *Observed and Projected Climate Shifts 1901-2100 Depicted by World Maps of the Köppen-Geiger Climate Classification*. Meteorologische Zeitschrift 19, pp 135-141. DOI: 10.1127/0941-2948/2010/0430, 2001
82. Rudland, D. J., Lancefield, R. M., Mayell, P. N., (2001). Contaminated Land Risk Assessment – A Guide to Good Practice, CIRIA Report: C552
83. Shi, C., (2004). Quality Assurance/Quality Control (QA/QC) for Waste Stabilisation/Solidification, in *Stabilisation and Solidification of Hazardous, Radioactive, and Mixed Wastes*, (R. D. Spence and C. Shi, Eds.). CRC Press, Boca Raton, FL.
84. STARNET: Stabilisation/Solidification Treatment and Remediation Network: http://www-starnet.eng.cam.ac.uk. Accessed 18.8.12
85. Sweeting, M.M., (1960). The Caves of the Buchan Area, Victoria. *Zeitschrift fur Geomorphologie* 2, pp 81-91
86. Terzaghi, K., Peck, R.B. and Mesri, G., (1996). *Soil Mechanics in Engineering Practice*, John Wiley and Sons, 3rd Edition
87. The Single European Act: http://europa.eu/legislation_summaries/institutional_affairs/treaties/treaties_singleact_en.htm. Accessed 18.8.12
88. The Sixth Environment Action programme of the European Community 2002-2012: http://ec.europa.eu/environment/newprg/intro.htm. Accessed 18.8.12
89. Thornthwaite, C.W., (1948). An Approach toward a Rational Classification of Climate. *Geographical Review* 38, pp. 55-94

90. USACE (U.S. Army Corps of Engineers), (1995). Engineering and Design Treatability Studies for Treatability Studies for Solidification/Stabilisation of Contaminated Material. Technical Letter No. 1110-1-158
91. USEPA, (2001). Comprehensive Five-Year Review Guidance, OSWER 9355.7-03B-P, EPA 540-R-01-007, June 2001, http://www.epa.gov/superfund/accomp/5year/index.htm
92. USEPA, (1989). Stabilisation/Solidification of CERCLA and RCRA Wastes: Physical Tests, Chemical Testing Procedures, Technology Screening, and Field Activities, Office of Research and Development. EPA/625/6-89/022
93. USEPA, (1992). Guidance for Conducting Treatability Studies under CERCLA. EPA/540/R- 92/071a, 1992,United States Environmental Protection Agency, Office of Research and Development
94. USEPA, (1993). Technical Resource Document: Solidification/Stabilisation and Its Application to Waste Materials. Office of Research and Development. EPA/530/R-93/012. June 1993, http://tinyurl.com/pc42ksj
95. US Department of Energy, (2000). Ancient Cementitious Materials. Available at: http://www.wipp.energy.gov/PICsProg/documents/Ancient%20Cementitious%20Materials.pdf Accessed 14.08.12
96. USEPA, (1986). *Handbook for Stabilisation/Solidification of Hazardous Wastes*, Office of Research and Development, EPA/540/2-86/001, June 1986
97. USEPA, (2008). 2005 and 2008 Corrective Action Baselines - Final Results, Summary of Environmental Indicator and Final Remedy results: 2006-2008 (PDF), http://www.epa.gov/epawaste/hazard/correctiveaction/baseline.htm
98. USEPA, *Five Year Reviews Online*, undated, http://cumulis.epa.gov/fiveyear/index.cfm
99. USEPA, (1993). Guidance for Evaluating the Technical Impracticability of Groundwater Restoration, U.S. Environmental Protection Agency, OSWER Directive 9234.2-25, 1993, EPA/540/R/93/080
100. USEPA, (2014). National Priorities List (NPL), http://www.epa.gov/superfund/sites/npl/index.htm, accessed 01.07.2014
101. USEPA, (2005). *National Strategy to Manage Post Construction Completion Activities at Superfund Sites*, OSWER 9355.0-105 http://www.epa.gov/fedfac/pdf/pcc_strategy_final.pdf
102. USEPA, (2000). Solidification/Stabilisation Use at Superfund Sites, EPA-542-R-00-010, Page 5, http://www.clu-in.org/download/remed/ss_sfund.pdf
103. USEPA, (2011). Summary of the Resource Conservation and Recovery Act, March 2011, http://www2.epa.gov/laws-regulations/summary-resource-conservation-and-recovery-act
104. USEPA, (2010). Superfund Remedy Report, Thirteenth Edition, September 2010, EPA-542-R-10-004, September 2010, http://clu-in.org/download/remed/asr/13/SRR_13th_MainDocument.pdf
105. USEPA, (2013). Superfund Remedy Report, Fourteenth Edition, 2013, EPA 542-R-13-016, http://www.clu-in.org/download/remed/asr/14/SRR_14th_2013Nov.pdf
106. USEPA, (2007). Treatment Technologies for Site Clean up: Annual Status Report Twelfth Edition, 2007, EPA-542-R-07-012 http://clu-in.org/download/remed/asr/12/asr12_main_body.pdf
107. USEPA, (2009). Technology Performance Review: Selecting and Using Solidification/Stabilisation Treatment for Site Remediation, U.S. Environmental Protection Agency, National Risk Management Research Laboratory, Office of Research and Development, Cincinnati, OH, November 2009, EPA/600/R-09/148, http://www.epa.gov/nrmrl/pubs/600r09148.html
108. USEPA, (2011). Whitehouse Oil Pits, Site Summary Profile, updated on March 22, 2011 http://www.epa.gov/region4/superfund/sites/npl/florida/whthowsopfl.html
109. van Zomeren A., van Wetten, H., Dijkstra, J.J. van der Sloot, H.A. and Bleijerveld, R., (2003). Long Term Prediction of Release From a Stabilised Waste Monofill and Identification of Controlling Factors. In: Proceedings Ninth International Waste Management and Landfill Symposium, 6-10th October, 2003, S. Margherita di Pula, Cagliari, CISA, Italy
110. Wayman, E., (2011). The Secrets of Ancient Rome's Buildings: What is it about Roman concrete that keeps the Pantheon and the Colosseum still standing? Available at http://www.smithsonian.com. Accessed 14.08.2012
111. Wheeler, P., (1995). Leach Repellent. *Ground Engineering* 28, 5, pp 20-22
112. Yang, F., Zhang, B., Ma, Q., (2010). Study of Sticky Rice - Lime Mortar Technology for the Restoration of Historical Masonry Construction. *Accounts of Chemical Research*, 43, 6, pp 936-944

CONTRIBUTING AUTHOR BIOGRAPHIES

Ken Andromalos *BSc., MSc.*

Ken Andromalos has 35 years experience in engineering, construction management and operations in the specialty geotechnical and environmental remediation markets. Has acquired extensive project experience, across the United States and Canada. Remediation experience includes: off-site removal activities, in-place closures, sludge stabilization, dredging, slurry walls and other types of groundwater cut-off walls, in-place soil mixing, soil vapour extraction, groundwater pump and treatment systems, and in-situ solidification/ stabilization (ISS).

Edward Bates *BSc., MSc.*

Ed Bates retired from the U. S. Environmental Protection Agency in 2009 after 32 years of service, including over 20 years working as an expert technical advisor on site characterization, remedy design, and remedy construction for nearly 100 CERCLA and RCRA sites. He has extensive experience of the regulation of S/S sites, including 25 S/S, 22 S/S treatability studies and 20 site specifications for treatment by S/S. He was a contributor to the IRTC Guidance document, a member of the PASSiFy project team. He has won numerous awards including 9 USEPA bronze medals, two EPA Scientific and Technologies Achievement Award and the National Notable Achievement Team Award.

Steven R Birdwell *BSc., MSc.*

Mr. Birdwell has over 27 years of experience in the management of environmental remediation and geotechnical construction projects. He began his career in 1987 and then founded Remedial Construction Services, L.P. (RECON) in 1989 to meet the demand for environmental services created by new regulation. Mr. Birdwell has presented at 4 International conferences on environmental remediation technologies, including in-situ bioremediation, stabilization of ash products, and the remediation of chlorinated solvents in groundwater.

CONTRIBUTING AUTHOR BIOGRAPHIES

Donald Burke MSc.

Mr. Burke is an Executive Director with the Province of Nova Scotia with more than 20 years experience in administrative and project management in the government (both municipal & provincial) and private sector particularly in environmental remediation and civil construction. During his career he was involved in or responsible for schedule, budget, quality control, environmental management, regulatory compliance, stakeholder management and risk for large projects. Most recently Sydney Tar Ponds and Coke Ovens Cleanup Project and Cape Breton Regional Municipality.

Kate Canning PhD.

Kate Canning is a Geo-environmental consulting engineer specialising in brownfield development, with over 15 years experience. Following a PhD evaluating the use of synthetic zeolites for the treatment of heavy metal contaminated effluents, Kate undertook a research fellowship investigating the use of novel cement stabilisation methods for the remediation of contaminated soils and wastes at the University of Greenwich. Funded by Blue Circle Cement, the project included establishing field and bench-scale trials for a range of stabilisation/solidification techniques, and supporting a Mobile Plant Licence application.

Paula Carey PhD., BSc.

Paula Carey has 25 years as a university lecturer in a department of Geology/Environmental Science teaching up to Masters level. Subjects include: petrography, structural geology, metamorphic geology, waste management, contaminated land remediation, site investigation, and natural materials for the construction industry. Paula was co-director of the Centre for Contaminated Land Remediation at the University of Greenwich and carried out research in the durability and testing of construction materials and the use of accelerated carbonation in the treatment of waste and contaminated soils. She has over 11 publications on accelerated carbonation and was a contributor to the Science Review of Solidification and Stabilisation for the Environment Agency, accompanying national guidance on S/S, published by the Environment Agency.

Colin Dickson *DipEng., BEng.*

Colin is a certified Professional Engineer with the Association of Professional Engineers of Nova Scotia (Engineers Nova Scotia in 1991). Colin completed his Executive Master of Business Administration at Saint Mary's University in 2008 and received the Gold Medal of Academic Excellence for the highest academic achievement in the program. He has enjoyed continuing his professional development in fields related to marine, environmental, industrial, civil and structural engineering.

Cameron Ells *BSc.*

Cameron Ells is a Professional Engineer (P. Eng.); environmental consultant; and entrepreneur, based in Halifax NS. He is a remediation specialist providing risk assessment, regulatory response support, and other services, to public and private sector clients. Cameron provided responsible party laboratory testing recommendations regarding Solidification/Stabilization and conceived and directed bench-scale laboratory testing programs, to identify preferred S/S mixes for site-specific project conditions. He also supported contractor teams in preparing competitive S/S application proposals.

Robert Garrett *BSc.*

Mr. Garrett is an experienced professional in the assessment and remediation of environmental contamination. He has served as a consultant since 1972, beginning his career in aquatic environmental assessments of both estuarine and fresh water ecosystems. He has both participated and managed multidisciplinary projects. His interests led him into the hazardous waste field where he has provided site environmental assessments, laboratory analyses, and development and implementation of remedial alternatives. Most recently, he has been responsible for project management of hazardous waste remediation projects, including bioremediation, soil fixation/solidification and building decontamination. Mr. Garrett formed Garrett Consulting, Inc. in 1990.

Peter Gunning PhD., MSc., BSc.

Dr Gunning joined Carbon8 in 2008 specialising in the treatment of contaminated soil and waste. Since his appointment, he has been responsible for laboratory research including product and process development on industrial waste recycling, treatment of contaminated materials, and mineral carbon capture. He has co-ordinated numerous pilot-scale proof-of-concept trials including the design and construction of novel carbon capture and waste/soil treatment plants. Dr. Gunning has continued to contribute to research at the University of Greenwich, and has an active role in the supervision of the research team at the Centre for Contaminated Land Remediation.

Colin D Hills CSci., BSc., MSc., PhD., DIC., MIMMM., FGS.

Colin Hills has over 30 years' experience in geo-materials and cement-based systems for the treatment of soil and waste, including working in Europe, the Middle East and Africa, primarily as a geologist. For the past 25 years he has been working extensively on S/S, has published over 100 papers, and has authored guidance on S/S for the Environment Agency (England and Wales). His work has attracted international recognition, has won a number of national and regional awards and led to innovative treatments for the management of difficult wastes.

Diane Ingraham PhD., BSc., SM.

Dr. Ingraham is a senior project manager with substantial (30 years) experience in management of project quality, schedule, budget, and risk for large projects, specifically those involving heavy civil construction and environmental regulatory compliance or remediation. Most recently Nalcor Energy's Lower Churchill Muskrat Falls Hydroelectric Project, Sydney Tar Ponds and Coke Ovens Cleanup Project, and Technical Training Consultant at Exxon/LearnCorp International.

Craig Lake *PhD., BEng.*

Craig Lake has 35 years experience in engineering, construction management and operations in the specialty geotechnical and environmental remediation markets. Extensive project experience across the United States and Canada. Remediation experience includes off-site removal activities, in-place closures, sludge stabilization, dredging, slurry walls and other types of groundwater cut-off walls, in-place soil mixing, soil vapour extraction, groundwater pump and treatment systems, and in-situ solidification/stabilization. This remediation experience includes working at various active and in-active facilities including steel making facilities, wood treating sites, manufactured gas plant (MGP) sites, Superfund sites, former Chemical Warfare facilities, petroleum refineries, manufacturing facilities, landfills, chemical plants, and numerous disposal sites.

Paul Lear *PhD., MSc., BSc.*

Dr. Lear has over 25 years of experience in hazardous waste treatment, laboratory management, and chemical process development. His experience includes selecting and evaluating treatment alternatives, providing data for preliminary design activities and project equipment specifications, assisting project design teams, and implementing the final design. Dr. Lear has hands-on experience with full-scale remediation activities and specializes in process troubleshooting. He has provided technical operational support to bioremediation, dewatering, soil washing, stabilization, thermal, and wastewater treatment activities at toxic, hazardous, and radioactive waste remedial sites.

Jerome MacNeil *BSc.*

Mr. MacNeil is a Senior Project Manager and Environmental Engineer with more than 17 years of experience. Over the course of his career he has developed specific expertise in the areas of contract management, environmental management and contaminated site remediation. Mr. MacNeil has managed in excess of 130 million dollars worth of environmental remediation and heavy civil projects. He also has experience in environmental site assessments, asbestos and mold abatement, air quality assessments, solid waste management, OH&S plan development and implementation, and development of environmental training programs.

Aiman Naguib *MSc., BSc.*

Mr Naguib has more than 24 years of experience in general remediation, and specialty environmental and geotechnical construction. He has worked extensively on complex environmental remediation and geotechnical projects involving in-situ S/S, slurry walls, permeable reactive barriers and cap/containment construction. As a Project Coordinator and Technical Advisor at ENTACT, he is responsible for managing and coordinating S/S projects. This includes cost estimates, proposal preparation, work planning, mix designs/treatability studies and the management of field operations, including resource allocations.

Stany Pensaert *MSc.*

Stany has 20 years service as head of the R&D Department at DEC-DEME Environmental Contractors NV. He is responsible for the scientific and process support for all environmental works at: site remediations, soil treatment, environmental dredging and sediment treatment. His work is world-wide supporting the DEME group. Has published over one hundred publications, case studies and technology appraisals with respect to soil remediation and environmental dredging.

Thomas Plante *MSc., BSc.*

Thomas Plante is an engineering consultant with over 25 years of experience in site characterization, remedy evaluation and selection, remedial design and remedial construction, primarily focused on municipal, industrial, and utility clients throughout the United States. His primary environmental remediation focus area is the investigation and remediation of former manufactured gas plant (MGP) sites containing coal tar, cyanide, ammonia, various metals, and other contaminants.

Daniel Ruffing *MSc., BSc.*

As a Project Manager at Geo-Solutions, Daniel is responsible for generating estimates, preparing proposals, managing bench-scale studies, and the management of projects, involving slurry trench cut-off walls, soil mixing, jet grouting, and other specialty geotechnical contracting techniques. Daniel's experience is founded in research conducted during his university education supplemented by field engineering experience, project supervision, and project management at Geo-Solutions. Daniel's field engineering and management experience on S/S projects spans 5+ years.

Bob Schindler *BSc.*

Mr. Schindler is President/CEO of Geo-Solutions Inc., an established international, environmental remediation and geotechnical contracting company specializing in slurry walls, bio-polymer drains, reactive barriers, in-situ soil solidification/stabilization, soil mixing, jet grouting and other related techniques. As President of Geo-Solutions, he oversees all company functions, operations, finance, administration, health and safety, quality assurance, marketing, and business development. He has managed numerous S/S projects and treatability studies. Quantities of in-situ S/S treated materials exceed one million cubic yards, using a variety of application methods and reagents.

Roy Wittenberg *MSc., BSc.*

Mr. Wittenberg has over 27 years of experience performing project engineering and management, technical supervision, design engineering and analysis, construction oversight and regulatory interface. His project management and construction experience includes a number of site remedial restoration projects along major waterways. His environmental experience includes conducting remedial alternatives evaluations, and development of risk-based clean-up approaches. His technical experience includes environmental engineering for soil/sediment and groundwater treatment, bench- and pilot-scale testing and civil engineering applications for site restoration and redevelopment.

APPENDIX A: REMEDIATED SITES EMPLOYING S/S

The following list of sites employing S/S in the USA and Europe was compiled from data supplied by S/S construction contractors and engineers.

The data is presented as it was received from these sources and has not been independently verified by the editors. However each project listed in the following Table identifies the person that submitted data along with the name and E-mail contact (or in a few cases the relevant publication), should additional information be required.

This extensive list of over 200 completed S/S projects illustrates the widespread successful application of S/S to a wide variety of contaminants and site types.

APPENDIX A: REMEDIATED SITES EMPLOYING S/S

Sites Outside The United States

Treatment Vendor/date of treatment	Site Name/Location	Quantity Media	Contaminant	In-situ/Ex-situ & Equipment	Purpose	Reagent Formula/ treat specs	Submitted by	Reference/Contact
ENTACT, LLC – 2013	Sydney Tar Ponds Solidification/Stabilisation – Sydney., Nova Scotia CANADA	Soil; 570,000 cubic meters of impacted sediment	Metals, PAH's, VOC's, PCB's, TPH	In-situ – Hydraulic Excavators	Remediation; treatment	Cement	ENTACT, LLC	Sydney Tar Ponds Agency www.tarpondscleanup.ca
DEC nv Belgium 2012	Hoedhaar Lokeren Belgium	10,000 tons of soil	Hg	Excavator	Comply to non-haz landfill	Fe° + iron hydroxides	Stany Pensaert	Stany Pensaert Pensaert.Stany@deme.be
DEC nv Belgium 2011	Obourg	170,000 tons sediment	none	Lime blending machine	Reuse as engineered backfill	Paper ash + CKD	Stany Pensaert	Stany Pensaert Pensaert.Stany@deme.be
DEC nv Belgium 2010	Söderhamn Sweden	31,000 tons soil	As/Cu/Cr	Mobile pug-mill mixer	Comply to non-haz landfill	Fe 2%	Stany Pensaert	Stany Pensaert Pensaert.Stany@deme.be
DEC nv Belgium 2007-2010	London Olympics	60,000 tons soil	Heavy metals, PAH, TPH	Mobile pug-mill mixer + allu bucket on excavator	Reuse as engineered backfill	Fe 2% Biochar 1%	Stany Pensaert	Stany Pensaert Pensaert.Stany@deme.be
DEC nv Belgium 2005-2008	Total Ertvelde Belgium	250,000 tons acid tar + soil	Various hydrocarbons	Fixed plant based on pug-mill mixer	Comply to haz landfill	Various fly ashes + cement	Stany Pensaert	Stany Pensaert Pensaert.Stany@deme.be

APPENDIX A: REMEDIATED SITES EMPLOYING S/S | 251

Sites Outside The United States

Treatment Vendor/date of treatment	Site Name/Location	Quantity Media	Contaminant	In-situ/Ex-situ & Equipment	Purpose	Reagent Formula/ treat specs	Submitted by	Reference/Contact
DEC nv Belgium 2007	Bekaert Belgium	50,000 tons soil	Cu/Pb/Cd/Zn	Mobile pug-mill mixer	Reuse as backfill	Paper ash 2%	Stany Pensaert	Stany Pensaert Pensaert.Stany@deme.be
DEC nv Belgium 2004	La Floridienne Belgium	60,000 tons waste	As	Mobile pug-mill mixers	Comply to haz landfill	Fe 2%	Stany Pensaert	Stany Pensaert Pensaert.Stany@deme.be
DEC nv Belgium 2004	La Floridienne Belgium	75,000 tons waste	Cyanides	Mobile pug-mill mixers	Comply to haz landfill	OPC + GGBS	Stany Pensaert	Stany Pensaert Pensaert.Stany@deme.be
DEC nv Belgium 2003	Guernsey UK	31,000 tons sediment	TBT	Allu rotary bucket	Reuse for port extension	Biochar 2%	Stany Pensaert	Stany Pensaert Pensaert.Stany@deme.be

APPENDIX A: REMEDIATED SITES EMPLOYING S/S

Sites Inside The United States

Treatment Vendor/date of treatment	Site Name/Location	Quantity Media	Contaminant	In-situ/ Ex-situ & Equipment	Purpose	Reagent Formula/ treat specs	Submitted by	Reference/Contact
WRScompass 2010, 2011, 2013	Wilmington Coal Gasification Site, Wilminton, DE	45,000 cy soil	PAHs, coal tar, DNAPL	Excavator	Redevelopment	1.5-3% Portland + 4.5-9% slag >50 psi UCS <1x10^{-6} cm/s permeability	Paul Lear	Paul.Lear plear@envirocon.com
Geo-Solutions 2012	Wood Preservers Remediation Warsaw, VA	58,454 cy	Wood Treating Chemicals	In-situ Auger Bucket	Stabilisation/ Solidification	Portland Cement	K. Andromalos	K. Andromalos kandromalos@geo-solutions.com
Geo-Solutions 2012	West End Remediation Cincinnati, OH	88,000 cy	MGP impacted soils	In-situ Auger	Stabilisation/ Solidification	Cement Bentonite	K. Andromalos	K. Andromalos kandromalos@geo-solutions.com
Geo-Solutions 2012	PG&E Front and T Street Site Sacramento, CA	42,500 cy	BTEX PAHs	In-situ Auger	Stabilisation	Portland cement Regenerated Granular Activated Carbon	K. Andromalos	K. Andromalos kandromalos@geo-solutions.com
Geo-Con (trade name of Geo-Solutions) 2012	Atlantic Wood Industries Superfund Site Portsmouth, VA	47,000 cy	DNAPLs	In-situ Excavator/ rotary blender system	Stabilisation/ Solidification	Portland Cement Slag Cement Organophilic Clay	K. Andromalos	M. Kitko mkitko@geo-solutions.com

Sites Inside The United States

Treatment Vendor/date of treatment	Site Name/Location	Quantity Media	Contaminant	In-situ/Ex-situ & Equipment	Purpose	Reagent Formula/treat specs	Submitted by	Reference/Contact
Geo-Con (trade name of Geo-Solutions) 2012	MGP Site Remediation Homer, New York	54,300 cy	PAHs, BTEX	In-situ Large diameter auger	remediation	1×10^{-6} cm/sec permeability UCS 50 psi	K. Andromalos	D. Payne dpayne@geo-solutions.com
Geo-Con (trade name of Geo-Solutions) 2012	Former Texarkana Wood Preserving Company Site Texarkana, Texas	42,000 cy	Creosote, PCPs	In-situ Large diameter auger	Superfund remediation	Cement Powdered activated carbon	K. Andromalos	G. Maitland gmaitland@geo-solutions.com
Geo-Solutions 2012	Indian Head Naval Base Indian Head, MD	1,300 cy	Chlorinated Solvents	In-situ Auger	Treatment/ Chemical Reduction		K. Andromalos	K. Andromalos kandromalos@geo-solutions.com
Geo-Solutions 2012	Former Hanley Area Ordnance Site St. Louis, MO	1,400 cy	Impacted Soils	In-situ Auger	Stabilisation		K. Andromalos	K. Andromalos kandromalos@geo-solutions.com
Geo-Solutions 2012	Former Columbus Wood Treaters Columbus, IN	4,500 cy	Creosote	In-situ Auger	Stabilisation	Powdered Activated Carbon	K. Andromalos	K. Andromalos kandromalos@geo-solutions.com
Geo-Solutions 2012	Former Municipal Wastewater Treatment Lagoon	7,500 m^3	Chlorinated Solvents	In-situ Auger	Chemical Reduction	ZVI	K. Andromalos	K. Andromalos kandromalos@geo-solutions.com

Sites Inside The United States

Treatment Vendor/date of treatment	Site Name/Location	Quantity Media	Contaminant	In-situ/Ex-situ & Equipment	Purpose	Reagent Formula/treat specs	Submitted by	Reference/Contact
WRScompass, 2012	Valero NRP Closure Project, Paulsboro, NJ	18,000 cy, sludge/sediment	TPH, Metals	In-situ Excavator	Lagoon Closure, Redevelopment	2% Portland + 6% slag + 2% Ferrous sulfate, >50 psi UCS <1x10^{-6} cm/s permeability	Paul Lear	Paul Lear plear@envirocon.com
WRScompass 2012	PSEG, Camden, NJ	19,000 cy soil	PAHs, coal tar, DNAPL	In-situ Excavator	Redevelopment	3% Portland + 9% slag, >50 psi UCS <1x10^{-6} cm/s permeability	Paul Lear	Paul Lear plear@envirocon.com
RECON 2012	Pond 9 Marcus Hook, PA	48,000 cy tarry sludge	Acid tar, with H$_2$S and SO$_2$, benzene, asbestos	In-situ bucket mixing	Cap	Patented LSS	Remedial Construction Services, L.P.	S. Birdwell steven.birdwell@reconservices.com
RECON 2012	Bayou Trepagnier Remediation, St. Charles Parish, LA	50,000 cy sediments	Hydrocarbons	In-situ bucket mixing	Strengthen for cap construction	Patented LSS	Remedial Construction Services, L.P.	S. Birdwell steven.birdwell@reconservices.com

APPENDIX A: REMEDIATED SITES EMPLOYING S/S

Sites Inside The United States

Treatment Vendor/date of treatment	Site Name/Location	Quantity Media	Contaminant	In-situ/ Ex-situ & Equipment	Purpose	Reagent Formula/ treat specs	Submitted by	Reference/Contact
RECON 2012	Refinery Pond Closure, Norco, LA	40,000 cy sludge	Hydrocarbons	In-situ bucket mixing	Impoundment closures	Patented LSS	Remedial Construction Services, L.P.	S. Birdwell steven.birdwell@reconservices.com
RECON 2012	Pond 9 Marcus Hook, PA	48,000 cy tarry sludge	Acid tar, with H_2S and S_O2, benzene, asbestos	In-situ bucket mixing	Cap	Patented LSS	Remedial Construction Services, L.P.	S. Birdwell steven.birdwell@reconservices.com
RECON 2012	Bayou Trepagnier Remediation, St. Charles Parish, LA	50,000 cy sediments	Hydrocarbons	In-situ bucket mixing	Strengthen for cap construction	Patented LSS	Remedial Construction Services, L.P.	S. Birdwell steven.birdwell@reconservices.com
RECON 2012	Refinery Pond Closure, Norco, LA	40,000 cy sludge	Hydrocarbons	In-situ bucket mixing	Impoundment closures	Patented LSS	Remedial Construction Services, L.P.	S. Birdwell steven.birdwell@reconservices.com
Envirocon, 2011	Manufactured Gas Plant Site, Ohio	35,000 cy	Coal Tar (BTEX, PAHs)	In-situ/ excavator bucket	Site remediation, risk reduction	Cement 2% GBFS 6% UCS 50 psi $\leq 1 \times 10^{-6}$ cm/sec permeability	T. Plante	T. Plante tplante@haleyaldrich.com 207-482-4622

Sites Inside The United States

Treatment Vendor/date of treatment	Site Name/Location	Quantity Media	Contaminant	In-situ/ Ex-situ & Equipment	Purpose	Reagent Formula/ treat specs	Submitted by	Reference/Contact
ENTACT, LLC - 2011	WE Energies Gaslight Point MGP Site Remediation – Racine, WI	33,361 cy of MGP impacted soils	Benzo(a)pyrene (PAHs); Coal; Tar; BTEX	In-situ – 10 ft diameter mixing auger mounted on drill rig.	Remediation	Cement-Slag grout	ENTACT, LLC	Wisconsin Department of Natural Resources
WRScompass, 2011	NYSEG Elmira MGP Site, Elmira, NY	10,000 cy soil	PAHs, coal tar, DNAPL	In-situ auger	Containment	4% Portland + 0.5% bentonite >50 psi UCS <1x10^{-6} cm/s permeability	Paul Lear	Paul.Lear plear@envirocon.com
WRScompass 2011	Sanford Gasification Facility, Sanford, FL	125,000 cy soil	PAHs, coal tar, DNAPL	In-situ auger	Redevelopment	2% Portland + 6% slag >50 psi UCS <1x10^{-6} cm/s permeability	Paul Lear	Paul.Lear plear@envirocon.com
Geo-Con (trade name of Geo-Solutions) 2011	F-Area Seepage Basin Aiken, South Carolina	1,400 linear feet	Tritium	In-situ Multiple Auger	Groundwater barrier	Impermix 1x10^{-6} cm/sec UCS 50 psi	K. Andromalos	G. Maitland gmaitland@geo-solutions.com

Sites Inside The United States

Treatment Vendor/date of treatment	Site Name/Location	Quantity Media	Contaminant	In-situ/Ex-situ & Equipment	Purpose	Reagent Formula/treat specs	Submitted by	Reference/Contact
Geo-Con (a trade name of Geo-Solutions) 2011	Rosemont Levee Improvements Lowell, MA	4,900 cy	N/A	In-situ Auger	Reinforcement	50 psi UCS Cement	K. Andromalos	G. Maitland gmaitland@geo-solutions.com
Geo-Con (a trade name of Geo-Solutions) 2011	Perimeter Wall Stabilisation Kingston, TN	560,000 cy	Coal Fly Ash	In-situ Excavator	Containment	250 psi UCS Cement Bentonite	K. Andromalos	S. Artman sartman@geo-solutions.com
Geo-Solutions 2011	MW-250 Site Remediation East Rutherford, NJ	7,626 cy	TCE	In-situ Auger	Oxidation	Potassium Permanganate/Cement	K. Andromalos	K. Andromalos kandromalos@geo-solutions.com
Geo-Solutions 2011	OMC Plant Site Remediation Waukegan, IL	8,900 cy	TCE	In-situ Auger	Chemical Reduction/Treatment	Bentonite	K. Andromalos	K. Andromalos kandromalos@geo-solutions.com
Geo-Solutions 2011	Inner Slip Site Remediation New Bedford, MA	6,457 cy	MGP impacted sediment	In-situ Auger	Solidify		K. Andromalos	K. Andromalos kandromalos@geo-solutions.com
Geo-Solutions 2011	Former Miller Chemical Site Robbinsville, NJ	2,778 cy	Pesticide Xylene	In-situ Auger	Chemical Oxidation/Stabilisation	Sodium Persulfate/Lime	K. Andromalos	K. Andromalos kandromalos@geo-solutions.com

APPENDIX A: REMEDIATED SITES EMPLOYING S/S

Sites Inside The United States

Treatment Vendor/date of treatment	Site Name/Location	Quantity Media	Contaminant	In-situ/Ex-situ & Equipment	Purpose	Reagent Formula/treat specs	Submitted by	Reference/Contact
RECON 2010	Cell 11 Port Arthur, TX	250,000 cy soil and sludge	Various hydrocarbons	In-situ bucket mixing	Stabilisation for construction foundation	Cement/ash blend	Remedial Construction Services, L.P.	S. Birdwell steven.birdwell@reconservices.com
RECON 2010	Fly Ash Ponds Delaware City, DE	115,000 cy fly ash	Petroleum coke and vanadium	In-situ bucket mixing	Impoundment closure	Lime kiln dust, Portland cement and soil	Remedial Construction Services, L.P.	S. Birdwell steven.birdwell@reconservices.com
RECON 2010	Soil Mixing Demo Project, Kingston, TN	1,500 cy fly ash	Fly ash	In-situ soil mixing	Construct barrier	289 pounds of Portland cement per cubic yard	Remedial Construction Services, L.P.	S. Birdwell steven.birdwell@reconservices.com
Geo-Solutions 2010	SAR Levee Repair Newport Beach, California	6,100 cy		In-situ Auger	Stabilisation		K. Andromalos	K. Andromalos kandromalos@geo-solutions.com
Geo-Solutions 2010	MW-520 Site Remediation East Rutherford, NJ	7,626 cy	TCE	In-situ Auger	Oxidation/Solidification	Potassium Permanganate /Portland Cement	K. Andromalos	K. Andromalos kandromalos@geo-solutions.com
Geo-Con (trade name of Geo-Solutions) 2010	Former Mfg Facility Parsippany, NJ	15,000 cy	Chlorinated solvent	In-situ Excavator mounted blender head	Oxidation Solidification	Potassium Permanganate /Portland Cement	K. Andromalos	G. Maitland gmaitland@geo-solutions.com

Sites Inside The United States

Treatment Vendor/date of treatment	Site Name/Location	Quantity Media	Contaminant	In-situ/Ex-situ & Equipment	Purpose	Reagent Formula/treat specs	Submitted by	Reference/Contact
Geo-Con (trade name of Geo-Solutions) 2010	Shermerhorn Creek Schenectady, NY	10,800 cy	PCBs Chlorinated Solvents	In-situ Backhoe	Stabilisation	Portland Cement	K. Andromalos	M. Kitko mkitko@geo-solutions.com
WRScompass, 2010	NYSEG Norwich MGP site, Norwich, NY	25,000 cy soil	PAHs, coal tar, DNAPL	In-situ auger	Containment	8% Portland + 0.75% bentonite >50 psi UCS <1×10^{-6} cm/s permeability	Paul Lear	Paul.Lear plear@envirocon.com
WRScompass, 2010	Newell-Rubbermaid Palmieri Site, Monaca, PA	32,000 cy soil and glass	Lead, arsenic, cadmium	Ex-situ pug-mill	Remediation	1.5% - 4.5% EnviroMag	Paul Lear	Paul.Lear plear@envirocon.com
WRScompass, 2010	United Metals Inc, Marianna, FL	36,000 cy soil	Lead	Ex-situ Pug-mill	Remediation	8% Portland cement + 8% TerraBond SPLP lead < 0.015 mg/L >50 psi UCS <1×10^{-6} cm/s permeability	Paul Lear	Paul.Lear plear@envirocon.com

Sites Inside The United States

Treatment Vendor/date of treatment	Site Name/Location	Quantity Media	Contaminant	In-situ/ Ex-situ & Equipment	Purpose	Reagent Formula/ treat specs	Submitted by	Reference/Contact
WRScompass 2009	ConEd, White Plains NY	30,000 cy soil	PAHs, coal tar, DNAPL	In-situ auger	Re-development	2.5% Portland + 7.5% slag >50 psi UCS <1x10^{-6} cm/s permeability	Paul Lear	Paul.Lear plear@envirocon.com
WRScompass, 2009	Orkin Atlanta Site, Atlanta, GA	5,000 cy soil	Chlordane	In-situ excavator	Closure	5% Portland cement + 0.05% carbon	Paul Lear	Paul.Lear plear@envirocon.com
WRScompass, 2009	West Doane Lake IRAM Project	18,000 cy Sludge/ sediment	Pesticides, metals	In-situ Excavator	Impoundment Closure	18% Portland + 3% Bentonite + 3% organoclay + 3% carbon >50 psi UCS <1x10^{-6} cm/s permeability	Paul Lear	Paul.Lear plear@envirocon.com
WRScompass, 2009	ArvinMeritor Sludge Lagoon closure Project, Granada, MS	17,500 cy sludge	TPH, metals	In-situ Excavator	Impoundment Closure	12% cement kiln dust >50 psi UCS	Paul Lear	Paul.Lear plear@envirocon.com

APPENDIX A: REMEDIATED SITES EMPLOYING S/S

Sites Inside The United States

Treatment Vendor/date of treatment	Site Name/Location	Quantity Media	Contaminant	In-situ/ Ex-situ & Equipment	Purpose	Reagent Formula/ treat specs	Submitted by	Reference/Contact
Geo-Con (trade name of Geo-Solutions) 2009	Vandenberg Air Force Base, California	22,000 vertical wall square feet	Chlorinated Solvents	In-situ Auger	Permeable Reactive Barrier	BOS-100	K. Andromalos	G. Maitland gmaitland@geo-solutions.com
Geo-Con (trade name of Geo-Solutions) 2009	MGP Site Macon, GA	16,290 cy	BTEX PAHs	In-situ Large Diameter Auger	Stabilisation	Slag Cement Portland Cement Bentonite	K. Andromalos	M. Kitko mkitko@geo-solutions.com
RECON 2009	Site Preparation Port Arthur, TX	425,000 cy dredge spoil and marsh sediment	Trace metals	In-situ bucket mixing	Stabilisation for construction foundation	Patented LSS/25 psi UCS	Remedial Construction Services, L.P.	S. Birdwell steven.birdwell@reconservices.com
RECON 2009	Site Preparation, Sabine Pass, TX	650,000 cy dredge spoils	Trace metals	In-situ excavator	Stabilisation for construction foundation	Patented blend of fly ash	Remedial Construction Services, L.P.	S. Birdwell steven.birdwell@reconservices.com
RECON 2009	Cell 9 Port Arthur, TX	172,000 cy sludge	Various hydrocarbons	In-situ bucket mixing	Impoundment closures	Cement/ash blend	Remedial Construction Services, L.P.	S. Birdwell steven.birdwell@reconservices.com

APPENDIX A: REMEDIATED SITES EMPLOYING S/S

Sites Inside The United States

Treatment Vendor/date of treatment	Site Name/Location	Quantity Media	Contaminant	In-situ/ Ex-situ & Equipment	Purpose	Reagent Formula/ treat specs	Submitted by	Reference/Contact
RECON 2009	Soil Stabilisation, Tuscumbia, AL	21,000 cy soil	Tetrachloro-ethene	In-situ soil mixing	Site closure	Granular ground blast furnace slag and Portland cement	S. Birdwell	S. Birdwell steven.birdwell@reconservices.com
RECON 2009	Bayou Stabilisation, Port Arthur, TX	675,000 cy sediments	Hydrocarbons	In-situ bucket mixing	Cap	Cement and fly ash blend/25 psi UCS	Remedial Construction Services, L.P.	S. Birdwell steven.birdwell@reconservices.com
GCI 2008	Brunswick Wood Preserving Site Brunswick, GA	100,000 tons	Wood Preserving Waste	Ex-situ ARAN Pug-mill	Produce Soil Cement	Cement 10% Fly ash 10%	R. Garrett	Robert Garrett bobgarrett417@bellsouth.net
WRScompass, 2008	Central Hudson Gas & Electric MGP, Poughkeepsie, NY	5,000 cy soil	PAHs, coal tar, DNAPL	In-situ Excavator	Containment	3% Portland + 9% slag >50 psi UCS <1x10⁻⁶ cm/s permeability	Paul Lear	Paul Lear plear@envirocon.com
Geo-Con (trade name of Geo-Solutions) 2008	Camp Lejeune, North Carolina	30,000 cy	TCE/PCE	In-situ Large diameter auger	Remediation	ZVI	K. Andromalos	G. Maitland gmaitland@geo-solutions.com

Sites Inside The United States

Treatment Vendor/date of treatment	Site Name/Location	Quantity Media	Contaminant	In-situ/Ex-situ & Equipment	Purpose	Reagent Formula/treat specs	Submitted by	Reference/Contact
Geo-Solutions 2008	MGP Site Remediation Rochester, New York	13,000 cy	BTEX, napthalene	In-situ Large diameter auger	Stabilisation	Cement/Slag / Bentonite UCS 50 psi $1.0E^{-6}$ cm/sec permeability	K. Andromalos	K. Andromalos kandromalos@geo-solutions.com
Geo-Con (trade name of Geo-Solutions) 2008	Linskey Way Cambridge, MA	1,200 cy	MGP impacted soils	In-situ Jet Grouting	Stabilisation	Cement/ Bentonite	K. Andromalos	M. Kitko mkitko@geo-solutions.com
Geo-Con (a trade name of Geo-Solutions) 2008	Hunter Ferrell Landfill Irving, TX	8,428 ft²	N/A	In-situ Auger	Seepage Barrier	1.00×10^{-6} cm/sec permeability Bentonite	K. Andromalos	M. Kitko mkitko@geo-solutions.com
Geo-Con (a trade name of Geo-Solutions) 2008	ISS Perimeter Wall Sag Harbor, NY	7,200 cy	BTEX PAHs	In-situ Auger	Barrier Wall Excavation Support	1.00×10^{-6} cm/sec permeability 50 psi UCS Cement	K. Andromalos	M. Kitko mkitko@geo-solutions.com
Geo-Solutions 2008	Former Rocky Mountain Arsenal Closure Commerce City, CO	1,917 linear feet	CWM impacted soils	In-situ Multiple auger	Barrier Wall	Bentonite	K. Andromalos	K. Andromalos kandromalos@geo-solutions.com

Sites Inside The United States

Treatment Vendor/date of treatment	Site Name/Location	Quantity Media	Contaminant	In-situ/Ex-situ & Equipment	Purpose	Reagent Formula/treat specs	Submitted by	Reference/Contact
WRScompass 2008	Manufactured Gas Plant Site, Waterville, ME	22,000 cy	Coal Tar (BTEX, PAHs)	In-situ/ excavator bucket	Site remediation, risk reduction	Interior- 7.4% ; Exterior - 20%. Reagent consisted of 1 part Type I/II Portland cement to 3 parts slag. UCS 30 psi, $\leq 1 \times 10^{-6}$ cm/sec permeability	T. Plante Paul Lear	T.Plante tplante@haleyaldrich.com 207-482-4622 Paul Lear plear@envirocon.com
Geo-Solutions 2007	Utica, New York	15,000 cy	PCBs	In-situ Large diameter auger	Stabilisation	Cement	K. Andromalos	K. Andromalos kandromalos@geo-solutions.com
Geo-Con (a trade name of Geo-Solutions) 2007	CKD Groundwater Remediation Metaline Falls, WA	31,400 ft^2	Cement Kiln Dust	In-situ Excavator	Containment	1.00×10^{-6} cm/sec 50 psi	K. Andromalos	S. Artman sartman@geo-solutions.com

Sites Inside The United States

Treatment Vendor/date of treatment	Site Name/Location	Quantity Media	Contaminant	In-situ/ Ex-situ & Equipment	Purpose	Reagent Formula/ treat specs	Submitted by	Reference/Contact
Geo-Con (a trade name of Geo-Solutions) 2007	Lowes Heidelberg Carnegie, PA	7,134 cy	Saturated Unconsolidated Soils	In-situ Excavator	Soil Improvement	100 psi UCS Cement	K. Andromalos	M. Kitko mkitko@geo-solutions.com
Geo-Con (a trade name of Geo-Solutions) 2007	Stabilised Soil Barrier Plattsburgh, NY	45,750 ft^2	Coal tar VOCs	In-situ Auger	Barrier Wall	1.00×10^{-6} cm/sec permeability 40 psi UCS Cement Bentonite	K. Andromalos	M. Kitko mkitko@geo-solutions.com
WRScompass 2007	Niagara Mohawk, Sarasota Springs, NY	45,000 cy soil	PAHs, coal tar, DNAPL	In-situ auger	Containment	8% Portland + 1% bentonite >50 psi UCS $<1 \times 10^{-6}$ cm/s permeability	Paul Lear	Paul.Lear plear@envirocon.com
WRScompass, 2007	Cambridge Creek MGP, Cambridge, MD	20,000 cy soil	PAHs, coal tar, DNAPL	In-situ auger	Containment	2% Portland + 6% slag >50 psi UCS $<1 \times 10^{-6}$ cm/s permeability	Paul Lear	Paul.Lear plear@envirocon.com
WRScompass, 2007	Foote Minerals site, Exton, PA	220,000 cy	Lithium	In-situ Auger	Redevelopment	7-10% slag >50 psi UCS $<1 \times 10^{-6}$ cm/s permeability	Paul Lear	Paul.Lear plear@envirocon.com

Sites Inside The United States

Treatment Vendor/date of treatment	Site Name/Location	Quantity Media	Contaminant	In-situ/Ex-situ & Equipment	Purpose	Reagent Formula/treat specs	Submitted by	Reference/Contact
WRScompass, 2007	Motiva Crude Expansion Project, Port Arthur, TX	120,000 cy, sludge	TPH	In-situ Excavator	Refinery Expansion	Portland cement or cement kiln dust >10 psi UCS	Paul Lear	Paul Lear plear@envirocon.com
RECON 2007	Site Remediation, Tarrant City, AL	15,000 cy soil	Chlorinated solvents, hydrocarbons & trace metals	In-situ soil mixing	Risk reduction	10% cement, UCS of 50 psi and 1×10^{-7} cm/sec permeability	Remedial Construction Services, L.P.	S. Birdwell steven.birdwell@reconservices.com
RECON 2007	Section 7 Port Arthur, TX	1,550,000 cy sludge	Various hydrocarbons	In-situ bucket mixing	Impoundment closures	Fly ash/20 psi UCS	Remedial Construction Services, L.P.	S. Birdwell steven.birdwell@reconservices.com
RECON 2007	Site Preparation Cameron Parish, LA	825,000 cy dredge spoil	Trace metals	Ex-situ bucket mixing	Stabilisation for construction foundation	Patented LSS/25 psi UCS	Remedial Construction Services, L.P.	S. Birdwell steven.birdwell@reconservices.com
RECON 2007	Site Preparation Sabine Pass, TX	800,000 cy dredge spoils	Trace metals	In-situ excavator	Stabilisation for construction foundation	Patented blend of fly ash	Remedial Construction Services, L.P.	S. Birdwell steven.birdwell@reconservices.com
RECON 2007	Purity Superfund Site, Fresno, CA	45,000 cy soil	Low pH, hydrocarbons	In-situ with rotating mixing head	Superfund remediation	Cement/25 psi UCS and leachability	Remedial Construction Services, L.P.	S. Birdwell steven.birdwell@reconservices.com

APPENDIX A: REMEDIATED SITES EMPLOYING S/S

Sites Inside The United States

Treatment Vendor/date of treatment	Site Name/Location	Quantity Media	Contaminant	In-situ/ Ex-situ & Equipment	Purpose	Reagent Formula/ treat specs	Submitted by	Reference/Contact
RECON 2006	Oxidation Basin Baytown, TX	35,000 cy soil and sludge	Hydrocarbons	In-situ excavator	Strengthening for cap construction	Fly ash	Remedial Construction Services, L.P.	S. Birdwell steven.birdwell@reconservices.com
RECON 2006	Cell 4 Port Arthur, TX	52,000 cy sludge	Various hydrocarbons	In-situ bucket mixing	Stabilisation for construction foundation	Cement/ash blend	Remedial Construction Services, L.P.	S. Birdwell steven.birdwell@reconservices.com
Geo-Con (Geo-Solutions) 2006	Peerless Photo Shoreham, NY	10,500 cy	Silver Cadmium	In-situ Jet Grouting	Stabilisation	Cement/ Bentonite	K. Andromalos	M. Kitko mkitko@geo-solutions.com
Geo-Con (trade name of Geo-Solutions) 2006	MGP Site Nyack, NY	11,400 cy	BTEX PAHs	In-situ Large diameter auger	Solidification	Cement/ Bentonite 1×10^{-5} cm/sec permeability 50 psi UCS	K. Andromalos	M. Kitko mkitko@geo-solutions.com
Geo-Con (a trade name of Geo-Solutions) 2006	Grey's Landfill Sparrows Point, MD	11,702 ft^2	N/A	In-situ Auger	Ground Improvement	85 psi UCS Cement	K. Andromalos	M. Kitko mkitko@geo-solutions.com

Sites Inside The United States

Treatment Vendor/date of treatment	Site Name/Location	Quantity Media	Contaminant	In-situ/ Ex-situ & Equipment	Purpose	Reagent Formula/ treat specs	Submitted by	Reference/Contact
Geo-Con (a trade name of Geo-Solutions) 2006	Walmart Supercenter Scarborough, ME	8,400 ft^2	N/A	In-situ Auger	Sub-grade Improvement	194 psi UCS Cement	K. Andromalos	G. Maitland gmaitland@geo-solutions.com
WRScompass, 2006	West side and Racine MGP sites, Racine, WI	32,000 cy soil	PAHs, coal tar, DNAPL	In-situ auger	Re-development	2% Portland + 5% slag + Rheomag >50 psi UCS <1x10^{-6} cm/s permeability	Paul Lear	Paul.Lear plear@envirocon.com
Shaw, 2006	Goodfellow Air force Base Small Arms Range, San Angelo, TX	11,000 cy	Lead, PAHs	En situ excavator	Remediation	5% Portland cement TCLP lead <5 mg/L	Paul Lear	Paul.Lear plear@envirocon.com
WRScompass, 2006	Kane County Remediation & Storm Water Improvement, St Charles, IL	19,500 cy, sludge	Lead, copper	In-situ Excavator	Impoundment closure	5-10% lime kiln dust TCLP lead <5 mg/L	Paul Lear	Paul.Lear plear@envirocon.com
WRScompass, 2006	Larimer County Landfill Firing Range remediation, Fort Collins, CO	3,000 cy soil	Lead	Ex-situ Pug-mill	Closure	3% Triple Superphosphate TCLP Lead <5 mg/L	Paul Lear	Paul.Lear plear@envirocon.com

Sites Inside The United States

Treatment Vendor/date of treatment	Site Name/Location	Quantity Media	Contaminant	In-situ/ Ex-situ & Equipment	Purpose	Reagent Formula/ treat specs	Submitted by	Reference/Contact
WRScompass, 2006	Walter J Heinrich Training Facility, Tampa, FL	1,000 cy soil	Lead	Ex-situ Excavator	Closure	3% Triple Superphosphate TCLP lead <5 mg/L	Paul Lear	Paul.Lear plear@envirocon.com
WRScompass, 2006	Normandy Park, River Hills & Residential Cleanup, Tampa, FL	24,000 cy soil and sediment	Lead	Ex-situ excavator	Remediation	3% Triple Superphosphate TCLP lead < 5 mg/L	Paul Lear	Paul.Lear plear@envirocon.com
Shaw, 2006	FMC Atvex Fibers Fine Fractions Stabilisation, Front Royal, VA	7,000 cy soils, sludges	Lead, arsenic, antimony	Ex-sitJ excavator	Closure	3% Triple Superphosphate + 0.5% ferrous sulfate	Paul Lear	Paul.Lear plear@envirocon.com
Shaw, 2006	City of Arlington Skeet Range, Arlington, WA	8,000 cy	Lead, PAHs	Ex-sitJ excavator	Closure	7.5% cement kiln dust TCLP lead <075 mg/L TCLP PAHs <0.010 mg/L	Paul Lear	Paul.Lear plear@envirocon.com
Shaw, 2005	FUSRAP Colonie Site Remediation Project, Colonie, NY	75,000 cy soil	Lead	Ex-situ pug-mill	Closure	1% phosphoric acid TCLP lead <5 mg/L	Paul Lear	Paul.Lear plear@envirocon.com

APPENDIX A: REMEDIATED SITES EMPLOYING S/S

Sites Inside The United States

Treatment Vendor/date of treatment	Site Name/Location	Quantity Media	Contaminant	In-situ/ Ex-situ & Equipment	Purpose	Reagent Formula/ treat specs	Submitted by	Reference/Contact
WRScompass, 2005	Augusta 15th Street, Augusta, GA	125,000 cy soil	PAHs, coal tar, DNAPL	In-situ auger	Redevelopment	2% Portland + 6% slag, >50 psi UCS <1x10^{-6} cm/s permeability	Paul Lear	Paul.Lear plear@envirocon.com
WRScompass, 2005	Americus MGP Site, Americus, GA	28,000 cy soil	PAHs, coal tar, DNAPL	In-situ auger	Redevelopment	8% Portland + 1% bentonite >50 psi UCS <1x10^{-6} cm/s permeability	Paul Lear	Paul.Lear plear@envirocon.com
WRScompass, 2005	Vandalia Road Facility, Pleasant Hill, IA	29,000 cy soil	PAHs, coal tar, DNAPL	In-situ Excavator	Containment	3% Portland + 9% slag >50 psi UCS <1x10^{-6} cm/s permeability	Paul Lear	Paul.Lear plear@envirocon.com
Shaw 2005	Fort McClellan Iron Mountain Ranges 12 and 13, Anniston, AL	21,000 cy soil	Lead	Ex-situ excavator	Closure	5% Portland cement TCLP lead <5 mg/L	Paul Lear	Paul.Lear plear@envirocon.com
WRScompass, 2005	Former Scrap Yard Lead Removal, Benton Harbor, MI	52,000 cy soil	Lead	Ex-situ Excavator	Redevelopment	5% Portland cement TCLP lead <5 mg/L	Paul Lear	Paul.Lear plear@envirocon.com

Sites Inside The United States

Treatment Vendor/date of treatment	Site Name/Location	Quantity Media	Contaminant	In-situ/ Ex-situ & Equipment	Purpose	Reagent Formula/ treat specs	Submitted by	Reference/Contact
RECON 2005	Impoundment Closure, Houston, TX	20,000 cy sludge	RCRA metals	Ex-situ bucket mixing	Impoundment Closure	Fly ash	Remedial Construction Services, L.P.	S. Birdwell steven.birdwell@reconservices.com
RECON 2005	Tex Tin Superfund, Texas City, TX	167,000 cy sludge, sediment, soil	Acid and heavy metals	In-situ and ex-situ bucket mixing	Superfund remediation	Fly ash/ 20 psi UCS	Remedial Construction Services, L.P.	S. Birdwell steven.birdwell@reconservices.com
RECON 2005	Landfill Emergency Action Southern, CA	67,000 cy sludge	Various hydrocarbons, drilling muds, tank bottoms	Ex-situ bucket mixing	Stabilise existing berms around lagoons	Strength and paint filter	Remedial Construction Services, L.P.	S. Birdwell steven.birdwell@reconservices.com
RECON 2005	SWB Closure Nederland, TX	125,000 cy soil and sediment	hydrocarbons	Excavator bucket mixing	Impoundment closures	Fly ash and Portland cement	Remedial Construction Services, L.P.	S. Birdwell steven.birdwell@reconservices.com
RECON 2005	Site Preparation Cameron Parish, LA	1,172,000 cy dredge spoils	Trace metals	In-situ excavator	Stabilisation for construction foundation	Patented blend of fly ash	Remedial Construction Services, L.P.	S. Birdwell steven.birdwell@reconservices.com
RECON 2005	Cell 3 Port Arthur, TX	80,000 cy sludge	Various hydrocarbons	In-situ bucket mixing	Stabilisation for construction foundation	Cement/ash blend	Remedial Construction Services, L.P.	S. Birdwell steven.birdwell@reconservices.com

Sites Inside The United States

Treatment Vendor/date of treatment	Site Name/Location	Quantity Media	Contaminant	In-situ/ Ex-situ & Equipment	Purpose	Reagent Formula/ treat specs	Submitted by	Reference/Contact
Geo-Con (trade name of Geo-Solutions) 2005	Refining Facility Lima, OH	20,000 cy	Petroleum	In-situ/ Backhoe	Stabilisation	Bentonite UCS 100 psi	K. Andromalos	B. Buccille bbuccille@geo-solutions.com
Geo-Con (a trade name of Geo-Solutions) 2005	Funnel & Gate Muskegon, MI	9,921 ft²	Impacted Groundwater	In-situ Excavator	Funnel System	1.00×10^{-7} cm/sec permeability Bentonite	K. Andromalos	M. Kitko mkitko@geo-solutions.com
RECON 2004	Basin 15 Closure Roxana, IL	300,000 cy sludge	hydrocarbons	In-situ bucket and rake mixing	Impoundment closure	Class C fly ash, cement and bed ash	Remedial Construction Services, L.P.	S. Birdwell steven.birdwell@reconservices.com
RECON 2004	Laydown Facility, Richmond, CA	85,000 cy soil and sludge	Hydrocarbons	In-situ excavator	Strengthening for cap construction	Fly ash and cement	Remedial Construction Services, L.P.	S. Birdwell steven.birdwell@reconservices.com
RECON 2004	Remedial Action, Buffalo, NY	5,400 cy soil	Nitrobenzene	In-situ jet grouting	Remedial action	Potassium Permanganate 600 points per column	Remedial Construction Services, L.P.	S. Birdwell steven.birdwell@reconservices.com
IT Corporation, 2004	B.F. Shaw Connex Facility, Troutville, VA	5,000 cy	lead	In-situ excavator	Remediation	3% Triple Superphosphate TCLP lead <5 mg/L	Paul Lear	Paul Lear plear@envirocon.com

Sites Inside The United States

Treatment Vendor/date of treatment	Site Name/Location	Quantity Media	Contaminant	In-situ/ Ex-situ & Equipment	Purpose	Reagent Formula/ treat specs	Submitted by	Reference/Contact
WRScompass, 2004	Augusta MGP Off-sites, Augusta, GA	15,000 cy soil	PAHs, coal tar, DNAPL	In-situ auger	Re-development	2% Portland + 6% slag >50 psi UCS <1x10^{-6} cm/s permeability	Paul Lear	Paul.Lear plear@envirocon.com
IT Corporation, 2004	Sunflower Army Ammunition Depot SWMU 22, De Soto, KS	15,000 cy soil and sludges	Lead, explosives, propellants	Ex-situ excavator	Closure	5% Portland cement TCLP lead < 5 mg/L, No reactivity	Paul Lear	Paul.Lear plear@envirocon.com
IT Corporation, 2004	Former Volunteer Army Ammunition Plant CFI Lease Site, Chattanooga, TN	13,000 cy	Lead	Ex-situ excavator	Remediation	5% Portland cement TCLP lead <5mg/L	Paul Lear	Paul.Lear plear@envirocon.com
IT Corporation, 2004	Texas American Oil Site, Midlothian, TX	13,000 cy sludge	Lead, TPH	In-situ excavator	Closure	10% EnviroBlend SPLP lead <0.015 mg/L >15 psi UCS <1x10^{-6} cm/s permeability	Paul Lear	Paul.Lear plear@envirocon.com
IT Corporation, 2004	Former Volunteer Army Ammunition Plant scrap Yard, Chattanooga, TN	16,000 cy soil	Lead	Ex-situ excavator	Remediation	6% Portland cement TCLP lead <5 mg/L	Paul Lear	Paul.Lear plear@envirocon.com

Sites Inside The United States

Treatment Vendor/date of treatment	Site Name/Location	Quantity Media	Contaminant	In-situ/Ex-situ & Equipment	Purpose	Reagent Formula/treat specs	Submitted by	Reference/Contact
IT Corporation, 2004	Fernald Silos 1 & 2 Stabilisation, Fernald, OH	4,000 cy sludge	Lead, radionuclides	Ex-situ pug-mill	Closure	8% Portland cement Pass paint filter test	Paul Lear	Paul.Lear plear@envirocon.com
WRS Infrastructure & Env. 2003	Lake Wire Lakeland, FL	Lake Sediment, 9,000 cy	Pb	Ex-situ Excavator mixing	State Action Risk Reduction	EnviroBlend 80/20 3% ZapSorb 1%	D. Wheeler	Dave Wheeler dwhee87@yahoo.com
WRScompass, 2003	Augusta MGP North Parcel, Augusta, GA	25,000 cy soil	PAHs, coal tar, DNAPL	In-situ auger	Re-development	2% Portland + 6% slag >50 psi UCS <1x10^{-6} cm/s permeability	Paul Lear	Paul.Lear plear@envirocon.com
WRScompass, 2003	Appleton MGP Fox River Canal Riverbank, Appleton, WI	32,000 cy soil	PAHs, coal tar, DNAPL	In-situ auger	Re-development	2% Portland + 6% slag >50 psi UCS <1x10^{-6} cm/s permeability	Paul Lear	Paul.Lear plear@envirocon.com
WRScompass, 2003	Sumter MGP Site, Sumter, SC	22,000 cy soil	PAHs, coal tar, DNAPL	In-situ auger	Re-development	3% Portland + 9% slag >50 psi UCS <1x10^{-6} cm/s permeability	Paul Lear	Paul.Lear plear@envirocon.com

APPENDIX A: REMEDIATED SITES EMPLOYING S/S

Sites Inside The United States

Treatment Vendor/date of treatment	Site Name/Location	Quantity Media	Contaminant	In-situ/Ex-situ & Equipment	Purpose	Reagent Formula/treat specs	Submitted by	Reference/Contact
IT Corporation, 2003	Andersen Air Force Base, Guam	13,000 cy soil	Lead, antimony	In-situ Tiller	Remediation	2% - 4% Triple Superphosphate TCLP lead <5 mg/L	Paul Lear	Paul Lear plear@envirocon.com
IT Corporation, 2003	Yorktown Naval Air Station Site 4 Burn Pit Remediation, Yorktown, VA	6,000 cy	Lead, cadmium	In-situ tiller	Remedaition	5% Portland cement TCLP lead <5 mg/L	Paul Lear	Paul Lear plear@envirocon.com
IT Corporation, 2003	Former Lee Field Naval Air Station Small Arms range, Green Cove, FL	7,500 cy soil	Lead	In-situ excavator	Closure	4% Portland cement TCLP lead <5 mg/L	Paul Lear	Paul Lear plear@envirocon.com
RECON 2003	CAMU Construct El Paso, TX	125,000 cy soil	Various hydrocarbons	Bucket mixing	Leachability requirements	Fly ash/ 20 psi UCS	Remedial Construction Services, L.P.	S. Birdwell steven.birdwell@reconservices.com
RECON 2003	250 Canal, Richmond, CA	15,000 cy sludge	Hydrocarbons	In-situ excavator	Strengthening for cap construction	Fly ash and cement	Remedial Construction Services, L.P.	S. Birdwell steven.birdwell@reconservices.com

APPENDIX A: REMEDIATED SITES EMPLOYING S/S

Sites Inside The United States

Treatment Vendor/date of treatment	Site Name/Location	Quantity Media	Contaminant	In-situ/ Ex-situ & Equipment	Purpose	Reagent Formula/ treat specs	Submitted by	Reference/Contact
RECON 2002	Surface Impoundment Closure, Edgemoor, DE	70,000 cy sludge	Ferric chloride sludge	In-situ bucket mixing	Impoundment closures	Federal Safe Drinking Water Act/ 20% hydrated lime, 10% Portland cement and 10% water	S. Birdwell Remedial Construction Services, L.P.	S. Birdwell steven.birdwell@reconservices.com
RECON 2002	Oxidation Pond, Richmond, CA	110,000 cy sludge	Hydrocarbons	In-situ excavator	Strengthening for cap construction	Fly ash and cement	S. Birdwell Remedial Construction Services, L.P.	S. Birdwell steven.birdwell@reconservices.com
Geo-Con 2002	Flyash Pond, Yarmouth, ME	Over 10,000 cy	Metals	In-situ, long-reach excavator	Flyash pond closure	Cement 10%	T. Plante	T.Plante tplante@haleyaldrich.com 207-482-4622
ENTACT, LLC - 2002	Browns Battery NPL Superfund Site – Pennsylvania	43,000 cubic yards of impacted soil	Lead	Ex-situ – Pug-mill	Superfund remediation; risk reduction	Phosphate based reagent	ENTACT, LLC	USEPA Region 3, PADEP, US Department of the Interior Fish & Wildlife Services
IT Corporation, 2002	Docklands Redevelopment, Melbourne, Victoria, Australia	60,000 cy Sludge, tar, soils	PAHs	In-situ excavator	Re-development	10% Portland cement + 10% cement kiln dust or 10% fly ash >25 psi UCS	Paul Lear	Paul.Lear plear@envirocon.com

Sites Inside The United States

Treatment Vendor/date of treatment	Site Name/Location	Quantity Media	Contaminant	In-situ/ Ex-situ & Equipment	Purpose	Reagent Formula/ treat specs	Submitted by	Reference/Contact
WRScompass, 2002	Athens Riverbank, Athens, GA	60,000 cy soil	PAHs, coal tar, DNAPL	In-situ auger	Re-development	2% Portland + 6% slag >50 psi UCS <1×10^{-6} cm/s permeability	Paul Lear	Paul.Lear plear@envirocon.com
IT Corporation, 2002	Chevron Perth Amboy Tank Cleaning Impoundments, Perth Amboy, NJ	45,000 cy of tank bottoms	TPH, lead	In-situ, Excavator	Remediation	5%–10% Portland cement >50 psi UCS TCLP lead <5 mg/L	Paul Lear	Paul.Lear plear@envirocon.com
WRScompass, 2002	Macon MGP Site, Macon, GA	125,000 cy soil	PAHs, coal tar, DNAPL	In-situ auger	Re-development	2% Portland + 6% slag >50 psi UCS <1×10^{-6} cm/s permeability	Paul Lear	Paul.Lear plear@envirocon.com
IT Corporation, 2002	American Home Products Impoundment 26 Closure, Boundbrook, NJ	24,000 cy Tar, soil	Tar	Ex-situ pug-mill	Closure	15% Portland cement >50 psi UCS <5% strain	Paul Lear	Paul.Lear plear@envirocon.com
IT Corporation, 2002	Sunflower Army Ammunition Depot SWMU 10 & 11, De Soto, KS	55,000 cy soil and sludges	Lead, explosives, propellants	Ex-situ pug-mill	Closure	5% Portland cement TCLP lead < 5 mg/L, No reactivity	Paul Lear	Paul.Lear plear@envirocon.com

Sites Inside The United States

Treatment Vendor/date of treatment	Site Name/Location	Quantity Media	Contaminant	In-situ/Ex-situ & Equipment	Purpose	Reagent Formula/treat specs	Submitted by	Reference/Contact
WRScompass, 2001	Savannah MGP Site, Savannah, Georgia	25,000 cy soil	PAHs, coal tar, DNAPL	In-situ auger	containment	2% Portland + 6% slag >50 psi UCS <1x10^{-6} cm/s permeability	Paul Lear	Paul.Lear plear@envirocon.com
WRScompass, 2001	Waycross Canal, Waycross, GA	50,000 cy soil	PAHs, coal tar, DNAPL	In-situ auger	Re-development	3% Portland + 9% slag >50 psi UCS <1x10^{-6} cm/s permeability	Paul Lear	Paul.Lear plear@envirocon.com
WRScompass, 2001	Former Police Pistol Range, Tampa, FL	32,000 cy soil	Lead	Ex-situ Pug-mill	Closure	3% Enviroblend TCLP lead < 5 mg/L	Paul Lear	Paul.Lear plear@envirocon.com
IT Corporation, 2001	Former Henry Woods & Sons Paint Factory Paint Pigment Disposal Area, Wellesley, MA	20,000 cy soil	Chromium	In-situ excavator	Closure	2% ferrous sulfate + 2% calcium polysulfide TCLP Cr <5 mg/L	Paul Lear	Paul.Lear plear@envirocon.com
IT Corporation, 2001	Former Kaiser Steel Plant Tar Pits Remediation, Fontana, CA	9,000 cy	TPH, sulfuric acid	In-situ excavator	Remediation	5% Portland cement + 10% Class C fly ash >25 psi UCS	Paul Lear	Paul.Lear plear@envirocon.com

APPENDIX A: REMEDIATED SITES EMPLOYING S/S

Sites Inside The United States

Treatment Vendor/date of treatment	Site Name/Location	Quantity Media	Contaminant	In-situ/ Ex-situ & Equipment	Purpose	Reagent Formula/ treat specs	Submitted by	Reference/Contact
IT Corporation, 2001	York Oil Superfund Site OU-2 Wetlands Soils, Moira, NY	12,000 cy	PCBs	Ex-situ excavator	Remediation	25% Portland cement >50 psi UCS	Paul Lear	Paul.Lear plear@envirocon.com
IT Corporation, 2001	Barbers Point Scrap Yard Remediation, Oahu, HI	41,000 cy soil	Lead	In-situ Dozer	Closure	1-4% Triple Superphosphate TCLP lead <5 mg/L	Paul Lear	Paul.Lear plear@envirocon.com
IT Corporation, 2001	Camp Allen Naval Station Salvage Yard, Norfolk, VA	9,000 cy	Cadmium	In-situ excavator	Remediation	5% Portland cement TCLP cadmium <1 mg/L	Paul Lear	Paul.Lear plear@envirocon.com
RECON 2001	Peak Oil and Bay Drums Superfund, Tampa, FL	55,000 cy soil	Lead and chlordane	Ex-situ pug-mill	Superfund remediation	Portland cement and TSP 282 g/l SPLP and 1×10^{-5} cm/sec permeability	Remedial Construction Services, L.P.	S. Birdwell steven.birdwell@reconservices.com
RECON 2001	Power Plant Stabilisation, Denton, TX	2,100 cy sludge	Wastewater and cooling tower sludge	In-situ bucket mixing	Disposal as Class I	Portland cement and quicklime	Remedial Construction Services, L.P.	S. Birdwell steven.birdwell@reconservices.com

Sites Inside The United States

Treatment Vendor/date of treatment	Site Name/Location	Quantity Media	Contaminant	In-situ/ Ex-situ & Equipment	Purpose	Reagent Formula/ treat specs	Submitted by	Reference/Contact
RECON 2001	SWIB/WWTS Cleanout Project, Norco, LA	35,000 cy sludge	Hydrocarbons	In-situ bucket mixing	Levee construction	Patented	Remedial Construction Services, L.P.	S. Birdwell steven.birdwell@reconservices.com
RECON 2000	Chemical Oxidation, Aiken, SC	10,000 cy soil and sludge	Radionuclides, heavy metals, organic compounds and inorganic compounds	In-situ soil mixing	Risk reduction	Cement, fly ash, bentonite, and zeolite	Remedial Construction Services, L.P.	S. Birdwell steven.birdwell@reconservices.com
Geo-Solutions, Four Seasons 2000	South 8th St, West Memphis Arkansas	40,000 cy	Pb, acid oily sludge	In-situ/auger Ex-situ/ bucket mix	Superfund remediation risk reduction	limestone 16% cement 13%, Fly ash 6.5%, UCS ≥ 100 psi ≤ 1X10^{-6} cm/sec SPLP Pb≤ 15 ug/L	E. Bates	USEPA 2009
IT, 2000	Seneca Army Depot Firing Range Remediation, Romulus, NY	6,000 cy Soils and sediment	Lead	Ex-situ excavator	Remediation	5% Portland cement TCLP lead <5 mg/L	Paul Lear	Paul Lear plear@envirocon.com

Sites Inside The United States

Treatment Vendor/date of treatment	Site Name/Location	Quantity Media	Contaminant	In-situ/ Ex-situ & Equipment	Purpose	Reagent Formula/ treat specs	Submitted by	Reference/Contact
IT/OHM 2000	American Creosote, Jackson Tennessee	45,000 cy	creosote, PCP, dioxins, furans	Ex-situ/ pug-mill	Superfund remediation risk reduction	cement 5%, fly ash 4.5 %, carbon 1.3% UCS ≥ 100 psi ≤ 1X10^{-6} cm/sec SPLP PCP ≤ 200 ug/L, SPLP Dioxin≤30 pg/L	E. Bates	USEPA 2009
IT Corporation, 2000	Drake Chemical Superfund site Incinerator Ash Stabilisation, Lock Haven, PA	1,000 cy ash	Arsenic, barium, cadmium, chromium, lead	Ex-situ excavator	Waste treatment	5% Portland cement + 2% ferrous sulfate TCLP metals < Drinking Water Standards	Paul Lear	Paul Lear plear@envirocon.com
IT Corporation, 2000	Roma Street Parklands Redevelopment, Brisbane, Queensland, Australia	16,000 cy soil, coke, ash, cinders	PAHs, TPH	Ex-situ excavator	Redevelopment	5% Portland cement >50 psi UCS, <1x10^{-6} cm/s permeability, S PLP PAHs <0.01 mg/L	Paul Lear	Paul Lear plear@envirocon.com

Sites Inside The United States

Treatment Vendor/date of treatment	Site Name/Location	Quantity Media	Contaminant	In-situ/ Ex-situ & Equipment	Purpose	Reagent Formula/ treat specs	Submitted by	Reference/Contact
IT Corporation, 2000	ABEX Superfund Site, Portsmouth, VA	20,000 cy	Lead	Ex-situ pug-mill	Remediation	5% Portland cement TCLP lead <0.75 mg/L	Paul Lear	Paul Lear plear@envirocon.com
WRScompass, 2000	Ashley River Site, Charleston, SC	14,000 cy, sediment	PAHs, creosote	Ex-situ Excavator	Remediation	Portland cement >50 psi UCS TCLP lead <5 mg/L	Paul Lear	Paul Lear plear@envirocon.com
IT Corporation, 2000	Fort Benjamin-Harrison Small Arms Firing Ranges, Indiana	15,000 cy	Lead	In-situ tiller	Remediation	4% Triple Superphosphate TCLP lead <5 mg/L	Paul Lear	Paul Lear plear@envirocon.com
Earth Tech 1999	Southwire, Carrolton, GA	Soil, 15,000 cy	Pb	Ex-situ Excel Port-a-Pug	State Action, risk reduction	EnviroBlend 80/20, 3% TCLP Pb≤ 5 mg/L	D. Wheeler	Dave Wheeler dwhee87@yahoo.com
IT Corporation, 1999	Willow Run Creek Site Remedial Action, Washetaw County, MI	380,000 cy	TPH, PCBs	Ex-situ Pug-mill	Closure	5% Cement kiln dust + 10%-15% Class F fly ash >50 psi UCS <1x10⁻⁶ cm/s permeability	Paul Lear	Paul Lear plear@envirocon.com

APPENDIX A: REMEDIATED SITES EMPLOYING S/S

Sites Inside The United States

Treatment Vendor/date of treatment	Site Name/Location	Quantity Media	Contaminant	In-situ/ Ex-situ & Equipment	Purpose	Reagent Formula/ treat specs	Submitted by	Reference/Contact
OHM, 1999	Exxon Bayway Refinery Impoundment Closure, Linden, NJ	120,000 cy	TPH	In-situ excavator	Closure	5% Portland cement + 5% - 10% cement kiln dust >10 psi UCS	Paul Lear	Paul Lear plear@envirocon.com
IT Corporation, 1999	Former Blackhills Army Ammunition Plant, Igloo, SD	15,000 cy soil/sludge	Chromium	Ex-situ excavator	Closure	1% ferrous sulfate, + 1% calcium polysulfide + 5% Portland cement TCLP Cr <5 mg/L	Paul Lear	Paul Lear plear@envirocon.com
RECON 1999	Section 9 Port Arthur, TX	126,000 cy soil and sludge	Various hydrocarbons	In-situ bucket mixing	Strengthen for cap construction	Patented LSS/25 psi UCS	Remedial Construction Services, L.P.	S. Birdwell steven.birdwell@reconservices.com
RECON 1999	Track F Port Arthur, TX	300,000 cy soil and sludge	Various hydrocarbons	In-situ bucket mixing	Strengthen for cap construction	Patented LSS/25 psi UCS	Remedial Construction Services, L.P.	S. Birdwell steven.birdwell@reconservices.com
RECON 1998	Pond Closure, El Paso, TX	23,000 cy sludge	Heavy metals	In-situ bucket mixing	Risk based closure	Fly ash and Portland cement	Remedial Construction Services, L.P.	S. Birdwell steven.birdwell@reconservices.com

Sites Inside The United States

Treatment Vendor/date of treatment	Site Name/Location	Quantity Media	Contaminant	In-situ/Ex-situ & Equipment	Purpose	Reagent Formula/treat specs	Submitted by	Reference/Contact
ENTACT, LLC - 1998	Schuylkill Metals NPL Superfund Site - Plant City Florida	265,000 tons of impacted soils and sediments; battery casings	Lead; metals; battery casings	Ex-situ - Pug-mill	Superfund remediation; risk reduction	cement 10% TSP 2% UCS ≥ 50 psi ≤ 1X10^{-6} cm/sec SPLP Pb≤ 1 mg/L TCLP Pb≤ 5 mg/L	E. Bates	USEPA 2009
Earth Tech 1998	ByPass 601 NPL Concord, NC	10,420 cy	Pb, from battery recycling	Ex-situ Excel Port-a-pug	Superfund remediation risk reduction	Cement 5% EnviroBlend 80/20 5%	D. Wheeler	Dave Wheeler dwhee87@yahoo.com
OHM, 1998	NL Taracorp Residential Cleanup, Granite City, IL	4,000 cy	Lead	Ex-situ pug-mill	Remediation	5% Portland cement TCLP lead <5 mg/L	Paul Lear	Paul Lear plear@envirocon.com
OHM, 1998	Double Eagle Refinery Acid Tar Ponds, Muskogee, OK	60,000 cy acid tar sludge	TPH, sulfuric acid, hydrogen sulfide, sulfur dioxide	In-situ excavator	Closure	15% Class C fly ash + 5% Portland cement >50 psi UCS pH>7	Paul Lear	Paul Lear plear@envirocon.com

Sites Inside The United States

Treatment Vendor/date of treatment	Site Name/Location	Quantity Media	Contaminant	In-situ/Ex-situ & Equipment	Purpose	Reagent Formula/treat specs	Submitted by	Reference/Contact
IT Corporation, 1998	Fort Gillem Small Arms Range Remediation. Atlanta, GA	5,000 cy soil	Lead	Ex-situ excavator	Remediation	5% Portland cement TCLP Pb <5 mg/L	Paul Lear	Paul.Lear plear@envirocon.com
OHM, 1998	Fayette Equipment and Salvage Site, Uniontown, PA	10,000 cy soil	Lead	Ex-situ pug-mill	Remediation	8% lime kiln dust TCLP lead <5 mg/L	Paul Lear	Paul.Lear plear@envirocon.com
OHM, 1997	Camp Pendleton IR Sites 3 and 6, Oceanside, CA	44,000 cy soil	DDT, DDD, DDE, dioxins, PCBs, TPH, lead, cadmium	Ex-situ Pug-mill	Closure	5% Class C fly ash + 1% carbon SPLP organochlorine pesticides <0.001 mg/L SPLP PCBs <0.001 mg/L	Paul Lear	Paul.Lear plear@envirocon.com
IT Corporation, 1997	Shaler/JTC Soil Remediation Project, Bruin, PA	60,000 cy soil	BTEX	In-situ auger	Remediation	10% Portland cement, >50 psui UCS, <1x10^{-6} cm/s permeability	Paul Lear	Paul.Lear plear@envirocon.com
RECON 1996	French Limited Superfund Site, Crosby, TX	30,000 cy sludge	Organics - biosolids	In-situ bucket mixing	Stabilisation	Soil and fly ash	Remedial Construction Services, L.P.	S. Birdwell steven.birdwell@reconservices.com

Sites Inside The United States

Treatment Vendor/date of treatment	Site Name/Location	Quantity Media	Contaminant	In-situ/ Ex-situ & Equipment	Purpose	Reagent Formula/ treat specs	Submitted by	Reference/Contact
GNB Env. Svcs. 1996	Yellow Water Rd NPL Baldwin FL	4,472 cy soils	PCBs	Ex-situ, Excel Port-a-Pug	Superfund remediation	Cement 27% UCS ≥ 50psi ≤ 1×10^{-6} cm/sec	D. Wheeler	Dave Wheeler dwhee87@yahoo.com
OHM, 1996	Hercules 009 Landfill, Brunswick, GA	30,000 cy	Toxaphene	In-situ excavator	Closure	15% Portland cement, >50 psi UCS, <1×10^{-6} cm/s permeability TCLP Toxaphene <0.050 mg/L	Paul Lear	Paul.Lear plear@envirocon.com
OHM, 1996	NL Taracorp Superfund Site, Granite City, IL	15,000 cy	Lead	Ex-situ pug-mill	Remediation	5% Portland cement TCLP lead <5 mg/L	Paul Lear	Paul.Lear plear@envirocon.com
IT Corporation, 1995	MOTCO Facility Impoundment Closure, LaMarque, TX	14,000 cy sludge, salts	Mercury	In-situ Excavator	Closure	15% lime kiln dust TCLP Hg <0.02 mg/L, >25 psi UCS	Paul Lear	Paul.Lear plear@envirocon.com

Sites Inside The United States

Treatment Vendor/date of treatment	Site Name/Location	Quantity Media	Contaminant	In-situ/ Ex-situ & Equipment	Purpose	Reagent Formula/ treat specs	Submitted by	Reference/Contact
IT Corporation, 1995	Mobil Refinery Stormwater Pond Closures, Beaumont, TX	600,000 cy	TPH	In-situ excavator	Closure	Portland cement, cement kiln dust, fly ash, bed ash (mix design varied throughout project) >10 psi UCS	Paul Lear	Paul.Lear plear@envirocon.com
OHM, 1995	Johnston Atoll Ash Pit Stabilisation Project, Johnston Island	12,000 cy soil, ash, metallic debris	Lead	Ex-situ Pug-mill	Closure	5% hydrated Lime TCLP lead <5 mg/L SPLP lead <5 mg/L	Paul Lear	Paul.Lear plear@envirocon.com
IT Corporation, 1994	BabCOCk & Wilcox Koppell Facility, Koppel, PA	46,000 cy Electric arc furnace dust, soil, sludge	Metals	Ex-situ pug-mill	Remediation	15% Portland cement TCLP Pb <5 mg/L >50 psi UCS	Paul Lear	Paul.Lear plear@envirocon.com

Sites Inside The United States

Treatment Vendor/date of treatment	Site Name/Location	Quantity Media	Contaminant	In-situ/Ex-situ & Equipment	Purpose	Reagent Formula/treat specs	Submitted by	Reference/Contact
Chemical Waste Management /OHM, 1994	GM Fisher Guide Landfarm Closure, Flint, MI	370,000 cy Soil, sludge, sediment	Metals	Ex-situ Pug-mill	Closure	10% Lime Kiln Dust + 5% Class C Fly Ash TCLP Cd <0.11 mg/L, Cr <0.6 mg/L, Pb <0.75 mg/L, Ni < 11 mg/L, Ag <0.14 mg/L >50 psi UCS	Paul Lear	Paul.Lear plear@envirocon.com
Chemical Waste Management 1994	DOE LEHR Facility, Davis, CA	2,500 cy soil, sludges, lab residuals	Radionuclides	Ex-situ pug-mill	Remediation	20% Portland cement	Paul Lear	Paul.Lear plear@envirocon.com
OHM, 1994	Umatilla Army Ammunition Depot Burn Pits, Hermiston, OR	25,000 cy soil	Lead, TNT, DNT, TNB, RDX, HMX	Ex-situ pug-mill	Closure	5% Portland cement TCLP lead <5 mg/L, TNT < 0.20 mg/L, RDX < 0.13 mg/L, TNB < 0.18 mg/L, RDX <0.2 mg/L, HMX <18 mg/L	Paul Lear	Paul.Lear plear@envirocon.com

Sites Inside The United States

Treatment Vendor/date of treatment	Site Name/Location	Quantity Media	Contaminant	In-situ/ Ex-situ & Equipment	Purpose	Reagent Formula/ treat specs	Submitted by	Reference/Contact
Geo-Con 1993	Georgia Power Gas Plant, Columbus Georgia		PAHs, BTEX, cyanide	In-situ/auger	Remediation Site reuse	cement 10-25% UCS 60 psi ≤ 1×10^{-5} cm/sec Interior and ≤ 1×10^{-6} cm/sec edge columns	E. Bates	USEPA 2009
OHM, 1993	Chevron Pollard Pond Remediation, Richmond, CA	150,000 cy acid sludge tar, bay mud	TPH, sulfuric acid, hydrogen sulfide, sulfur dioxide	Ex-situ Pug-mill	Redevelopment	10% calcium carbonate + 5% Portland cement + 5% Class C fly ash >50 psi UCS 1% alkalinity reserve 5>pH <12	Paul Lear	Paul.Lear plear@envirocon.com
Chemical Waste Management, 1993	Selma Pressure Treating Superfund Site Remediation, Selma, CA	11,500 cy soil	Arsenic, copper, chromium, pentachlorophenol, dioxins	Ex-situ pug-mill	Remediation	12% Portland cement + 4% organoclay >50 psi UCS	Paul Lear	Paul.Lear plear@envirocon.com
OHM, 1993	Beals Battery Site	15,000 cy soil	Lead	Ex-situ pug-mill	Remediation	15% Portland cement TCLP Pb <5 mg/L >50 psi UCS	Paul Lear	Paul.Lear plear@envirocon.com

Sites Inside The United States

Treatment Vendor/date of treatment	Site Name/Location	Quantity Media	Contaminant	In-situ/ Ex-situ & Equipment	Purpose	Reagent Formula/ treat specs	Submitted by	Reference/Contact
Silicate Technology Corp. 1993	Selma Pressure Treating, Selma California	13,000 cy	PCP, As, Cu, Cr, dioxins furans	Ex-situ paddle mixer	Superfund remediation risk reduction	cement activated carbon	E. Bates	USEPA 2009 USEPA 2004
OHM, 1993	Kassouf-Kimerling Battery Disposal Superfund Site, Tampa FL	14,000 cy	Lead	Ex-situ pug-mill	Remediation	10% Portland + 1% hydrated lime + 10% class C fly ash TCLP <5 mg/L >50 psi UCS <1x10^{-6} cm/s permeability	Paul Lear	Paul.Lear plear@envirocon.com
Vendor unknown 1993	Pepper Steel, Medley Florida	85,000 cy soils	Pb, As, PCBs	Ex-situ equipment unknown	Superfund remediation risk reduction	cement fly ash UCS ≥ 21psi ≤ 1X10^{-6} cm/sec	E. Bates	USEPA 2009

Sites Inside The United States

Treatment Vendor/date of treatment	Site Name/Location	Quantity Media	Contaminant	In-situ/ Ex-situ & Equipment	Purpose	Reagent Formula/ treat specs	Submitted by	Reference/Contact
GCI 1992	Private Client Baltimore, MD	20,000 tons	Sewage Cake	Ex-situ ARAN Pug-mill	Produce Potting Soil	Lime 20% Soil 20%	R. Garrett	Robert Garrett bobgarrett417@bellsouth.net
OHM, 1992	Henkel Corporation Impoundment Closure, Carlstadt, NJ	52,000 cy	Lead, arsenic, TPH, PAH, PCBs	In-situ excavator	Closure	5% Portland cement + 15% fly ash >50 psi UCS <1×10^{-6} cm/s permeability	Paul Lear	Paul.Lear plear@envirocon.com
OHM, 1992	Lee's Farm Superfund Site, Woodville, WI	12,000 cy soil	Lead	Ex-situ pug-mill	Remediation	15% Portland cement TCLP Pb <5 mg/L >50 psi UCS <1×10^{-6} cm/s permeability	Paul Lear	Paul.Lear plear@envirocon.com
Chemical Waste Management, 1992	Portsmouth Gaseous Diffusion Plant Site X-231, Portsmouth, OH	1,000 cy soil	TCE, radionuclides	In-situ auger	Remediation	TCLP TCE <0.5 mg/L >50 psi UCS <1×10^{-6} cm/s permeability ANS 16.1 Leach Indices >10 for radionuclides	Paul Lear	Paul.Lear plear@envirocon.com

APPENDIX A: REMEDIATED SITES EMPLOYING S/S

Sites Inside The United States

Treatment Vendor/date of treatment	Site Name/Location	Quantity Media	Contaminant	In-situ/Ex-situ & Equipment	Purpose	Reagent Formula/treat specs	Submitted by	Reference/Contact
Chemical Waste Management, 1991	Port of Los Angeles Berths 212 and 215 Remediation, Los Angeles, CA	75,000 cy soil	Lead	Ex-situ pug-mill	Re-development	15% limestone + 5% sodium polycarbonate Cal WET lead <5 mg/L TCLP lead <5 mg/L	Paul Lear	Paul.Lear plear@envirocon.com
OHM, 1991	New Hampshire Plating Superfund Site, Merrimack, NH	27,000 cy Sludge and soil	metals	Ex-situ excavator	Remediation	15% Portland cement TCLP Cd <0.11 mg/L, Cr <0.6 mg/L, Pb <0.75 mg/L, Ni < 11 mg/L, Ag <0.14 mg/L	Paul Lear	Paul.Lear plear@envirocon.com
WRScompass, 1991	Columbus MGP FCRC, Columbus, GA	80,000 cy soil	PAHs, coal tar, DNAPL	In-situ auger	Re-development	3% Portland + 9% slag >50 psi UCS <1x10⁻⁶ cm/s permeability	Paul Lear	Paul.Lear plear@envirocon.com
Chemical Waste Management, 1991	Solvay Animal Health RCRA Impoundment Closure, Charles city, IA	5,000 cy	Arsenic	In-situ excavator	Closure	10% Class c fly ash TCLP As <5 mg/L >50 psi UCS	Paul Lear	Paul.Lear plear@envirocon.com

APPENDIX A: REMEDIATED SITES EMPLOYING S/S

Sites Inside The United States

Treatment Vendor/date of treatment	Site Name/Location	Quantity Media	Contaminant	Ir-situ/Ex-situ & Equipment	Purpose	Reagent Formula/treat specs	Submitted by	Reference/Contact
GCI 1991	Johnson Controls Atlanta, GA	Soil 3,000 tons	Pb	Ex-situ ARAN Pug-mill	State Action Property Transaction	CKD 15%	R. Garrett	Robert Garrett bobgarrett417@bellsouth.net
GCI 1991	BFI Atlanta, GA	10,000 tons	Liquid Waste	Ex-situ ARAN Pug-mill	Produce Solids	Lime 15% Soil 20%	R. Garrett	Robert Garrett bobgarrett417@bellsouth.net
GCI 1990	Johnson Controls Atlanta, GA	Soil 12,000 tons	Pb	Ex-situ ARAN Pug-mill	State Action Property Transaction	Proprietary 12%	R. Garrett	Robert Garrett bobgarrett417@bellsouth.net
GCI 1990	U.S. Navy San Francisco, CA	10,000 tons	Pb	Ex-situ ARAN Pug-mill	Shooting Range Remediation	CKD 15%	R. Garrett	Robert Garrett bobgarrett417@bellsouth.net
Chemical Waste Management, 1990	Lack Industries Impoundment Closure, Grand Rapids, MI	20,000 cy sludge	Metals	Ir-situ Excavator	Closure	15% Lime Kiln Dust TCLP Cd <0.11 mg/L, Cr <0.6 mg/L, Pb <0.75 mg/L, Ni <11 mg/L, Ag <0.14 mg/L	Paul Lear	Paul.Lear plear@envirocon.com

Sites Inside The United States

Treatment Vendor/date of treatment	Site Name/Location	Quantity Media	Contaminant	In-situ/Ex-situ & Equipment	Purpose	Reagent Formula/treat specs	Submitted by	Reference/Contact
Geo-Con (a trade name of Geo-Solutions)	Dundalk Marine Terminal Baltimore, MD		Chromium Ore Processed Residue3	In-situ Auger	Strain Relief	Horticultural Grade Peat Moss	K. Andromalos	M. Kitko mkitko@geo-solutions.com
Unknown unknown	New York Harbor and Port Authority, New York	over 145,000 cy	metals, dioxins, PAHs, PCBs	Ex-situ/ pug-mill	dredge spoils disposal, Brownfield re-development	cement 8 %	E. Bates	USEPA 2009
Greenleaf Env. Svcs	Edgewood Ave. Site Atlanta, GA	Soil 4,500 cy	Pb	Ex-situ Excavator mixing	State Action Property transaction	EnviroBlend 80/20 3% <5 mg/L TCLP Pb	D. Wheeler	Dave Wheeler dwhee87@yahoo.com

Appendix A References

USEPA, Technology Performance Review: Selecting and Using Solidification/Stabilisation Treatment for Site Remediation, EPA/600/R-09/148, November 2009, http://www.epa.gov/nrmrl/pubs/600r09148.html

USEPA, Remedial Action Report Soil Remedy: Selma Pressure Treating Site, Selma, California, EPA CERCLIS ID Number CAD029452141 USEPA Region 9, San Francisco, California, September 29, 2004.

APPENDIX B: CASE STUDIES EMPLOYING S/S

Appendix B presents a collection of S/S case studies from the USA and Europe. Each case study is presented as a separate file named for the location of the project. The files are presented in alphabetical order. These case studies are presented as received from leading S/S vendors and the editors have not independently verified the information. However for each case study the name of the person submitting the case study and their E mail address is provided should anyone wish to obtain more information regarding that project.

No.	Case Study	Author	Contact
1	Abex Superfund, VA	Paul Lear	plear@envirocon.com
2	American Creosote Superfund Site	Paul Lear	plear@envirocon.com
3	Bayou Trepagnier, LA	Steven R. Birdwell	steven.birdwell@reconservices.com
4	BHAD Chromium Pond, SD	Paul Lear	plear@envirocon.com
5	Cambridge, MA	Mark Kitko	mkitko@geo-solutions.com
6	Former Camden Gas Works Site, NJ	Paul Lear	plear@envirocon.com
7	Camp Pendleton Scrap Yard, CA	Paul Lear	plear@envirocon.com
8	Chevron Pollard Landfill Site	Paul Lear	plear@envirocon.com
9	Carnegie, PA	Mark Kitko	mkitko@geo-solutions.com
10	Columbus, IN	Ken Andromalos	kandromalos@geo-solutions.com
11	CSX Benton Harbor, MI	Paul Lear	plear@envirocon.com
12	Double Eagle Refinery Site, OK	Paul Lear	plear@envirocon.com
13	Dundalk, Baltimore, MD	Mark Kitko	mkitko@geo-solutions.com
14	East Rutherford, NJ	Ken Andromalos	kandromalos@geo-solutions.com
15	Foote Minerals Superfund Site, PA	Paul Lear	plear@envirocon.com
16	Ghent, Belgium	Stany Pensaert	pensaert.stany@deme.be
17	Guernsey, United Kingdom	Stany Pensaert	pensaert.stany@deme.be

No.	Case Study	Author	Contact
18	Hercules 009 Landfill Site, GA	Paul Lear	plear@envirocon.com
19	Hoedhaar Lokeren, Belgium	Stany Pensaert	pensaert.stany@deme.be
20	Irving, Tx	Mark Kitko	mkitko@geo-solutions.com
21	Johnston Atoll	Paul Lear	plear@envirocon.com
22	Kingston, TN	Steve Artman	sartman@geo-solutions.com
23	London, United Kingdom	Stany Pensaert	pensaert.stany@deme.be
24	Martinsville, VA	Mark Kitko	mkitko@geo-solutions.com
25	Milwaukee, Wisconsin	Roy E. Wittenberg	rwittenberg@naturalrt.com
26	Nederland, TX	Steven R. Birdwell	steven.birdwell@reconservices.com
27	New Bedford, MA	Ken Andromalos	kandromalos@geo-solutions.com
28	Nyack, NY	Mark Kitko	mkitko@geo-solutions.com
29	NYSEG Norwich NY MGP Site	Paul Lear	plear@envirocon.com
30	Obourg, Belgium	Stany Pensaert	pensaert.stany@deme.be
31	Perth Amboy, NJ	Paul Lear	plear@envirocon.com
32	Portsmouth, VA	Mark Kitko	mkitko@geo-solutions.com
33	Rieme, Belgium	Stany Pensaert	pensaert.stany@deme.be
34	Roma Street Station	Paul Lear	plear@envirocon.com
35	Sag Harbor, NY	Mark Kitko	mkitko@geo-solutions.com
36	Sanford MGP Superfund Site	Paul Lear	plear@envirocon.com
37	Soderhamn, Sweden	Stany Pensaert	pensaert.stany@deme.be
38	Southeast Wisconsin	Roy E. Wittenberg	rwittenberg@naturalrt.com
39	St Louis, MO	Ken Andromalos	kandromalos@geo-solutions.com
40	Sunflower Army Ammunition Depot	Paul Lear	plear@envirocon.com
41	Sydney Tar Ponds	Jerome MacNeil Diane Ingraham Donnie Burke	jerome@practicalenvironmental.com diane.ingraham@stantec.com donnie@tarpondscleanup.ca

No.	Case Study	Author	Contact
42	Umatilla Ammunition Depot	Paul Lear	plear@envirocon.com
43	Valero Paulsboro Refinery	Paul Lear	plear@envirocon.com
44	Waukegan, IL	Ken Andromalos	kandromalos@geo-solutions.com
45	West Doane Lake Site	Paul Lear	plear@envirocon.com
46	X-231B Pilot Study	Paul Lear	plear@envirocon.com
47	Zwevegem, Belgium	Stany Pensaert	pensaert.stany@deme.be

CASE STUDY: 1
Abex Superfund Site

Location:	Portsmouth, VA, USA
History:	Former foundry
Contaminants:	Lead, cadmium, zinc
Scale:	28,354 tons (25,722 tonnes)
Reagents:	Paper sludge ash
Method:	Ex-situ
End-Use:	Housing

Site Description

This project included a multi-faceted removal action at Former Abex Foundry and the adjacent Portsmouth Redevelopment Housing Authority's (PRHA's) Washington Park Homes area. The project had extensive regulatory oversight, and involved:

- Excavation, on-site staging, and classification of 85,000 tons (77,100 tonnes) of lead-impacted soil and debris
- Pug-mill stabilization using lime/Portland cement for 28,354 tons (25,722 tonnes) of hazardous lead-contaminated soil to UTS Standards
- Removal and disposal of 6,000 tons (5,443 tonnes) of construction debris
- Transportation and disposal of 87,600 tons (79,469 tonnes) of non-hazardous waste and 1,200 tons (1,088 tonnes) of hazardous waste

Excavation included removal of 1 to 5 ft (0.3-1.5 m) of soils determined by future land usage as mandated by the Consent Decree involving Pnuemo Abex and the USEPA. Relocation of the residents and securing of the PRHA apartments was performed prior to excavation. In addition, a video survey was performed to capture the pre-remediation condition of structures and ground cover.

Areas were excavated and direct-loaded into 16 yd³ (12 m³) dump trucks. Excavation grade was maintained utilizing a laser level operated by a ground man who worked closely with the equipment operator. Materials were transported and staged on a soil storage and treatment pad in 100 ton (98 tonnes) delineated bins for waste classification.

The classification phase was performed to define which materials could be disposed of as a non-hazardous waste without first requiring stabilization. The material requiring stabilization (greater then 5 ppm TCLP lead) was then treated using a Portec 53 Pug-mill. The reagent admixture rate was defined through treatability testing and applied to 23,354 tons (22,985 tonnes) of material which achieved the criteria of 0.75 ppm TCLP lead (UTS Standard) at a 99 % efficiency pass rate. Following post-treatment verification sampling, the material was loaded and transported and disposed of at a local non-hazardous landfill.

Drivers

Public housing was constructed on top of soil contaminated from the past foundry activities. The contaminated soils required removal to prevent contact with the contaminated soil by the residents.

Objectives

- Remove the contaminated soil
- Treat (if applicable) the soil to remove its RCRA hazardous characteristic D008 (lead), dispose of the soil at an off-site landfill
- Restore the site

Method

Material with greater than 5 mg/L TCLP-leachable lead was treated with Portland cement to reduce the TCLP-leachable lead concentrations to less than 0.75 mg/L.

A pug-mill-based system, consisting of a feed hopper with conveyor, pug-mill, reagent silo, and stacking conveyor was used to treat 150-200 tons (148-197 tonnes) of contaminated soil per hour. The treated soil was segregated into 200 ton (197 tonne) stockpiles for confirmation testing.

CASE STUDY: 2
American Creosote Superfund Site

Image: Google

Location:	Jackson, TN
History:	Former wood preserving site
Contaminants:	PAHs, PCP, arsenic, dioxins
Scale:	22,500 tons (22,144 tonnes)
Reagents:	Cement, fly ash & activated carbon
Method:	Ex-situ
End-Use:	Brownfield redevelopment

Site Description

The site consists of a former wood preserving site where wood was pressure-treated with creosote and pentachlorophenol. Wastewater sludges from the wood preserving operations were land applied, contaminating the site with polyaromatic hydrocarbons (PAHs), pentachlorophenol (PCP), dioxins, and arsenic. The primary goals of the remediation were to limit the solubility of the chemicals of concern to minimize their leachability. The treated waste was then to be buried in an on-site excavation area, covered with a clay liner and topsoil, and seeded with grass

OHM Corporation (OHM) completed remediation work at the American Creosote Site in Jackson, Tennessee under contract with the State of Tennessee and the US EPA Region IV. The project scope involved the remedial design, construction of a slurry cut-off wall around the impacted soils, excavation of the impacted soils, collection and treatment of perched water and DNAPL, stabilization treatment of the impacted soils, compaction and placement of the treated soil back into the excavation area, and capping the treated soil with clay and topsoil.

A 2,400 ft (731 m) slurry cut-off wall was constructed around the perimeter of the impacted soil. This cut-off wall was filled with a soil-bentonite mixture which had a permeability of less than 1×10^{-9} m/s and was keyed into the native clay layer underlying the site at a depth of 3 to 5 ft (0.3-1.5 m) below ground surface.

The impacted sandy soil inside of the slurry wall was excavated to the underlying clay layer. Visibly stained clay material was also removed during the excavation. The perched water and DNAPL in the excavation area were collected and removed by a series of drainage trenches. The perched water and DNAPL were treated on-site in a wastewater treatment plant designed, built, and operated by Shaw. In the wastewater treatment plant, an oil/water separator was used to separate the water from the DNAPL. The water was then passed through a sand filter and activated carbon before it was discharged to a local POTW. The collected DNAPL was containerized and shipped off-site for treatment and disposal.

The excavated material was stockpiled and screened through a screen plant consisting of a grizzly screen and a screen deck. The oversized material (>2 in/5 cm) from the screening operation was stockpiled for placement with the treated soil. The material passing through the screen deck was suitably sized for treatment and was subsequently transported to the pug-mill feed hopper for stabilization treatment. The soil was mixed with the reagents and water inside the pug-mill. The pug-mill system was operated 12 hours per day, 5 day per week, treating 1,200 tons (1,181 tonnes) of soil per day.

The treated soil exited the pug-mill on a conveyor, and was placed in temporary stockpiles until daily sampling was complete. The treated material exiting the pug-mill had a moist soil-like consistency. Confirmatory testing for the leachability of PAHs, PCP, dioxins, and arsenic, permeability, and unconfined compressive strength was performed on the sample from each temporary stockpile. All of the confirmatory samples met the performance requirements for leachability, permeability, and unconfined compressive strength. After confirmatory sampling results were received for a treated soil stockpile, the treated soil was placed back into the excavation area in 12 in (30 cm) lifts and compacted. Daily surveys of the excavated area defined the placement position of each stockpile of the treated soil.

After placement of all the treated soil, OHM graded the placed and compacted soil to establish the required 1-2 % slopes for the clay/topsoil cap. The graded soil was covered by a Claymax geosynthetic clay liner (GCL) in accordance with the manufacturer's instructions. The GCL was then covered with 18 in (45 cm) of common fill and 6 in (15 cm) of topsoil. The common fill was placed in 1 ft (0.3 m) loose lifts and track-walked into place. The topsoil was then hydromulched and seeded.

Risk Drivers

The site was slated for brownfield redevelopment by a local utility. The contaminated soils required treatment to prevent off-site migration of the contaminants into the surficial aquifer.

Objectives

- Remove the contaminated soil
- Treat to reduce the leachability of the contaminants and produced a high strength, low permeable treated material
- Place and compact the treated material into the excavation area
- Restore the site

Method

The contaminated soil was treated with a mixture of Portland cement, Class F fly ash, and activated carbon to produce a treated material which met the performance criteria listed in Table 1.

Table 1: SPLP Results for the Wood Preserving Site Soil

Portland Cement		0.05
Class F Fly Ash		0.06
Activated Carbon	Untreated	0.02
Water		0.05
Parameter	SPLP (g/L)	SPLP (g/L)
Pentachlorophenol	9,600	<100
Benzo(a)pyrene Potency Equivalence	155	<1
2,3,7,8 –TCCD Toxicity Equivalence	0.25	<0.03

Validation

Temporary 500 ton (492 tonne) stockpiles of the treated soil were created. Confirmatory testing for the leachability of PAHs, PCP, dioxins, and arsenic, permeability, and unconfined compressive strength was performed on the sample from each temporary stockpile. All of the confirmatory samples met the performance requirements for leachability, permeability, and unconfined compressive strength.

Equipment Used

The stabilization treatment involved the use of a pug-mill system to mix the soil with activated carbon, Class F fly ash, and Portland cement. OHM's pug-mill system consisted of a variety of feeders, conveyors, silos, and a pug-mill mixer integrated into a complete system for the continuous mixing of wastes and reagents. The screened material was fed to an 8 yd³ (6 m³) feed hopper. The hydraulically-driven belt, located in the bottom of the feed hopper, fed the material onto a 40 ft (12 m) long by 24 in (60 cm) wide belt conveyor. The conveyor belt was equipped with a Ramsey single idler belt scale for the accurate, real-time determination of the rate, in tons per hour, of soil being treated. The conveyor belt conveyed the material into the pug-mill for blending with the stabilization additives. The pug-mill mix box was 4 ft (1.2 m) wide by 3 ft (0.9 m) high by 10 ft (3 m) long. Paddles were bolted onto structural steel shafts with replaceable shafts flanged on both ends for ease of maintenance. The mixer was V-belt driven by a 125-hp motor.

Dry stabilization additives were stored on-site in vertical cement silos. The silos were self-leveling and had a capacity of 200 barrels of material. The silos were also equipped with a top mount baghouse for dust control during silo filling. The feed from each silo was controlled by a 14 in (35 cm) diameter screw auger powered by a 5 hp motor. The dry stabilization additives were introduced from silo feeders which are attached to the pug-mill through a central feed auger. The soil was mixed with the reagents and water inside the pug-mill. The pug-mill system was operated 12 hours per day, 5 day per week, treating 1,200 tons (1,181 tonnes) of soil per day.

Specific issues

For this site, the EPA Superfund Technical Assistance and Response Team (START) promulgated risk-based SPLP leaching criteria for PAHs, PCP, and dioxins. Based on the successful completion of the American Creosote Site remediation, the US EPA used these leachability criteria for PAHs, PCP, and dioxins for use at other wood-preserving sites contaminated with creosote.

CASE STUDY: 3
Bayou Trepagnier

Location:	St. Charles Parish, Louisiana
History:	Petrochemical plant
Contaminants:	Hydrocarbons
Scale:	50,000 yd³ (1,134 m³)
Reagents:	Paper sludge ash
Method:	In-situ bucket mixing
End-Use:	N/A

Site Description

The Bayou Trepagnier was the original drainage canal of two local petrochemical plants, sandwiched in between the Mississippi River and Lake Pontchartrain. Over the course of many years, residual petroleum impacted the sediment and banks of the Bayou. The scope of this remediation project was to immobilize the petroleum impacted sediment in the first 6,000 ft (1,828 m) of the Bayou and build a cap over it, protecting human health and receptors in the wetland/swamp.

In addition, the work included removal and chipping of trees, access road construction and installation of a FlexiFloat Bridge over the new drainage canal, construction of storm water holding cells for bayou dewatering activities, solidification of bayou sediments, placement of a clean clay cap over solidified sediments, and seeding of disturbed areas. A "clean zone" was also required that would entail excavating stabilized sediments in a 500 ft (152 m) segment, and reuse these to bulk-out impacted sediment downstream that required stabilization, and backfill the "clean zone" with imported fill material so a future diversion canal could be installed through non-impacted soil.

Characterisation

RECON was involved early in the design phase, undertook treatability studies and constructability reviews, and a field pilot demonstration project. The final design was completed after incorporating data generated from the treatability study in the Feasibility Study Phase, and a successful pilot field study project. During the design phase of the project, RECON ensured that the data gathered and lessons learned (during the Feasibility Study) were incorporated into the successful "full-scale" project. Waste composition, including reagents were as follows:

Sediment ~ 20 %
Moisture Content ~ 62-67 %
Oil & Grease ~ <3 %
Cement/Lime ~ 10-15 %

Risk Drivers

Unstable petroleum-impacted sediment was at risk of cross-contaminating the LaBranche Wetlands. The site was already impacted by saltwater intrusion from Lake Pontchartrain, as a result of storm surge flooding, leading to die-back of the wetland and swamp vegetation.

Objectives

The objectives of the project were to:

- Access the site, minimizing the disturbance of wetlands
- Install dams throughout to divide the bayou into several 'reaches'
- Stabilize the sediment in the bottom of the bayou to be supportive of a cap
- Flip and reuse the access road material as the cap
- Vegetate the cap and access with native plants

Method

Stabilization, not simply solidification, was the chosen remedy to mitigate risk of chemically-impacted sediment transfer off-site. As with any chemically-impacted material that is left in-place, stabilization binds both chemical and sediment together to form a matrix that greatly reduces the leachability of the chemicals into the environment. Therefore, no waste material required off-site disposal.

RECON utilized hydraulic excavators to perform mixing of reagents into the sediment. Tandem dumps were used to import access road/cap material. Tracked dump trucks were used to transport bank spoils to reaches throughout the bayou for incorporation into the mix. A flexi-float bridge was installed at the egress point in Engineers Canal to allow for one-way traffic of dump trucks. Long-reach excavators were used to take advantage of reaching across and mixing grids the width of the bayou. Other ancillary equipment included equipment mats, pumps, etc.

Validation

RECON used a pocket penetrometer to gauge the strength of stabilized material, and performed Unconfined Compressive Strength (UCS) testing to verify. It was determined that by creating a stabilized matrix, it would not require testing for chemicals.

Issues

The project was a significant challenge and a great accomplishment, as the stabilized cap over sediment within a bayou that meandered through a pristine swamp, was deemed to have returned all project access to pre-existing conditions.

The Bayou Trepagnier Team worked closely together with regulatory and other agencies, including the U.S. Army Corps of Engineers (USACE), State of Louisiana Department of Environmental Quality (LDEQ), Pontchartrain Levee District (PLD), State of Louisiana Coastal Protection Restoration Authority (CPRA) and its division the Office of Coastal Protection and Restoration (OCPR), the Louisiana Department of Wildlife and Fisheries (LDWF), and the Louisiana Department of Natural Resources (LDNR) Office of Coastal Management.

Most risks were mitigated in the Feasibility and Design phases of the project by having Owner, Engineer, Consultants and Contractor-RECON working together. The project ended up approximately $1,500,000 under budget, and was on schedule despite being shut down 4 weeks during high water levels in the Mississippi River that required the USACE to open the Bonnet Carre Spillway gates.

CASE STUDY: 4
BHAD Chromium Pond

Location:	Igloo, SD
History:	Ordnance manufacture
Contaminants:	Hexavalent chromium
Scale:	13,400 tons (13,188 tonnes)
Reagents:	Reducing agent + alkaline reagent
Method:	Ex-situ
End-Use:	N/A

Site Description

The former Blackhill Army Depot (BHAD) is located in Fall River County, Igloo, South Dakota. The BHAD consists of approximately 21,100 acres (4,046 m²) and was a former defence site, at which many forms of ordnance were stored, manufactured, and destroyed from 1942, until its closure in 1967. Since closure, much of the property has been used for livestock grazing and most of the former facilities show varying stages of decay.

The chromium pond site was located in the ammunition workshop area and was adjacent to the clean and paint building (Building 3038). The soil contained up to 8,000 mg/kg of chromium, primarily in the form of hexavalent chromium.

The remediation of the chromium pond areas entailed:

- Conducting a soil stabilization treatability study
- Work plan preparation
- Mobilization of necessary personnel and equipment
- Site setup including placement of project offices and installation of process equipment and staging areas
- Monitoring well abandonment
- Excavation and treatment of 15,000 tons (14,763 tonnes) on chromium-contaminated soil

- Transportation and disposal of the wastes
- Confirmatory sampling and analysis
- Surveying
- Restoration, re-vegetation, and site clean-up

Soil exceeding the risk-based remediation criteria for hexavalent chromium of 5.2 mg/kg for soil for 0 to 1 ft (0-0.3 m) below ground surface, and 7.0 mg/kg for soil below 1 ft (0.3 m) below ground surface, was excavated and stockpiled. Processing took place in two lined 150 ft (45 m) x 100 ft (31 m) bermed areas: the first used for reduction and interim storage, and the second for stabilization, and final storage during confirmatory testing.

Each bermed area was graded to a sump to collect storm and decontamination water for re-use in the stabilization treatment process. The process area contained two 50 yd^3 (38.2 m^3) mixing boxes, interim storage of sixteen 40 ton (39 tonne) batches awaiting stabilization, and storage for thirty-six 40 yd^3 (30 m^3) batches awaiting confirmatory testing and off-site disposal.

Characterisation

The soil was a silty clay soil with up to 8,000 mg/kg hexavalent chromium.

Risk drivers

The soil contained up to 8,000 mg/kg of hexavalent chromium. Crystals containing hexavalent chromium were also found on the surface of the chromium pond. The contaminated soils required removal to prevent contact with workers under the potential commercial/industry redevelopment of the site.

Objectives

- Remove the contaminated soil
- Treat (if applicable) the soil to remove its RCRA hazardous characteristic D007 (chromium)
- Dispose of the soil at an off-site landfill
- Restore the site

Method

The treatment of the contaminated soil involved two stages: (1) mixing with reducing agents to reduce the chromium to its trivalent oxidation state; (2) then treatment with an alkaline reagent to immobilize the trivalent chromium. The stabilization treatment of the metals-contaminated soil involves four process steps:

- Reduction
- Interim Staging
- Stabilization
- Confirmatory Testing

For the reduction step, an excavator or loader transferred 40 tons (39 tonnes) of soil from the untreated stockpile to a mixing box, where 380 gallons (1,438 litres) of calcium polysulfide solution and 2,850 gallons of water were added. Two supersacks, each containing 1 ton (1 tonne) of ferrous sulfate, were then added and the reagents and water mixed into the soil.

The equipment operator mixed the reducing agents, water and soil until it was visibly determined that the mixture was homogeneous and was green in colour. The reduced soil was then removed from the mixing box and transferred to one of the interim storage bins. A sample of the treated soil from each treatment batch was evaluated on-site to ensure no hexavalent chromium was present.

For the alkaline stabilization step, a loader transferred soil from the interim storage bin to the stabilization mix box, where 2 tons (2 tonnes) of Portland cement was added. The equipment operator then used an excavator to produce a visibly homogeneous final product, which was removed to a final storage bin, prior to final sampling verification. Following verification, the treated soils were transported to a licensed disposal facility for disposal as non-hazardous waste/soils.

Validation

Samples of the treated soil were obtained at a frequency of every 40 tons (39 tonnes) treated and subjected to TCLP testing for metals. All of the treated samples met the performance criteria of less than 0.6 mg/L TCLP-leachable chromium on the first pass.

Specific Issues

Residents of the nearby community were concerned with the disposal of the treated soil in the local municipal landfill. Public meetings were held to explain how the treatment reduces both the toxicity and mobility of the chromium contamination.

CASE STUDY: 5
Cambridge

Image: Google

Location:	Cambridge, Massachusetts
History:	Manufactured gas plant
Contaminants:	MGP Waste (BTEX, PAHs)
Scale:	1,200 yd^3 (917 m^3)
Reagents:	Portland Cement, Bentonite
Method:	In-situ, Jet Grouting
End-Use:	Not Disclosed

Site Description

Geo-Con was contracted by BMR-Kendall Development to stabilize approximately 1,200 yd^3 (cy) (917 m^3) of MGP impacted soils underlying Linskey Way in Cambridge, Massachusetts by the jet grouting method.

The impacted soils were the result of a residual contamination plume left over from the MGP site that occupied the parcel from the mid-1800's. The plume could not be stabilized during the in-situ remediation of the main site in 2000, because of the number of major utilities existing beneath Linskey Way (this earlier work received several awards).

Jet grouting uses the injection of grout at ultra-high (3,000 to 5,000 psi) (21 to 35 MPa) pressures to effectively cut and mix the soil, thereby creating soil-mix columns, forming a "monolith" in which soils are solidified and stabilized.

The jet grout column locations were drilled with a non-destructive vacuum method and cased to safely bypass the utilities located in the upper 6 ft (1.8 m) of the treatment area. A total-station-based movement monitoring system was implemented to monitor and protect utilities and adjacent structures. Vapour handling equipment with carbon treatment was utilized to minimize odors from the grouting operation, supported by real-time monitoring of air quality.

A 2-phase test program established the jet grouting parameters, column spacing, and reagent addition rates to meet the project solidification goals. Completed test-column locations were cored to verify effective column diameter, and sampled and subjected to centrifuge testing to verify that the cement addition was sufficient to immobilize contaminants within the soil-cement matrix.

This project represents the first time jet grouting has been used for the remediation of an MGP site. The nature of the site has presented numerous challenges to the project team including:

- Utilities – Numerous, highly sensitive, subsurface utilities including an antiquated, but active 36 in (90 cm) natural gas main, 18 in (45 cm) high pressure natural gas force main, 6-in water line, electrical duct bank, storm sewer lines, and overhead power lines
- Public Protection – The site is located in a very congested, high-traffic area. Site is in the heart of Kendall Square in downtown Cambridge Massachusetts bordered by business offices, parking garages, and a public skating rink
- Project Coordination – There were numerous entities involved in the project including Public Utilities, City Departments, adjacent building owners, multiple oversight engineers and other contractors. Permitting and coordination of all interests involved has been paramount throughout the course of the project
- Site Logistics – Maintaining a full-scale jet grouting operation including batch plant, cement/bentonite deliveries, jet grouting rig, vacuum pre-drilling, and spoils solidification and handling, on a 7500 ft^2 (697 m^2) parcel with as little impact to the public's day-to-day activities as possible proved to be a major challenge

Geo-Con installed approximately 680 columns (approximately 7200 ft^2) (669 m^2) to complete the stabilization of the contaminated soil under Linskey Way. The columns were approximately 18 ft (5.5 m) deep and were grouted from the top of clay later at approximately 18 ft (5.5 m) to 13 ft (4 m) below ground surface. Geo-Con used a cement-bentonite grout mix to complete the jet grouting. Grouting was completed using Geo-Con's C-7 jet grout rig and jet pump.

CASE STUDY: 6
Former Camden Gas Works Site

Location:	Camden, NJ
History:	Gas works
Contaminants:	PAHs, DNAPL
Scale:	24,000 yd^3 (18,349 m^3)
Reagents:	Portland Cement & blast furnace slag
Method:	In-situ
End-Use:	N/A

Site Description

The former Camden Gas Works is a former MGP facility located on five parcels of land (Parcels 1 through 5) comprising approximately 12.5 acres (50,585 m^2) in Camden, New Jersey. The parcels are bounded by Front, Second, Locust, Spruce, Cherry, Walnut, and Chestnut Streets in the City of Camden.

This project addressed soils on Parcel 2 (approximately 1.5 acres/6,070 m^2 in size), which is currently an unused lot owned by the City of Camden. Historical information indicated that Parcel 2 was used for MGP operations from 1891 to the late 1950s. Former MGP structures on Parcel 2 included 2 former gas holders (Gas Holder Nos. 3 and 4), tar and oil storage tanks and appurtenances, a governor house/laboratory and a valve house.

Risk drivers

A utility acquired the site from the City of Camden and planned to construct a 69-13 kV Class H substation on the property. The contaminated soils required removal/treatment to prevent contact with the contaminated soil by workers and to prevent migration of PAHs off-site.

Objectives

- Removal and disposal of the unsaturated soil
- Treatment of the saturated soil down to 30 ft (9.1 m) bgs
- Restore the site

The work performed by WRScompass consisted of the following activities:

- Site preparation activities were performed that included pre-work topographic surveying and existing conditions surveys of adjacent facilities, locating existing utilities using a combination of utility locator services and pre-trenching the site perimeter using soft dig techniques since the site was surrounded by public streets with numerous known and suspect active and abandoned utilities, installing erosion controls, removing the existing perimeter fence where necessary and installing temporary fence and concrete barriers around the site perimeter, setting up a support zone and contamination reduction zone, performing clearing and grubbing, constructing a material staging area, and installing a 200 gallon/min (757 litre/min) water treatment system
- Demolition activities were completed to remove existing above ground structures at the site. Most notable was the demolition of a large steel gas holder and demolition of two small structures (including the associated asbestos abatement). Concurrently, any abandoned utilities identified via the trenching discussed above were removed and capped and any concrete pads, structures, and other obstructions were removed from the work areas for ISS. WRScompass also removed the gas holder walls and floor slab and a notable quantity of wood piles used to support these structures. All resulting concrete and wooden debris was loaded out and disposed of off-site
- Steel sheeting was installed around the perimeter of the site once the pre-trenching was completed to clear the sheeting alignment and once certain active overhead utilities were relocated. The sheets were installed as permanent sheeting to protect a fragile ductile iron gas main located along the south side of the project site
- Excavation of contaminated soil for off-site disposal was performed in three areas of concern. Area 1 was only excavated to a depth of 3 ft (0.9 m). Since the planned construction of the site required imported fill with certain soil qualities to a depth of 12 ft (3.7 m) below ground, Area 2 was excavated that depth. In Area 3, the ISS-treated material was required to be no higher than 12 ft (3.7 m) below ground after treatment. Area 3 was excavated to between 13 and 14 ft (4-4.3 m) to accommodate the swell resulting from ISS treatment. This minimized the disposal of the denser ISS-treated material. Excavated soil was loaded into trucks and transported to the pre-approved thermal disposal facilities
- ISS was performed in Area 3 to the required depth of 30 ft (9.1m) using an excavator-mixing method. A grout plant was set up and the required reagent grout was produced in an on-site batch plant then conveyed to the ISS treatment cell, where it was added on a per weight basis using a pre-determined mix design of 9 % by weight for granulated blast furnace slag and 3 % by weight Portland cement. ISS was performed on over 24,000 yd^3 (18,349 m^3) of soil
- Once ISS was completed, the site was backfilled as required, using primarily an engineered fill material with the properties specified for the ensuing substation construction project. A 4-inch layer of crushed stone surfacing was installed above the engineered fill. Backfill material was installed in controlled lifts and compacted
- Other site restoration activities included the removal of temporary fence and barriers and installation of permanent fence where specified as well as replacement of concrete sidewalks and curbing. All equipment and temporary facilities were decontaminated and removed from the site and support areas were restored

Method

Portland cement and ground granulated blast furnace slag were used to chemically immobilize the PAH contamination and to improve the unconfined compressive strength to >50 psi (>0.3Mpa) and reduce the hydraulic conductivity to <1×10^{-9} m/sec.

Validation

Samples of the treated soil were obtained at a frequency of every 250 tons (246 tonnes) treated and subject to SPLP, UCS and permeability testing. The ISS performance criteria were : PAHs concentrations in the SPLP leachate less than 10 times the New Jersey Class II Groundwater Quality Standards, UCS > 50 psi (0.3 MPa), and hydraulic conductivity <1×10^{-9} m/sec. All of the treated samples met the performance criteria on the first pass.

Equipment

The ISS operations were conducted using an excavator and a batch plant. The batch plant was used to prepare the reagent grout. The appropriate amount of water was metered into an initial 5 yd^3 (3.8m^3) batch tank (equipped with a high-speed, high-shear mixer) and recorded. The reagents (Portland cement and ground granulated blast furnace slag) were transferred from the silos to the batch tank using the internal screw conveyor to deliver the specified volume of reagent. Each reagent was added separately to the mix tank and the scales on which the mix tank sets were tared before each reagent was added to verify that the correct amount of reagent had been added. When the correct grout composition was achieved, the blended grout was transferred to the excavator. The batch plant was also equipped with a second storage tank to allow for temporary storage of a blended batch to allow for uninterrupted production of batches. A high speed mixer in this second tank was to ensure the blended batch does not separate. The pre-determined grout volume was pumped to the treatment area based on the soil density, reagent admixture ratio, and the treatment cell dimensions.

Specific Issues

This was the first ISS project at an MGP site in New Jersey. WRScompass and the client met with the NJ Department of Environmental Protection and negotiated leachability performance criteria for the site. The performance standards were the leachate criteria determined as part of the risk-based development of site-specific impact to groundwater standards and were equal to 10 (default dilution-attenuation factor (DAF)) times the Class II NJ Groundwater Quality Standards for the PAHs. This manner of calculating ISS leachate performance criteria has now been used at multiple sites in New Jersey, though the DAF has been increased to 20.

CASE STUDY: 7
Camp Pendleton Scrap Yard

Location:	San Diego County, CA
History:	Scrap yard
Contaminants:	PAHs, PCP, arsenic, dioxins
Scale:	55,000 tons (54,131 tonnes)
Reagents:	Fly ash & activated carbon
Method:	Ex-situ
End-Use:	Restoration

Site Description

Marine Corp Base Camp Pendleton Site 6 was previously referred to as the DPDO scrap yard. The scrap yard operated from the early 1950s to 1979 as a storage, processing, and disposal area for scrap metal, salvage items, hazardous materials, and transformer fluids. The yard was divided into four areas:

- Polychlorinated biphenyls (PCB) spreading area
- Road burning area
- Battery electrolyte disposal area
- Hazardous waste drum storage area

Approximately 1,000 to 2,000 gallons (3,785 to 7,570 litre) of dielectric fluid from transformers was reportedly spread in the area for dust control. Immediately east of this area, flammable liquids such as fuels, solvents, and paint thinners were used as igniter fluids for burning wood debris during the 1950s and 1960s.

The scope of work issued by the Navy included the following requirements pertaining to remediation at Site 6:

Demolition and Excavation. Demolition was required to remove existing pavements and light structures and clearing and grubbing of vegetation in the area of contamination.

Soil Treatment System Site Development. Site development included construction of the treatment system's concrete foundation; utility connections (i.e. water and electrical); support facilities; and shelter structures. Adequate laydown, storage, and equipment access areas were constructed to ensure efficient operation of the treatment system.

Transportation. Transportation was required to haul contaminated soil from excavation areas to the treatment system. Upon satisfactory treatment, treated soil was transported to an on-base disposal site.

Soil Treatment System Pilot Test. Pilot testing with 250 tons (246 tonnes) of contaminated soil from Site 6 was performed to optimize the treatment system.

Operation of the Soil Treatment System. The soil treatment system was being operated to remediate contaminated soil from Site 6 to achieve nondetectable values for pesticides, PAHs, PCBs, and dioxins in the leachate from the Synthetic Precipitation Leaching Procedure (SPLP). OHM treated soils with a PORTEC Pug-mill Model 53 using Carbon and Type C fly ash to meet required specifications.

Sampling and Analysis. Confirmation sampling was performed to verify that all contaminated soil was removed from Site 6 and remaining soils met or exceeded cleanup criteria.

Disposal. Prior to disposal, the Navy approved an on-base disposal location which was prepared by clearing an appropriately sized area, excavating that area to a uniform depth below existing grade, and stockpiling those excavated soils for later use as a cover material for the treated waste. The treated waste was placed in the excavation and covered with a minimum of 1 ft of the stockpiled material.

Soil Treatment System Demobilization. Upon completion of soil treatment operations, all equipment and temporary facilities were removed from the site.

Site Restoration. Site 6 was restored to accommodate a wetlands habitat.

Risk drivers for the remediation

A preliminary human health risk assessment and ecological risk assessment was conducted and indicated that chemicals of concern at Site 6 included:

- Antimony
- Arsenic
- Beryllium
- Chromium (as Cr VI)
- Zinc
- dichloro-diphenyl-trichloroethane (DDT)
- DDE

- DDD
- Arochlor 1260 (PCB)
- Polyaromatic hydrocarbons (PAH)
- OCDD (dioxin)

Objectives

- Remove the contaminated soil
- Treat to reduce the leachability of the contaminants and produced a high strength, low permeable treated material
- Place and compact the treated material into the excavation area
- Restore the site

Method

The contaminated soil was treated with a mixture of Class C fly ash, and activated carbon to produce a treated material with nondetectable values for pesticides, PAHs, PCBs, and dioxins in the leachate from the Synthetic Precipitation Leaching Procedure (SPLP).

Validation

Temporary 500 ton (492 tonnes) stockpiles of the treated soil were created. Confirmatory testing for the leachability of pesticides, PAHs, PCBs, and dioxins was performed on the sample from each temporary stockpile. All of the confirmatory samples met the performance requirements for leachability.

Equipment

The stabilization treatment involved the use of a pug-mill system to mix the soil with Class C fly ash activated carbon. OHM's pug-mill system consisted of a variety of feeders, conveyors, silos, and a pug-mill mixer integrated into a complete system for the continuous mixing of wastes and reagents. The screened material was fed to an 8 yd³ feed hopper. The hydraulically-driven belt, located in the bottom of the feed hopper, fed the material onto a 40 ft (12.2 m) long by 24 in (60 cm) wide belt conveyor. The conveyor belt was equipped with a Ramsey single idler belt scale for the accurate, real-time determination of the rate, in tons per hour, of soil being treated. The conveyor belt conveyed the material into the pug-mill for blending with the stabilization additives. The pug-mill mix box was 4 ft (1.2 m) wide by 3 ft (0.9 m) high by 10 ft (3 m) long. Paddles were bolted onto structural steel shafts with replaceable shafts flanged on both ends for ease of maintenance. The mixer was V-belt driven by a 125-hp motor.

Dry stabilization additives were stored on-site in vertical cement silos. The silos were self-leveling and have a capacity of 200 barrels of material. The silos were also equipped with a top mount baghouse for dust control during silo filling. The feed from each silo was controlled by a 14 in (35 cm) diameter screw auger powered by a 5-hp motor. The dry stabilization additives were introduced from silo feeders which are attached to the pug-mill through a central feed auger. The soil was mixed with the reagents and water inside the pug-mill. The pug-mill system was operated 12 hours per day, 5 day per week, treating 1,200 tons (1181 tonnes) of soil per day.

CASE STUDY: 8
Chevron Pollard Landfill Site

Location:	Richmond, CA
History:	Landfill site
Contaminants:	Pb, H_2S, sulphuric acid
Scale:	150,000 yd³ (114,683 m³)
Reagents:	Portland Cement & limestone
Method:	Ex-situ
End-Use:	Restoration

Site Description

OHM developed, designed, and operated an on-site neutralization and stabilization process to treat 150,000 yd³ (114,683 m³) (more than double the amount originally slated for stabilization) of waste at a former landfill located at a major northern California refinery. The landfill was located immediately adjacent to Richmond Bay.

The waste consisted primarily of the following three materials:

Acid Sludge Tar (AST): A layer of soft, acidic tar produced as a sulfonated lubrication oil by-product.

Dredged Bay Mud (DBM): A layer of natural, dredged soil originally placed on top of the AST material. Over time, the DBM settled below the AST and transferred residual sulfuric acid from the AST to the DBM.

Interface Layer: This was the most variable of the three wastes in its chemical and physical characteristics, as there was little correlation between the vertical thickness of the mixed layer, the relative amounts of AST or DBM, and either total depth or location.

Characterisation

The AST was the waste by-product of sulfuric acid treatment of refinery stocks. The material had a pH range of 0.5 to 1.5 and a total acidity of 12 % to 25 %. The DBM was dredged material that was placed in the landfill as a matter of convenience and had become commingled with the AST. Substantial portions of all three waste groups were considered hazardous due to their corrosive characteristics. The DBM and the interface layer contained significant levels of sulfur compounds, which were a health and safety concern during excavation and processing. Additionally, the surface of the landfill was unstable and would not support heavy equipment.

Objectives

- Remove the AST and DBM from the landfill area
- Neutralize the AST
- Solidify the DBM and neutralized AST to produce a suitable structural fill material
- Place and compact the treated material on an adjacent parcel as structural fill
- Backfill the landfill area

Method

A 2-step neutralization/solidification process was utilized. Limestone was used to neutralize the AST. The DBM and neutralized AST were solidified using Portland cement.

Validation

Samples of the treated soil were obtained at a frequency of every 100 tons (98 tonnes) treated. The samples were tested to ensure that the treated material met:

- Final pH – 5 to 12
- 96-Hour Bioassay Test – pass
- Excess Alkalinity – greater than 1 wt%
- Geotechnical – less than 1 % creep in 48 hours at 500 psf applied pressure (0.02MPa)
- Particle size – greater than 97 % passing a No. 4 mesh
- Paint Filter Test – pass
- TCLP – RCRA non-hazardous

Over 96 % of the treated samples met the performance criteria on the first pass. Those 100 ton (98 tonnes) stockpiles not meeting the performance criteria were retreated.

The client also implemented a rigorous QA/QC program on the stabilized material that included on-site testing for pH, paint filter test, particle size, and creep/bearing.

Equipment

The stabilization process started with the excavation and transport of the DBM and AST from the landfill. The DBM was pumped from the landfill to a holding tank, which was designed to hold enough DBM for 2 days of processing. The AST was transported by truck to a clay pit, which also held a 2-day supply of material. The AST was fed to a hopper that continuously supplied a conveyor. The AST was fed from the conveyor to a crusher, designed to accept sticky, plastic material. Crushed rock was fed continuously into the crusher with the AST to act as an abrasive agent to keep the crusher clear. The crusher produced an AST product that was 97 % passing the No. 4 screen. This material was then fed into a pug-mill. In the pug-mill, the AST was mixed with limestone at a ratio of approximately 33 % of the total acidity. This resulted in AST particles encapsulated with limestone to meet the neutralization requirements. The treated AST material was conveyed to a second pug-mill unit.

Concurrent with AST treatment, DBM was pumped over a shaker screen to remove any AST particles greater than No. 4 mesh in size. The screened-out AST particles fell into the AST day pit. The DBM was fed into a mixing tank for equalization. The final stabilization step was performed in the second pug-mill. DBM, treated AST, and cement were mixed in the second pug-sealer and the final stabilized product was transported to a second landfill. The stabilized material was spread and allowed to cure. Once the material had cured, it was compacted to 90 % standard proctor density. The total system throughout was approximately 60 tons (59 tonnes) per hour.

CASE STUDY: 9
Carnegie

Location:	Carnegie, Pennsylvania
History:	Natural soil
Contaminants:	Saturated Unconsolidated Soils
Scale:	7,134 yd³ (5454 m³)
Reagents:	Portland cement
Method:	Ex-situ
End-Use:	N/A

Site Description

Geo-Con was contracted by Frank J. Zottola Construction (Zottola) to solidify 7,134 yd³ (5454 m³) of organic, saturated, unconsolidated soils at the site of the future Heidelberg Lowes Home Improvement store in Carnegie, Pennsylvania.

The problematic soils were located under a crucial corner of the structure foundation that would eventually house the electrical service for the building. The soils extended from the existing ground surface to a clay/till-confining layer located as deep as 18 ft (5.5 m) below ground surface. The difficulty of the project was increased as the majority of the unsuitable soil was beneath the groundwater table.

The objective of the project was to solidify the unsuitable soils beneath the site and provide a suitable foundation for building floors, roadways and parking areas.

CASE STUDY: 10
Columbus

Location:	Columbus, Indiana
History:	Wood treating plant
Contaminants:	Creosote
Scale:	4,500 yd^3
Reagents:	Cement & activated carbon
Method:	In-situ
End-Use:	N/A

Site Description

This project was performed at the former Columbus wood treating plant in Columbus, IN.

In-Situ Stabilization (ISS) work was performed on soils that were impacted by creosote. The treatment zone covered approximately 12,000 ft^2 of (1,114 m^2) surface area with depths ranging from 3 to 17 ft (0.9-5.2 m).

The project goal was to mix the contaminated soils in-place with a cement and powered activated carbon grout, with the subsequent mixture meeting the treatment criteria of a minimum 50 psi (0.3 MPa) unconfined compressive strength @ 14 days. Laboratory testing indicated that all of the soils mixed columns were in compliance with the treatment goals.

Geo-Solutions utilized state-of-the-art equipment and materials including soil mixing using a rig with a 9 ft (2.7 m) diameter mixing tool, automated grout mixing plant, and all required carbon materials.

Quality control was the responsibility of Geo-Solutions and was largely provided through the electronic monitoring of the Soil Mixing rig and supplementary daily grout quality control reports, which documented that the specified carbon and cement addition rates were applied.

CASE STUDY: 11
CSX Benton Harbor Scrapyard

Location:	Benton Harbor, MI
History:	Battery recycling
Contaminants:	Lead
Scale:	99,800 tons (90,537 tonnes)
Reagents:	Reducing agent + Alkaline Reagent
Method:	Ex-situ
End-Use:	Restoration

Site Description

This 7-acre (28,328 m^2)site was adjacent to a CSX switchyard and the Paw-Paw River near downtown Benton Harbor, MI. Contamination at the site resulted from scrapped metal processed for resale and recycled batteries, from 1959 to the mid-1980s. Scrap metal and battery casings were used to fill in low-lying areas of the site. Lead was the primary contaminant found at the site in soil.

Characterisation

The waste as a combination of soil, battery casings, scrap metal and debris used to fill wetlands on the site.

Drivers

The site and adjacent areas were identified for redevelopment, and the contaminated soil was required to be removed from the site. Treatment of the excavated material was necessary to allow disposal in a municipal landfill.

Objectives

- Remove the contaminated soil
- Treat (if applicable) the soil to remove its RCRA hazardous characteristic D008 (lead)
- Dispose of the soil at an off-site landfill
- Restore the site

Method

The excavated contaminated soil and debris were mixed with Portland cement to immobilize the lead. Each soil stockpile was sized to require one load (nominally 25 tons or 22.7 tonnes) of Portland cement.

An excavator was used to mix the contaminated soil and debris with the Portland cement, normally over 2 hours, until visibly homogeneous. Sampling and analysis in support of off-site disposal of the treated soils as non-hazardous waste/soils was performed.

Once the analysis confirmed full-treatment, the treated soil was loaded onto trucks and transported to licensed disposal facility for disposal as non-hazardous waste/soils. Loaders removed soil from the designated final storage bins and transferred the soil into the transport trucks. Load-out skirts contained any soil that may be spilled during load-out.

An excavator was used to mix the contaminated soil and debris with the Portland cement.

Validation

A 5-point composite sample was obtained from each stockpile after treatment. The samples of the treated soil were subjected to TCLP testing for metals. Over 95 % of the treated samples met the performance criteria of less than 0.75 mg/L TCLP-leachable lead on the first pass. Those that did not were retreated until passing.

Issues

The contaminated material continued below the groundwater level. Since the site was adjacent to the Paw Paw River, dewatering was not possible and the excavation had to occur in the wet. Visible confirmation of the removal was required. Backfilling also occurred below the groundwater level.

CASE STUDY: 12
Double Eagle Refinery Site

Location:	Muskogee, OK
History:	Refinery
Contaminants:	PAHs, TPHs, hydrogen sulfide
Scale:	60,000 yd^3 (45,873 m^3)
Reagents:	Portland cement & fly ash
Method:	In-situ
End-Use:	Restoration

Site Description

Impoundments at the Double Eagle Refinery in Muskogee, Oklahoma, contained acid sludge tars. These tars, a byproduct of the sulfuric acid cracking of lube oils and asphaltenes have a low pH (<2) and contain hydrogen sulfide. The recommended remedial action was neutralization and solidification of these acid sludge tars.

Risk drivers

Residential housing was located at the limits of the refinery property. The remediation had to be conducted without perimeter air exceedences or odour complaints from the nearby residents.

Objectives

- Treat (if applicable) the tar to increase its pH and improve its physical properties
- Cap the treated material
- Restore the site

The full-scale neutralization was accomplished as a 2-step process. In the first step, the fly ash was added to the acid sludge tar and homogenized thoroughly using an excavator. The minimal area of acid sludge tar was disturbed during the neutralization and the neutralized material was covered with Rusmar foam, to minimize odour and hydrogen sulfide emissions. In addition, a perimeter misting line was suspended from existing power poles and an odour-neutralizing agent was misted to prevent any odour emissions to nearby residents. The next day, the neutralized material was mixed with the Portland cement and recovered with Rusmar foam. After 5 days, the excavator could operate on top of the neutralized and solidified material, allowing the work to progress from the circumference of each impoundment out toward the middle. The neutralized and solidified material was capped with fill, clean soils and topsoil. Final grade was established to allow proper drainage over the impoundments and the area was seeded.

The combination of the 2-step treatment and the use of odor/vapour suppression allowed WRScompass personnel to complete the impoundment solidification and capping work in a cost effective manner and within the designated schedule. Just as important, the odour and vapour suppression systems employed eliminated any perimeter air quality exceedences and complaints from nearby residents

Method

The tar was neutralized using Class C fly ash. The neutralized material was solidified using Portland cement.

Validation

Samples of the treated soil were obtained at a frequency of every 250 yd^3 (191 m^3) treated. The treated material was tested to meet pH (>7), hydrogen sulphide (<5 ppm) and unconfined compressive strength (>50 psi or 0.3 MPa at 28 days). All of the treated samples met the performance criteria on the first pass.

Equipment

An excavator was used for both the neutralization and solidification steps.

CASE STUDY: 13
Dundalk

Location:	Baltimore, Maryland
History:	Natural soil
Contaminants:	Chromium ore processed residues
Scale:	350 lineal feet (106 m)
Reagents:	Peat moss/bentonite slurry
Method:	In-situ
End-Use:	Marine terminal

Site Description

Geo-Con was contracted by CH2M Hill Constructors, Inc. to install approximately 433 lineal feet (132m) of Augered Strain Relief Trench (SRT) at the 1800 Area of the Dundalk Marine Terminals in Baltimore, Maryland. Chromium Ore Processed Residue (COPR) was used extensively as backfill for the Marine Terminals during their construction. During long-term exposure to water, the COPR exhibited significant expansion, resulting in problematic movement of the Marine Terminal soils and damage to utilities and surface structures. The five segments of SRT were intended to relieve lateral ground stress and provide strain relief by permitting lateral expansion of subsurface layers of COPR.

The SRT was installed as a series of overlapping 2 ft (0.6 m) diameter soil-mix columns, placed on 1.4 ft (0.4 m) centers to provide the specified continuity and effective trench thickness.

The in-place soils were mixed with horticultural-grade peat moss at a rate of 8-10 % by weight of soil. The peat moss was slurried in a 6 % bentonite/water solution in Geo-Con's custom high-speed/high-shear batch plant and delivered through the mixing rig's wet-kelly system and mixed with the in-place soils using Geo-Con's custom made Shallow Soil Mixing (SSM) auger. The columns ranged from 10 to 15 ft (3 to 4.6 m) in depth. The SRT was part of an on-going Pilot Study at the marine terminals. The project was completed within budget and with no health and safety related incidents. Geo-Con was also able to accelerate the SRT installation schedule and provide the client a valuable schedule gain.

APPENDIX B: CASE STUDIES EMPLOYING S/S

CASE STUDY: 14
East Rutherford

Location:	East Rutherford, New Jersey
History:	Petrochemical plant
Contaminants:	TCE
Scale:	7,626 yd³ (5,830 m³)
Reagents:	Potassium permanganate & Portland cement
Method:	In-situ
End-Use:	Not Disclosed

Site Description

The purpose of this project was to use single auger soil mixing to treat TCE contaminated soils, using a 2-step approach involving: 1) In-situ oxidation with Potassium Permanganate (PP), and 2) In-situ solidification with Portland Cement (PC).

Geo-Solutions (GSI) was contracted by ERM to undertake soil mixing, site preparation and clean-up and excavation, including stockpiling of 1500 yd³ (1148 m³) of clean overburden, installation of 14 trench bedding "plugs" (to minimize PP migration off-site), erosion and sediment control, installation of monitoring wells, utility abandonment and site landscaping.

Soil mixing involved the installation of 242 soil columns using a Delmag RH-18 drill rig, of 9 ft (2.75 m) diameter to depths of 17-19 ft (5.2–5.8 m) below work-pad elevation. A column layout for the soil mixing work is shown opposite.

In the course of the soil mixing work, GSI encountered over fifty 14 in (35 cm) diameter deep foundation concrete piles. Obstructed soil columns were "isolated" and the obstructions excavated to the maximum possible depth, followed by re-treatment. Pictures of the obstruction removal and obstructions are shown opposite and overleaf. This approach worked as planned and was executed without significant delay to the original schedule.

Column layout for the soil mixing work

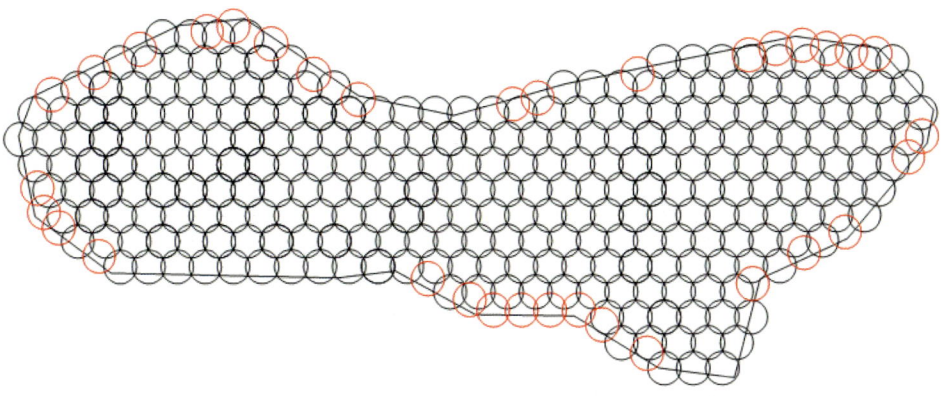

Excavated obstructions

Excavated obstructions

CASE STUDY: 15
Foote Minerals Superfund Site

Location:	Malvern, PA
History:	Lithium production
Contaminants:	Lithium tailings
Scale:	220,000 yd^3 (168,202 m^3)
Reagents:	Portland Cement & blast furnace slag
Method:	In-situ
End-Use:	Restoration

Site Description

The Foote Mineral Co. site is located on a 79 acre (319,702 m^2) property in East Whiteland Township, PA. Starting in 1941, the Foote Company operated a variety of process buildings for the manufacture of lithium metal and lithium chemicals and inorganic fluxes for the metal industry. Ores and minerals were also crushed and sized there. When the plant closed in 1991, the site had two quarries, a pit used to burn solvents, a lined basin and more than 50 buildings and process area.

The final site remedy included:
1) excavation and off-site removal of radiation-contaminated soils;
2) stabilization of process tailings located in one of the two on-site quarries;
3) consolidation of other site wastes, debris and contaminated soils into the quarry area;
4) capping of the quarry area at the surrounding grade.

WRScompass was contracted for the ISS of the process tailing in the on-site quarry.

Risk Drivers

Several nearby private wells were found to be impacted by contamination and a public water supply well about a mile downhill from the site had been shut down due to contamination.

Objectives

Approximately 220,000 yd³ (168,202 m³) of lithium tailings were stabilized in-situ. The overburden material from the ISS areas was excavated and hauled to the soil management area. Once the pre-ISS excavation elevations were created, a grout mix consisting of ground granular blast furnace slag (GGBFS) and water was mixed with the tailings. The mixing depth was up to 70 ft (21.3 m) with an average depth of 40 ft (12.2 m).

Prior to treated soil curing, representative samples were collected and placed into appropriate molds for quality control tests. Molded specimens were cured for the prescribed duration, then removed from their molds and tested.

Method

Portland cement and ground granulated blast furnace slag were used to improve the unconfined compressive strength to >50 psi (0.3 MPa) and reduce the hydraulic conductivity to <1×10^{-9} m/sec.

Validation

Prior to treated soil curing, representative samples were collected at a frequency of 1 sample per 500 yd³ (382 m³) treated and placed into appropriate molds for quality control tests. Molded specimens were cured for the prescribed duration, then removed from their molds and tested for UCS and permeability testing. Over 98 % of the treated samples met the performance criteria on the first pass. Those columns not meeting the performance criteria were retreated.

Equipment

The ISS operations within the site were conducted using a 4000-series Manitowoc crane equipped with an attached Hain Platform. The crane/platform assembly is supplemented with a swivel-mounted, top-feeding Kelly Bar capable of reaching a depth of 75 ft (22.9 m) bgs. Augers were attached to the bottom of the Kelly Bar. A 9 ft (2.7 m) diameter auger was utilized on this project.

A batch plant was used to blend prescribed proportions of reagent and water to produce a grout. This grout was pumped from the batch plant to the auger and mixed with the soil. Depending on the soil's physical properties, an excavator was used to loosen the soil and remove debris before treatment.

The tailings were treated by mixing grout into overlapping columns of soil. A prescribed quantity of grout was mixed into each column. Each column of soil was treated using the following procedure:

- When work began each morning and again whenever the grout storage/mixing tanks were empty, at least one batch of grout was prepared to create a small reservoir of grout for the work
- The horizontal coordinates of each column were located with surveying equipment and marked using grade stakes. Each column top elevation was also marked with the specified treatment depth. The required grout for each column was determined based upon the specified mix design and column volume
- The mixing drill rig was initially positioned atop the column to be treated. An operator lowered the Kelly bar and auger system through the soil and mixed the grout and soil in-situ to the required depth for three passes per column
- Immediately after the soil was mixed with grout, samples of the grouted soil were collected and placed into appropriate molds for testing

After completing one column, the crane crawled to treat another column, with adjacent columns overlapped to ensure complete coverage. This process was repeated until a strip of soil was treated. Actual locations of the panels were field verified and documented.

After completing ISS work, equipment that contacted contaminated soil was decontaminated and demobilized.

CASE STUDY: 16
Ghent

Location:	Ghent, Belgium
History:	Chemical/metallurgical industry
Contaminants:	Metals, cyanides
Scale:	135,000 tonnes (148,812 tons)
Reagents:	Zero-valent iron, cement/slag blend
Method:	Ex-situ
End-Use:	Not Disclosed

Site Description

The company La Floridienne was located in the port of Ghent (Belgium) and was active in the first half of the 20th century, producing metal salts and iron cyanide pigments from residues collected from other chemical and metallurgical industries.

The site was abandoned and contaminated, apart from the former production area, where two big stockpiles of production residue were found, called the 'red' waste and the 'grey' waste.

Characterisation

The so-called 'red' waste consisted of about 60,000 tons (54,431 tonnes) of roasted pyrite ashes. Consequently the material was very acidic. Its texture was silty but was not cohesive. The 'grey' waste consisted of 75,000 tons (68,038 tonnes) of very cohesive alkaline gypsum precipitate from water treatment operations employing excess lime. Both composition and leachability are outlined in Tables 1 and 2.

Table 1: Composition range of the 'red' and 'grey' waste materials

	Red waste tip	Grey waste tip
pH	3.8	12.5
As (mg/kg DM)	35,000 – 40,000	500 – 1000
Pb (mg/kg DM)	2000 – 10,000	10,000 – 25,000
Zn (mg/kg DM)	500 – 5000	500 – 2000
Cyanides (mg/kg DM)	100 - 500	500 – 8000

Table 2: Leaching of the 'red' and 'grey' waste (EN 12457-4 leaching test)

	Red waste tip	Grey waste tip
pH	3.8	12.5
As (mg/l)	6 – 8	0.01 – 0.05
Cd (mg/l)	0.1 -1	n.d.
Cu (mg/l)	0.1 – 8	n.d.
Pb (mg/l)	0.1 – 0.5	0.1 – 10
Zn (mg/l)	0.1 – 0.5	0.05 – 0.5
Cyanides (mg/l)	0.01 – 0.1	10 – 300

Due to the large volume of material to be treated, the option to keep the material on-site was preferred. Therefore, the wastes had to be treated to reduce leaching (and in order to fulfil the European Waste Acceptance Criteria for hazardous waste landfills) stored above groundwater and underneath an HDPE liner, to prevent rain percolation.

The chemical (leaching) objectives are given in Table 3.

As the site would later serve as a container terminal, the S/S material had to a high bearing capacity, i.e. a compressibility modulus of 17 MPa (2,465 psi) as measured in a plate test. Specific durability criteria were not pre-defined, but were provided later, based on a theoretical basis.

Table 3: Leaching criteria for hazardous waste landfills in Flanders (EN 12457-4 leaching test)

	Maximum leachability (mg/l)
pH	4 – 13
As	1
Cd	0.5
Cu	10
Pb	2
Zn	10
Cyanides	1

Drivers

Serious groundwater pollution from arsenic and cyanides resulted from the leaching of both waste piles. A first step in the site remediation was, therefore, the elimination of these piles as a pollution source, followed by the pumping-and-treating of impacted groundwater.

Method

The acidic 'red' waste had sufficient geotechnical properties, and only a reduction of arsenic leaching was required. This was achieved by adding 1 w/w% of zero valent iron. An alternative approach, based on cement-based solidification also decreased arsenic leaching, but strongly increased the release of cyanide to unacceptable levels.

The alkaline 'grey' waste had to be geotechnically improved, so a cementitious binder was required, but the use of cement and lime was not possible, and a strong impermeable matrix was achieved by adding 40 w/w% of a binder consisting of Portland cement and GGBS (ground granulated blast furnace slag). An interesting chemical stabilisation of cyanide in the form of 'Prussian Blue', was also investigated but did not work sufficiently.

The whole project had to be carried out in a very short time-frame of only 8 months. Therefore, both wastes were mixed in parallel in two different plants, each with a throughput of around 150 tons (136 tonnes) an hour.

Challenges

Dust control was extremely important, since the waste materials on-site were highly toxic, and the approach taken involved suppression spraying, using a minimum amount of water.

Validation

Both treated materials were validated after at least one month after treatment, by means of the EN 12457–4 leaching test, and checked against the criteria in Table 3.

APPENDIX B: CASE STUDIES EMPLOYING S/S | 337

Non-cohesive 'red' waste was mixed with the iron powder by means of a continuous liming mixer

Cohesive 'grey' waste was mixed in a batch concrete mixing plant (double axis pug-mill system

CASE STUDY: 17
Guernsey

Location:	Guernsey, Channel Islands, UK
History:	Harbour (ship maintenance)
Contaminants:	TBT (tributyltin)
Scale:	25,000 m³ (32,698 yd³)
Reagents:	Bio-char based reagent
Method:	Dredging & ex-situ
End-Use:	Marina

Site Description

The Port of St. Sampsons in the channel island of Guernsey was to be upgraded to a yacht marina.

The dry (at low tide) inner harbour required dredging by dry earth moving equipment followed by the construction of a sill and gate across the harbour entrance to impound water at +4.5 m above CD. Dredging was to +3.5 m above CD in order to maintain a constant water level of at least 1 m within the harbour.

The work involved the removal of 25000 m³ (32,698 yd³) of sediments contaminated with TBT (tributyltin) at concentrations between 640 µg/kg and 1770 µg/kg, making sea-dumping of these sediments impossible. The TBT resulted from ship hull maintenance, primarily sand blasting and painting, and work was successfully completed in 2003.

Characterisation

The sediments in the harbour contained a high percentage of sand and very little organic material so they were very prone to leaching (average concentrations in leachate were 1.8 µg/l).

Risk Drivers

Since the sandy material was suitable for land reclamation south of the port, the main risk involving re-use was the leaching of TBT to the groundwater.

Objectives

As no standards for TBT leaching existed at the time, a target of non-detectable levels of TBT (< 0,005 µg/l) in the leachate was preferred by the authorities.

Method

During the remedial design process, various laboratory-scale tests were carried out to determine the appropriate additive for immobilising the TBT. Some experience was already gained from testing of TBT contaminated sediments from the port of Zeebrugge, Belgium (Table 1).

The latter showed that stabilisation with cement is not working at all, as expected from literature as TBT becomes very soluble at high pH. A commercial product, E-clays (Envirotreat, UK), although being on previous TBT stabilisation projects, did not show any positive result.

The proprietary product Organodec, which is based on biochar, gave promising results, but when combined with cement the positive effect was diluted (Table 2). It was therefore decided to only use the addition of 2 % of Organodec in the Guernsey project.

The sediment was treated by means of a rotary mixing bucket (brand Allu) mounted on an excavator.

The production rate was about 30 tons (27 tonnes) an hour, and dosing was done by spreading 500 kg of Organodec (after soaking to prevent dust) over a batch of 16 m^3 (22 yd^3) (about 25 tons or 22.7 tonnes) of sediments. Mixing was carried out 3 times to ensure a sufficiently uniform distribution of additive.

Validation

Every batch of treated sediment (representing about 200 tons or 181 tonnes) was sampled and tested according to EN 12457-4, with the majority of results below the detection limit of 0.005 µg/l.

Table 1: Lab scale stabilisation tests on Zeebrugge sediment

Initial TBT leachability (μg/l)	Additives	TBT leachablity of treated material (μg/l)
4.5	5% OPC[2]	105.08
4.5	5% E-clays[3]	4.4
4.5	5% E-clays + 12%OPC	1,752.09
4.5	2% OrganoDEC[1]	0.11

[1] OrganoDEC = patented additive;
[2] OPC = Ordinary Portland Cement;
[3] E-clay = environmental clay

Table 2: Lab scale stabilisation tests on Guernsey sediment

Initial TBT leachability (μg/l)	Additive	TBT leachablity of treated material (μg/l)
1.8	2% OrganoDEC[1]	<0.005
1.8	2% OrganoDEC+2% OPC[2]	0.55
1.8	5% OrganoDEC+5% OPC	1.6

CASE STUDY: 18
Hercules 009 Landfill Site

Location:	Brunswick, GA
History:	Chemical waste landfill
Contaminants:	Toxaphene
Scale:	88,150 yd³ (67,395 m³)
Reagents:	Portland Cement
Method:	In-situ
End-Use:	Restoration

Site Description

The Hercules 009 Landfill Superfund Site is located on 16.5 acres near the City of Brunswick, Georgia. The property was used as a borrow pit during construction of Georgia State Highway 25 (Spur 25), which borders the property on the west. Hercules Incorporated was issued a permit in 1975 to use 7 acres at the northern end of the site, known as the 009 Landfill, to dispose of waste from the production of an agricultural pesticide, toxaphene.

Hercules Incorporated began producing toxaphene in 1948 and continued production through 1980. Toxaphene was one of the most heavily used insecticides in the United States until 1982, when the United States Environmental Protection Agency (EPA) cancelled the registrations of toxaphene for most uses. A registration is a license allowing a pesticide product to be sold and distributed for specific uses in accordance with specific use instructions, precautions, and other terms and conditions. All uses of toxaphene were banned by the EPA in 1990.

Between 1975 and 1980, Hercules Incorporated operated the 009 Landfill under a permit issued by the Georgia Environmental Protection Division (GEPD). The permit allowed the Brunswick, Georgia, Hercules plant to dispose of wastewater sludge from the production of toxaphene. Part of the Hercules 009 Landfill was also used for disposing empty toxaphene drums and toxaphene-contaminated glassware, rubble, and trash. The landfill was constructed as 6 cells divided by subsurface berms reportedly lined with a soil-bentonite clay mixture across the bottom and along the bermed walls. The landfill was closed in 1983. The EPA added the Hercules 009 Landfill to the Superfund National Priority List in 1984. Hercules designed and implemented the remedy for the site.

Objectives

- Treat the landfill contents and soil to remove its RCRA hazardous characteristic for toxaphene
- Cap the treated material
- Restore the site

Method

The remedy included Portland cement-based solidification/stabilization treatment. The contents of the landfill were treated by S/S while the material remained in-situ.

The performance standards for the S/S treated material were at least 50 psi (0.34 MPa) unconfined compressive strength (UCS) and toxaphene leaching of less than 0.5 mg/L, as determined using the Toxicity Characteristic Leaching Procedure (TCLP), on 28-day cured samples.

Validation

During treatment composite samples were made from S/S-treated material collected at one third and two thirds the total treatment depth of each subcell. Cement addition rates were verified based on written logs used for each subcell, which recorded subcell volume and weight of cement used. The average dosage of Portland cement applied to the 009 Landfill materials was approximately 14.8 %, judged within project tolerances. UCS was determined by pocket penetrometer on composite samples. Selected cylinders made from treated material from the remedial action start-up period were tested for UCS by ASTM D 2166 in order to correlate pocket penetrometer results to UCS. Compressive strength of the treated material increased over time and generally exceeded the 50 psi (0.34 MPa) requirement within 3 to 5 days cure time. TCLP testing was conducted on composite samples of blocks of four subcells. TCLP testing never revealed any presence of toxaphene in the leachate of the composite samples tested.

Equipment

An excavator was used to mix the landfill contents with the Portland cement. The landfill was divided into 25 x 25 ft (7.6 x 7.6 m) square cells for treatment. The total wet weight of the untreated soil and sludge in the cell was determined using a density of 100 lb/ft^3 (1600 kg/m^3) and the depth of untreated soil or sludge in the cell. Based on the total wet weight of the soil and sludge in the cell, the amount of Portland cement required for treatment was calculated. Up to 6 subcells were treated at one time. The remedial action contractor used an excavator to mix dry cement into the contaminated material while the material remained in place. Water for hydration of the cement was added as needed. Records were kept including depths of treatment of the "as-treated" (as-built) subcells. The depth of treatment extended below the bottom of the landfill contents sludge zone adding to the total volume of material to be treated. The treated depth of the majority of subcells extended into the regional groundwater table.

The remedial action construction was completed by regrading and revegetating the site. The primary intent of this activity was to establish an adequate vegetative cover over the soil-cement cap, the stabilized landfill contents, and other disturbed areas of the site resulting from remedial action activities. Rough grading involved adding some selected fill from a nearby borrow area. During this fill placement, these areas were rough graded and compacted to promote positive drainage. A vegetative cap was placed on top of the graded area comprised of 6 in (15 cm) of loose fill, which was fertilized and seeded.

CASE STUDY: 19
Hoedhaar Lokeren

(photo credit http://www.urbex.nl/site/hoedhaar/)

Location:	Hoedhaar Lokeren, Belgium
History:	Hat-making industry
Contaminants:	Mercury
Scale:	10,000 tonnes (11,023 tons)
Reagents:	Iron-based reagent
Method:	Ex-situ
End-Use:	Redevelopment/Residential

Site Description

During two centuries the city of Lokeren in Belgium was known for its production of high quality hats, made from rabbit fur felt. In addition to many hat-making workshops, the city had more than 40 felt production sites, using fur treatment by mercury nitrate brine.

Soil and groundwater pollution resulted from fur processing and the largest of these facilities in Lokeren was 'Hoedhaar' (Dutch for Hat-hair) and was situated near the city centre. When the City Council decided to re-develop the site as a residential area, the controlled demolition of the contaminated buildings, the removal of asbestos and soil from the site and excavation of a contaminated brook took place. From the 35,000 tonnes (38,580 tons) of soil excavated, 15,000 tonnes (16,534 tons) was soil-washed off-site, and 10,000 tonnes (11,023 tons) were chemically stabilized to immobilize mercury, prior to disposal to landfill. A further 10,000 tonnes (11,023 tons) were sent directly to landfill. The groundwater was pumped and treated by a physico-chemical treatment plant including microfiltration. The remediation took place from 2010 until 2012.

Characterisation

The soil consisted of loamy sand, contaminated with mercury, up to concentrations of 1000 mg/kg dry matter (DM); however, the average concentrations were between 100 and 300 mg/kg DM. As the soil was low in both clay and organic matter, mercury was highly leached, ranging between 0 and 40 mg/kg DM.

Risk Drivers

The presence of mercury in old buildings, soil and groundwater was the driver for the remediation of this derelict site.

Objectives

The site-specific remediation target was 120 mg/kg DM for mercury. This level is very high for a residential area, and so the site was covered with clean topsoil.

The target for the stabilisation of the soil was a residual leachability of 1 mg/kg DM of mercury, a criterion also used at the landfill where the soil was disposed of.

Method

Various chemical stabilisation trials, targeted at known mercury speciation (e.g. as a sulphide) were carried out, but without much success. A new approach employing zero-valent iron and ferric hydroxides was successful, and was limited to 5 %.

As the amount of soil was limited, the use of specialised mixing plant could not be justified. Therefore, a simple mixing approach using bucket mixing was used. The spreading of soil, additive dosing (one powder and one liquid), was followed by excavator mixing. During early stages of the work the quality of mixing, which was proportional to mixing time, was improved.

Validation

Batches of treated soil (representing about 250 tons or 226 tonnes) were sampled and tested according to EN 12457-4.

CASE STUDY: 20
Irving

Location:	Irving, Texas
History:	Landfill
Contaminants:	Mercury
Scale:	744 linear feet (226 m)
Reagents:	Slurried bentonite
Method:	In-situ
End-Use:	N/A

Site Description

Geo-Con was contracted by The City of Irving, Texas, to install a Soil Mixing Barrier Wall that tied into the existing slurry walls at the Hunter Ferrell Landfill. The soil-mix barrier wall was overlapped into the existing slurry walls on the 'Middle' and 'East' Tract cells. The site presented unstable/flowing sands and barrier wall installation necessitated using soil mixing, as traditional slurry-trenching was not possible.

Objectives

The barrier wall was installed to complete the seepage barrier along the south side of the future Middle Tract of the landfill. The work also included berm construction over the barrier wall alignment, to provide 25-year flood prevention.

Method

The length of this soil mixing wall was 744 linear feet (226 m) with an average depth of 45.1 ft (13.7 m) for a total of 8427.5 vertical ft^2 (782.9 m^2). The average depth to top of the shale key-in layer was 40 ft (12.2 m).

The Barrier Wall was mixed with a Delmag RH20 Drill Rig with a 4 ft (1.2 m) diameter soil mixing auger attached. The overlapping 4 ft diameter columns provided a minimum 36 in (90 cm) thick wall. Geo-Con keyed into the underlying shale layer approximately 6 ft (1.8 m) and no less than 2 ft (0.6 m) throughout production.

Slurried bentonite was used to create the soil mix material necessary to meet the permeability requirement of 1×10^{-10} m/sec. The slurried bentonite was pumped through the mixing tool while drilling and mixing the columns that created the continuous barrier wall.

Geo-Con exceeded every permeability requirement on all samples sent to the lab, with an average permeability being 1.79×10^{-8} cm/sec. Field samples of the freshly mixed material were taken once a day from differing depths of a finished column using Geo-Con's custom designed sampling device.

Samples were tested for permeability at Geotechnics Laboratory in Pittsburgh, Pennsylvania. All other quality control tests were done on-site by the Geo-Con Project Engineer. These tests included unit weight, viscosity, pH, and filtrate on the bentonite slurry.

Challenges

A significant challenge on the project was the presence of buried obstructions encountered during the mixing of the wall, including boulders, timbers, and steel cables. The obstructions required selective removed from the alignment with an excavator, so that all columns could be installed.

Ambient air in work zone air-monitoring was carried out on this site every one to two hours of production, because of the potential for encountering buried trash at the site. The project was completed in Level D at all times.

CASE STUDY: 21
Johnston Atoll Solid Waste Burn Pit

Location:	Pacific Ocean
History:	Hazardous waste incineration
Contaminants:	Lead
Scale:	22,000 tons (19,958 tonnes)
Reagents:	Portland Cement
Method:	Ex-situ
End-Use:	Restoration

Site Description

Johnston Atoll is located approximately 700 nautical miles (1,296 km) southwest of the Hawaiian Islands. It is an unincorporated territory of the United States under the operational control of the Defense Nuclear Agency (DNA). Johnston Atoll is also managed by the U.S. Fish and Wildlife Service (USFWS) as a National Wildlife Refuge. The atoll is comprised of four small islands: Johnston, Sand, North, and East Islands, which are surrounded by a coral reef.

The Solid Waste Burn Pit (SWBP), located on the northwest end of Johnston Atoll, was constructed in 1978 and utilized to burn refuse generated during the daily operation of the island. Hazardous materials, such as batteries, paints, and solvents, were burned previously in the inactive portion of the SWBP. The inactive portion of the SWBP contained ash material considered to be RCRA-characteristic hazardous waste, based on its TCLP lead levels. OHM Remediation Services (OHM) designed and mobilized an ash/residue treatment system to chemically stabilize and solidify the lead-contaminated SWBP ash and reduce the leachable lead levels to below the TCLP criteria.

The project was performed on an aggressive schedule due to the detection of lead in the adjacent lagoon and the potential for continued contaminant migration during the 1995 hurricane season. Mobilization of equipment to Johnston Atoll was initiated in April 1995, in advance of formal approval by EPA of the RCRA Class III Permit Modification. The permit modification request designated the SWBP and surrounding area as a Corrective Management Unit (CAMU), to allow treatment without triggering Land Ban Disposal Restrictions. The Johnston Atoll CAMU was the first CAMU approved by EPA Region IX.

This project required materials and equipment to be transported by ocean-going tug and barge from Seattle to Johnston Atoll via Honolulu. Over 1,000 tons (907 tonnes) of hydrated lime were containerized and transported, along with equipment including office, laboratory and decontamination trailers; 17 heavy construction vehicles and trucks; assorted materials handling equipment; and maintenance supplies and materials required for the entire project. Eighteen Shaw project personnel were assigned to the site for the entire project.

Risk drivers

Lead-contaminated particles from the SWBP were affecting the blood lead levels of monk seals and migratory fowl that frequented Johnston Atoll. The contaminated soils required treatment and capping to prevent contact with the contaminated soil by wildlife.

Objectives

- Remove the contaminated soil
- Treat the soil to remove its RCRA hazardous characteristic D008 (lead)
- Place and compact the treated material back into the excavation area
- Cap the treated material with 3 ft (0.9 m) of coral sand
- Restore the site

Method

Material with greater than 5 mg/L TCLP-leachable lead was treated with hydrated lime to reduce both the TCLP- and SPLP-leachable lead concentrations to less than 0.75 mg/L.

Validation

Samples of the treated soil were obtained at a frequency of every 200 tons (181 tonnes) treated and subject to TCLP and SPLP testing for metals. Over 99 % of the treated samples met the performance criteria on the first pass. Those stockpiles not meeting the performance criteria were retreated.

In 2003, samples of the treated material were collected by drilling through the cap. The treated material still had less than 0.75 mg/L in the TCLP and SPLP leachate. The treated material had less than 50 µg/L in the Physiology Based Extraction Test (PBET).

Equipment

A pug-mill system, consisting of a feed hopper with conveyor, pug-mill, reagent silo, and stacking conveyor were used to treat 150-200 tons (136-181 tonnes of contaminated soil per hour. The treated soil was segregated into 200 ton (181 tonnes) stockpiles for confirmation testing.

CASE STUDY: 22
Kingston

Location:	Kingston, Tennessee
History:	Natural soil
Contaminants:	Coal fly ash
Scale:	560,000 yd³ (428,151 m³)
Reagents:	Blast furnance slag, cement, bentonite
Method:	In-situ
End-Use:	Industrial spill amendment

Site Description

A dyke failure at the Kingston Fossil Plant in 2008 released an estimated 5.4 million cubic yards (4.12 million cubic metres) of coal ash from a dredge cell rupturing a natural gas line, and cutting power and transportation and contaminating the Emory River. Three homes were rendered uninhabitable, and a nearby residential area was evacuated.

The removal of coal ash from the Emory River was completed in May 2010. The reconstruction of the failed dyke, along with reinforcement of the remainder of the containment cells will prevent future movement of impounded ash. The installation of the Perimeter Wall Stabilization (PWS) is a key component of dike reinforcement.

The PWS is divided into 8 segments with differing design criteria in each. The design includes both continuous in- and out-board walls, constructed with a series of connecting shear walls perpendicular to the in-board wall lateral wall segments, designed to withstand shear loads.

In some segments, the shear walls extend beyond the out-board perimeter wall to form a buttress to the out-board wall. Each segment has different treatment widths and area-replacement ratios to account for differing failure modes in the most economical manner. The diagram opposite shows the typical design layout.

Typical Design Layout

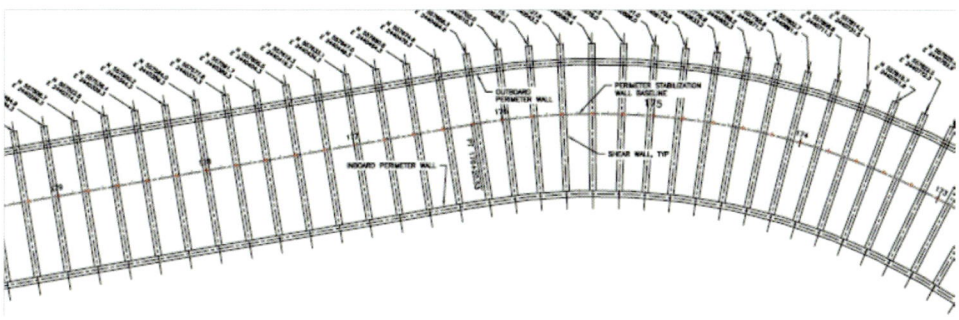

Method

Geo-Con is installing the PWS using the slurry wall trenching method. The slurry wall is installed using a custom designed long reach boom and long stick excavators fitted with 4 ft (1.2 m) wide trenching buckets. The trenching bucket is specifically designed for keying the wall into the shale bedrock formation at the site.

The trench is excavated using Cement-Bentonite (CB) slurry, which acts as hydraulic shoring during excavation and cures in the trench to become the permanent backfill material.

The self-hardening slurry includes a combination of Ground Granulated Blast Furnace Slag cement, Portland cement, and bentonite slurry. The proportions of the mixture were developed during an extensive laboratory testing program that evaluated over 70 candidate mixtures.

Multiple mix recipes have been used in production segments to accommodate different needs in the design. The cured CB slurry will reach an unconfined compressive strength (UCS) of 200-400 psi (1.4-2.8 MPa).

The CB slurry wall is keyed into the local shale bedrock formation which occurs at depths ranging from 45 to 65 ft (13.7 to 19.8 m) below ground surface (bgs) and bedding planes that dip from 15° to 20° from horizontal.

The self-hardening slurry wall trenching method employed on the PWS Project offers a number of distinct advantages over other methodologies including:

- A homogeneous wall is installed, creating a continuous perimeter wall system with more consistent strengths throughout
- No in-situ materials are used
- A high capacity batch plant produces a consistent product
- No cold joints: panels are overlapped for tight tie-in to previous work
- Able to follow sloping rock and more accurate ability to measure bedrock key-in than is inherent with in other geotechnical construction methods such as soil mixing
- The process is efficient with no wasted mixing: the entire width of the wall is the effective width of the wall.

The PW footprint will form a continuous boundary of approximately 11,500 ft (3,505 m) around the property, and when completed, Geo-Con will have installed more than 12 linear miles (560,000 yd^3 or 428,150 m^3) of trench, making it the largest Cement-Bentonite slurry wall installation in US history.

Geo-Con is working for TVA, which acts as the General Contractor. Construction is scheduled to be completed by 2014.

Issues

Key project challenges include concurrent design and construction.

Geo-Con works closely with the design engineer on upcoming segments of the design to ensure constructability and smooth transition between the segments.

An additional challenge is the strata through which excavation occurs. Generally, the top 15 ft of the soil profile consists of the coal ash which is highly liquefiable. This requires care in both the excavation and stability of work platform. The remainder of the profile consists of sand and silt underlain with the shale bedrock key-in layer. Because of the displacement during the ash slide, occasional non-native type materials have also been encountered.

CASE STUDY: 23
London Olympic Site

Location:	London, UK
History:	Former industry
Contaminants:	Metals, PAHs
Scale:	60,000 tonnes (66,138 tons)
Reagents:	Biochar & zero-valent iron
Method:	Ex-situ
End-Use:	2012 Olympic Games complex

Site Description

A large derelict site in East London was selected as the venue for the Olympic games of 2012. The site was a former marsh adjacent to the Lea River that had been reclaimed in Victorian times by backfilling of demolition waste and coal clinker. These and subsequent industrial activities caused contamination.

The soil was a mixture of natural and made ground, was found to be contaminated by heavy metals (e.g. arsenic), PAHs, TPH, sulphate, cyanide and ammonium. The remediation took place between 2007 and 2010.

Characterisation

Although the majority of the soil, over 1 million tons, was treated by soil washing, about 60,000 tons (66,138 tonnes) was too fine to treat with this technique. Therefore chemical stabilisation was proposed to the client. This soil varied from fine grained sandy loam to stiff clay. Some of the soils showed elevated leachabilities of heavy metals, in particular arsenic, others showed elevated leaching of organics (mainly PAHs), and for some soils both metals and organics were leachable.

Objectives

Tables 1 and 2 below show some typical site-specific leaching target values for reuse of soil on the site, for organic and inorganic contaminants, respectively. It should be noted that some of the target values, such as for PAH, were lower than the analytical methods' levels of detection, and it was, therefore, agreed with the Environment Agency to reduce the leaching to below the level of detection. No geotechnical improvement of the soil was required.

Table 1: Leachability (mg/L) according to EN 12457-4 of relevant organic pollutants before and after treatment by various additions of OrganoDEC for hydrocarbon immobilization

Contaminant	Input Average	1.5% OrganoDEC A	% Reduction	Target value
Naphthalene	0.050	<0.00092	>98	0.012
Acenaphthylene	0.0014	<0.00012	>91	0.00014
Acenaphthene	0.0068	<0.00036	>94	0.00012
Fluorene	0.0049	<0.0002	>96	0.00014
Phenanthrene	0.0026	<0.0003	>88	0.00011
Anthracene	0.0013	<0.00014	>89	0.00002
Fluoranthene	0.0019	<0.00038	>80	0.00002
Pyrene	0.0013	<0.00046	>64	0.0001
PAH (Sum of EPA 16)	0.072	<0.003	>96	-

Table 2: Leachability (mg/L) according to EN 12457-4 of relevant heavy metal pollutants before and after treatment by various additions of FeDEC, additive for heavy metal immobilization

Contaminant	Input Average	1% FeDEC	% Reduction	Target value
Arsenic	0.0082	<0.005	>39	0.05
Copper	0.0086	<0.005	>42	0.028
Zinc	0.088	0.018	79	0.125

Drivers

The Olympic site is criss-crossed by rivers and canals, and the environmental risk driver was the leaching of pollutants via groundwater and surface waters. This risk was translated into residual leachate concentrations of the treated soils, and although the leachate targets varied slightly over the site, they were all very low.

Method

Both for the organic and the metal contaminants, chemical stabilisation was applied. For the organics a proprietary biochar (organodec) was added to the soil at very low dosages (order 1 %). For the metals a zero valent iron powder was used, also at very low dosages.

Most of the soil was treated by means of a batch mixing plant with a throughput of about 120 tons (132 tonnes) an hour (a double axis pug-mill system). It was however, technically impossible to treat the very cohesive clay soils with this system, due to fouling, and for this reason a rotary bucket was applied.

Challenges

Due to the public interest in this site, and the fact that the Environment Agency was not familiar with chemical stabilisation, this project was seen as a demonstration project and was strictly followed-up.

Validation

All treated soils were validated immediately after treatment, by means of the EN 12457–4 leaching test, and checked against the site-specific leaching criteria.

356 | APPENDIX B: CASE STUDIES EMPLOYING S/S

Rotary bucket

Treated soil using batch mixing plant (double axis pug-mill system)

CASE STUDY: 24
Martinsville

Location:	Martinsville, Virginia
History:	Chemical Manufacturing
Contaminants:	Carbon Tetrachloride
Scale:	5,000 yd^3 (3,822 m^3)
Reagents:	ZVI, Kaolin Clay
Method:	In-situ, LDA
End-Use:	Not Disclosed

Site Description

Geo-Con was contracted by the DuPont de Nemours and Company, to stabilise contaminated soils at the DuPont Martinsville Unit I ISM Remediation site using In-situ Stabilization. Treatment involved the application of zero valent iron and Kaolin clay.

Zero valent iron was selected based on the primary contaminant of concern, carbon tetrachloride. Kaolin clay was used to minimize groundwater flow through the treatment zone. Elevated levels of carbon tetrachloride were identified on-site in the existing soil and groundwater with shallow soil concentrations as high as 30,000 mg/kg. Other contaminants detected included chloroform, methylene chloride, dichloroethylene, barium and chromium.

Geo-Con's scope consisted of in-situ treatment of 5,000 yd^3 (3822 m^3) of soil by the Shallow Soil Mixing technology (SSM), without excavation of removal.

In this application, Geo-Con used a Cassagrande Model CM15 crane-mounted drill rig to support the soil mixing operation. A wet or hollow Kelly bar connects the mix auger with the drill turntable and carries slurry from the plant to the auger.

An 8 ft (2.4 m) diameter mixing-auger was used to produce a homogeneous soil mix column. The zero valent iron/kaolin clay slurry was applied as specified to a depth of 35 ft (10.7 m). The 8 ft (2.4 m) diameter columns were spaced to provide complete coverage of the treatment areas.

Kaolin clay slurry was prepared in Geo-Con's on-site mixing plant and was injected as the mixing auger was advanced downward to the maximum treatment depth of 35 ft (10.7 m) to create the appropriate soil mix proportions within the column. The SSM columns were laid out to provide full coverage for the 3 rates of reagent dosage required at the site.

Once the Kaolin clay was injected and the column thoroughly mixed, iron was added by driving in a steel casing, filled with the prescribed amount of iron for the column, then pulling out leaving the sacrificial drive point and the iron in place. By proper sizing of the casing for the specific iron application rate, the iron was distributed evenly over the length of the column. The column was then remixed from top to bottom. Cement was incorporated into the upper 20 ft (6.1 m) of treated soil to improve the workability of the treated soil so that it could be graded and capped.

A total of 78 columns were successfully installed to treat the designated area. A sampling program was implemented to verify the amount of iron installed. Selected columns were sampled at 10 ft (3 m) intervals with Geo-Con's SSM sampling tool.

Soil mix samples were tested for iron content using a wet wash and iron separation test. Results of sampling indicated the average iron content of each column was greater than required and no individual samples were more than 20 % less than the required iron concentration.

Control of both carbon tetrachloride emissions and migrant nuisance odours to prevent activation of off-site alarms both real and perceived were a significant concern. Controls implemented included staged mixing and excavation, plastic liners, tarps and application of latex foam. Other work related to the Unit I ISM Remediation included demolition and removal of buried utilities and concrete structures, abandonment of monitoring wells and construction of an asphalt cap.

CASE STUDY: 25
Milwaukee, Wisconsin

Location:	Milwaukee, Wisconsin, USA
History:	Manufactured Gas Plant (MGP)
Contaminants:	MGP coal tar residuals
Scale:	25,000 yd³ (19,113 m³)
Reagents:	Portland cement and blast furnace slag
Method:	In-situ auger and bucket mixing
End-Use:	Property redevelopment and waterway restoration

Site Description

Natural Resource Technology supported all phases of planning and design through implementation and remedial construction of the former manufactured gas plant (MGP) facility located along a major waterway in Milwaukee, Wisconsin. Project planning required extensive permitting and coordination with the City of Milwaukee, Wisconsin Department of Natural Resources, state health department, property owner and local businesses. The site contains a number of historic brick MGP structures that have historic significance for the City, and are occupied by different business concerns.

This high profile project was conducted in 2 phases. Phase I consisted of the excavation and on-site thermal treatment or off-site disposal of approximately 58,000 tons (52,616 tonnes) of MGP-impacted soil and debris. Phase II involved the relocation of the main truck access ramp to the property, installation of an environmentally-sealed sheet pile system along the river, demolition of approximately 650 ft (198 m) of historic dock wall structure, in situ stabilisation/solidification (ISS) of approximately 25,000 yd³ (19,113 m³) of MGP-impacted soil, and restoration of the shoreline with riprap. ISS operations were performed directly along the river and in upland areas.

Remediation focused on maximizing removal or solidification of MGP impacted soil in areas previously identified to contain the heaviest MGP impacts. Thermally-treated material was beneficially reused for backfill in excavated areas. Excess ISS material (swell) was beneficially used on-site to rebuild the river bank and for site grading.

Characterisation

- Site Characterisation
- Remedial Alternatives Analysis
- Remedial Design
- Permitting/Agency Negotiation
- Bench- and Pilot-Scale Testing
- Construction Quality Assurance
- Construction Management

Risk Drivers

- Potential contributions of MGP residuals to groundwater and surface water
- Unstable historic dock wall structures and erosion of potentially MGP impacted materials to the river
- Protection of historic MGP structures

Objectives

- Minimize the potential for future contributions of MGP residuals to groundwater and surface water
- Restore the site for future property redevelopment
- Improve the waterfront to minimize potential for future erosion and degradation

Method

Extensive ISS bench-scale testing was conducted to evaluate a number of different mix designs. Final mix designs consisted of combinations of Portland cement and ground granular blast furnace slag. Specialized mix designs were developed to address specific strength and stability requirements for the river and roadway embankments. To reduce concerns for possible washout along the river, a commercial stiffening agent suitable for submerged concrete applications was also added. To access the MGP impacts, the existing truck access ramp was demolished and reconstructed in another area of the site. Reconstruction included partial demolition and reconstruction of a major city bridge that involved detailed coordination with city engineers. Demolition of historic dock wall structures was also required along the waterway to provide access for the ISS operations. Pilot- and full-scale ISS operations were completed by the contractor (WRScompass) using a combination of 8 and 12 ft (2.4 m and 3.6 m) diameter augers, and bucket mixing.

Validation

Discrete samples were collected from the top, middle or bottom of the selected ISS columns or grids by WRScompass according to the following frequency:

- One discrete sample collected every 1,000 yd^3 (764 m^3) or once per day of ISS production
- One discrete sample collected every 100 linear feet (30.5 m) around the perimeter of the ISS treatment area

Samples were tested for unconfined compressive strength and hydraulic conductivity to demonstrate compliance with the performance objectives.

Benefits of Solution

- Use of ISS substantially reduced concerns for potential direct contact exposure to the community and river during remediation
- Application of ISS reduced long-term liability and cost in lieu of off-site disposal of MGP impacted materials
- ISS was effective for river embankment restoration and reconstruction

CASE STUDY: 26
Nederland

Image: Google

Location:	Nederland, Texas
History:	Hazardous waste impounds
Contaminants:	Hydrocarbons
Scale:	65,000 yd³ (49,696 m³)
Reagents:	Fly ash
Method:	Excavator bucket mixing
End-Use:	N/A

Site Description

The project involved the closure of 4 hazardous waste impoundments of approximately 80 acres (323,749 m²), and the placement of over 300,000 yd³ (229,366 m³) of imported soils, 1.5m ft² of geosynthetic clay liner (GCL), 1.0m ft² (93,000 m²) of geogrid, and the stabilization of over 65,000 yd³ (49,696 m³) of excessively soft sediment. Work also included installation of a vinyl sheet pile wall to control the migration of NAPL-impacted groundwater from going off-site.

Other project features included performing a Resource Conservation and Recovery Act (RCRA) closure on one of the waste impoundments, installation of Articulated Block Mattress (ABM) at a storm water inlet that was constructed in the final erosion control phase, and creation of 20 acres (80,937 m²) of wetlands within the confines of the former Storm Water Impoundment (SWI).

Characterisation

The only sludge that required treatment via stabilization within the SWI was along the existing drainage sloughs. The sludge in most of the SWI was in solid form due to continual pumping of storm water from the basin, which allowed the sludge to build a crust. The sludge, through many years of investigation, was characterized to be non-hazardous, but still had a high hydrocarbon presence.

Waste composition of the stabilized material, including reagent, is shown below:

Solids ~ 35 % to 40 %
Moisture ~ 50 % to 55 %
Oil & Grease ~ <5 %
Fly Ash ~ 5 %

Risk Drivers

The risks associated with this project included the sludge and contaminant being transferred off-site as a result of groundwater migration, and flooding or a hurricane storm surge on the Neches River.

By stabilizing and capping the impoundment areas (including the installation of a RCRA cap and cutting-off the flow of groundwater with a barrier wall) these risks were mitigated requiring no further action from the agencies.

Objectives

The objectives of this remedial action were to:

1. Reduce sediment transfer off-site via surface water or groundwater
2. Reduce hydraulic conductivity from permeating through cap materials to get to the waste
3. Meet closure regulations stipulated by the Texas Commission on Environmental Quality (TCEQ)
4. Mitigate the damage to wetlands by creating wetlands on top of a portion of the cap
5. Enhance storm water quality that discharged to the Neches River

Method

It was not necessary to "stabilize" but rather "solidify" since the waste requiring strengthening was being classified as non-hazardous and groundwater was not at a high elevation. The sludge required little reagent dosage to gain the strength it would need to support a cap and eventual wetlands that was constructed on top the cap. Had there been any chance of the treated material being immersed in water, then stabilization would have been the preferred approach.

RECON utilized an excavator mounted on swamp pontoon tracks to install several drainage ditches in soft sediment/sludge. This, coupled with simple sump dewatering techniques, proved to lower the water elevation in the sludge, forming a 2 ft (0.6 m) thick crust.

This formation of the crust allowed low-ground pressure dozers to traverse most of the area. Other very excessively soft areas within the impoundment were stabilized using fly ash, a by-product generated from a local power plant. Long-stick excavators were used to mix the fly ash with the sludge. By excavating drainage ditches, RECON used 50 % less fly ash during the project. In addition, RECON was able to use a standard bi-axle geogrid in lieu of high-strength.

Sump dewatering techniques

Validation

RECON solidified the waste to support a cap and added the necessary small dosages of high available lime fly ash to reduce the water content in the sludge to produce a strength of approximately 10 lb/in^2 (psi) (0.07 MPa). This was monitored by use of pocket Penetrometer and Torvane field vane shear.

Issues

In order to speed-up the construction schedule and receive a minimum of 200 loads of soil per day, RECON developed and operated a borrow pit and purchased 17 dump trucks. With the help of a local hauler, RECON was able to haul in excess of 400 loads per day. This action allowed RECON to expedite the amount of GCL installed per day that required daily cover to protect it from wet weather.

CASE STUDY: 27
New Bedford

Location:	New Bedford, MA
History:	Manufactured gas plant
Contaminants:	MGP contaminated soil
Scale:	6,457 yd³ (4,936 m³)
Reagents:	Not specified
Method:	In-situ
End-Use:	N/A

Site Description

This project involved using a single auger soil mixing tool to solidify MGP impacted dredge sediments. Geo-Solutions (GSI) was subcontracted by Charter Environmental to complete all of the soil mixing work. The soil mixing required the installation of 178 x 8 ft (2.44 m) diameter and 32 x 3 ft (0.91 m) soil mixed columns to depths of 15-32 ft (4.6-9.6 m) below work pad elevation. A column layout for the soil mixing work is shown below. All of the soil mixing work was completed using GSI's Delmag RH-18 drill rig, batch plant, and silos. Prior to GSI's arrival, the dredge sediments were placed within sheet pile cells. The stabilization took place within the sheet pile cells using timber mats for rig stability.

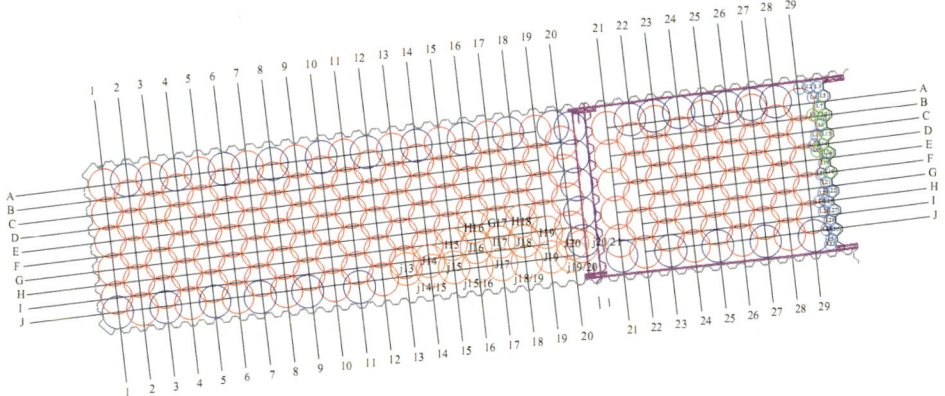

CASE STUDY: 28
Nyack

Location:	Nyack, New York
History:	Manufactured gas plant
Contaminants:	BTEX, PAHs
Scale:	11,400 yd³ (8,716 m³)
Reagents:	Portland Cement, Bentonite
Method:	In-situ
End-Use:	Not Disclosed

Site Description

A Manufactured Gas Plant (MGP) operated at this site from 1852 until 1965. The site covers a total land area of approximately 4 acres (16,187 m²), is in an urban setting, with adjacent properties used for a mix of commercial and residential purposes.

It is believed that gas was made from the coal carbonization process from 1852 until 1887, and then until 1889 the plant used oil instead of coal. From 1890 until 1938 the plant used both coal and oil as feedstock for the carbureted water gas (CWG) process. From 1938 until 1965 the site was used as an oil gas facility only during times of peak demand, a practice known as "peak shaving".

Coal tars are present at this site in the form of a dense oily liquid, which does not readily dissolve in water. Materials such as this are typically found at MGP sites, and are referred to as non-aqueous phase liquids or NAPL.

Specific volatile organic compounds of concern are benzene, toluene, ethylbenzene and xylenes. These are referred to collectively as BTEX.

The specific semivolatile organic compounds of concern in soil and groundwater are the following polycyclic aromatic hydrocarbons (PAHs):

- acenaphthene acenaphthylene
- anthracene benzo(a)anthracene
- benzo(a)pyrene benzo(b)fluoranthene
- benzo(g,h,i)perylene benzo(k)fluoranthene
- dibenzo(a,h)anthracene chrysene
- fluoranthene fluorene
- indeno(1,2,3-cd) pyrene 2-methylnaphthalene
- naphthalene phenanthrene
- pyrene

Excavation was used at the 'upper' terrace, whereas in-situ S/S was used for the 'lower' terrace of this site. Geo-Con was subcontracted by Sevenson Environmental (Sevenson) to perform the in-situ treatment of the lower terrace soils.

In-situ treatment involved Geo-Con's custom-made Calweld Drill Platform mounted on a Manitowoc 3900 crawler crane. The Calweld Drill is capable of over 400,000 ft-lbs (19.2 MPa) of torque and can turn a 10 ft (3 m) diameter mixing auger at up to 21 ft (6.4 m) below ground surface (fbgs) to the bedrock surface.

A mixture of water, Portland cement and Sodium Bentonite (Montmorillonite Clay) grout was prepared on-site and pumped through a hollow-stem Kelly bar to the mixing auger. Overlapping columns provided 100 % coverage of the treatment area involving 11,400 yd^3 (cy) (8,715 m^3). Treated soil was collected every 500 cy (382 m^3) and tested for Unconfined Compressive Strength and Flex-Wall Permeability to ensure immobilisation of contaminants. Treatment met a UCS of >50psi (0.3 MPa) and a permeability of approximately 1×10^{-9} m/s.

Geo-Con also performed high-pressure jet grout mixing to ensure the interface between the S/S columns and the site bedrock was sealed. The cement/bentonite grout was delivered through ports at 3500 psi (244.1 MPa) using a custom made sub-assembly on a 4 in (10 cm) drill string, and 755 lineal feet (230 m) of jet grouted columns were installed. An additional 311 cy (237 m^3) of soil were also stabilized in this project that was completed within budget and ahead of schedule and with no health and safety related incidents.

CASE STUDY: 29
NYSEG Norwich NY MGP Site

Location:	Norwich, NY
History:	Manufactured gas plant
Contaminants:	PAHs, DNAPL
Scale:	52,000 yd³ (39,756 m³)
Reagents:	Portland Cement, Bentonite
Method:	In-situ
End-Use:	Shopping plaza

Site Description

The NYSEG Norwich Former MGP site is located at 24 Birdsall Street, in the City of Norwich, Chenango County, New York. The exact starting date of MGP operations at the site is unknown. However, Sanborn fire insurance maps suggest that the plant operations started sometime between 1863 and 1887. By 1887 the site began supplying manufactured gas to the City of Norwich under the name "Norwich Gas Lighting Company". Little is known about the generation and disposal practices of residues from the MGP. However, 2 tar storage vessels existed in the subsurface prior to their removal in 1997. In addition, a potential purifier waste disposal area was identified in 1990 through an interview with a former employee of the MGP. Manufactured gas was produced at the site using the coal gasification and carbureted water gas processes. In 1892 the name of the facility operator was changed to "Norwich Illuminating Company", which was later changed to "Norwich Gas and Electric Company" in 1917. Coal gas was produced on-site until 1917 and then carbureted water gas produced from 1917 to 1953. NYSEG acquired the property in 1939.

The site previously occupied approximately 1 acre (4,046 m²) of land located at 24 Birdsall Street. In the years following cessation of gas production, former MGP structures were razed and subsequently NYSEG used the site for equipment storage. Presently, much of the property is paved with asphalt or covered with compacted gravel. A NYSEG electric substation exists on the eastern portion of the site.

The northern part of the site has been developed as a shopping plaza with retail shops. NYSEG purchased the former Aero Products facility located to the south and used the building for storage for several years. During the summer of 2006, NYSEG demolished the former Aero Products building. The off-site area that extends to the south of the former Aero Products building is comprised of mostly residential housing. NYSEG has purchased property at 37 and 41 Front Street and razed the structures located on these properties to allow for the ISS of the underlying soils.

WRScompass performed this in-situ stabilization (ISS) project for NYSEG at the Norwich Former MGP Site located in Norwich, New York. The scope of services consisted primarily of ISS treatment of 52,000 yd³ (cy) (39,756 m³) of soil using an auger mixing method, with other activities necessary to prepare for ISS work and to restore the site. The site was adjacent to an active shopping area that remained open during remediation – thus the work was highly visible to the public.

Risk Drivers

The contaminated soils required removal/treatment to prevent contact with the contaminated soil by the residents and to prevent migration of PAHs off-site toward the Chenango River.

Objectives

- **Removal and disposal of the unsaturated soil**
- **Treatment of the saturated soil down into the clay aquitard**
- **Restore the site**

Site preparation included abandoning monitoring wells located in the work area, performing a detailed utility location survey, exposing and retiring known gas mains in the work area, installing erosion controls, removing the existing perimeter fence where necessary and installing a temporary fence with privacy screen around the site perimeter, setting up a support zone and contamination reduction zone, and establishing a remote stockpiling area.

The site was pre-excavated in preparation for ISS work such that the treated material would be at least 4 ft (1.2 m) below the final grade as required by the NYSDEC. This included the removal of asphalt pavement and concrete slabs, removing of over 6,000 cy (4,587 m³) of expectedly non-impacted soil, and removal and off-site disposal of known hot spots of impacted soil. The soil that was potentially non-impacted was transported to a stockpiling site several miles from the site and at another property owned by the client, where it was sampled and verified to either be suitable for reuse as backfill material at the site or disposed of off-site.

ISS was performed using a 10 ft (3 m) diameter auger. A grout plant was set up and the required reagent admixture was produced on-site then conveyed to the auger rig, where it was added on a per weight basis using a pre-determined mix design of 8 % by weight for Portland cement and 1 % by weight for bentonite. As the work progressed, the reagent admixture was refined to reduce the amount of bentonite required as the bentonite addition rate was hampering the ability to productively complete the work under adverse winter weather conditions. ISS was first performed at a 10 ft wide perimeter that was keyed 4 ft (1.2 m) into the clay layer, followed by the interior ISS keyed 2 ft (0.6 m) into the clay layer.

Once ISS was completed, the site was backfilled to the required elevations. Suitable excavated and stockpiled soil was returned from the remote stockpile location and was supplemented by imported clean fill soil. Backfill material was installed in controlled lifts and compacted. Once the required grades were established a demarcation layer was installed atop the backfill material. Then the site restoration was completed by installing paving sub-base stone and pavement in a majority of the site, and topsoil and seeding in other specified areas.

Method

Portland cement and bentonite were used to improve the unconfined compressive strength to >50 psi (0.3 MPa) and reduce the hydraulic conductivity to <1×10^{-9} m/sec.

Validation

Samples of the treated soil were obtained at a frequency of every 500 cy (382 m^3), treated and subject to UCS and permeability testing. Over 98 % of the treated samples met the performance criteria on the first pass. Those columns not meeting the performance criteria were retreated.

Equipment

The ISS operations within the site were conducted using a 4000-series Manitowoc crane equipped with an attached Hain Platform. The crane/platform assembly is supplemented with a swivel-mounted, top-feeding Kelly Bar capable of reaching a depth of 75 ft (22.9 m) bgs. ISS was performed using a 10 ft (3 m) diameter auger attached to the bottom of the Kelly Bar.

Specific Issues

The ISS work was conducted adjacent to an operating shopping plaza, with ISS columns installed within 1 ft (0.3 m) of the footer for a grocery store. Extensive vibration, odor, and dust control was required to eliminate any impact of the operation on the shopping plaza.

APPENDIX B: CASE STUDIES EMPLOYING S/S

CASE STUDY: 30
Obourg

Location:	Obourg, Belgium
History:	Former Manufactured gas plant
Contaminants:	Coal Tar (PAHs, DNAPLs, LNAPLs)
Scale:	110,000 m³ (143,874 yd³)
Reagents:	Paper sludge ash
Method:	Soil liming equipment
End-Use:	Sediment treatment centre

Site Description

The Walloon authority SPW (Service Publique de Wallonie) wanted to install a new sediment treatment centre along the Canal du Centre, which is known for its four large ship elevators (UNESCO world heritage). Two sediment storage lagoons that were present on the site located near Obourg, were filled with soft poorly consolidated sediment.

The proposal was to solidify the soft sediments (110,000 m³ or 143,874 yd³ in total) and use the product in the construction of base layers and in new dykes. The picture above shows the excavation works in progress (top middle) and the two newly constructed lagoons (bottom middle) being lined with freshly solidified sediment.

Objectives

The objective of the work was to solidify the soft sediments and turn it into a construction material that could achieve a compressibility modulus of 11 MPa or 1595 psi (or 17 MPa/2465 psi at surface layer). The execution method chosen by the contractor, DEC imposed an extra boundary condition: after treatment with binder the material had to be stockpiled for several weeks before being compacted in its final place, ruling out the use of cement as an additive.

Formulation

The clayey-silty sediment present at the site varied between 40 % dry matter content (very soft) to 65 % dry matter content (plastic).

The sediment was not solidified with a traditional cementitious binder, as this would result in a short term curing. Therefore a specific additive was applied, which absorbs the sediment's pore water, improves the structure, and flocculated the clay particle. These properties result from using paper sludge ash, a meta-kaolin based fly ash, produced from incineration of paper mill waste. This product can be found in sufficient large quantities in most countries. On this site an average dosing of 20 % w/w was used, resulting in the use of about 35,000 tons (31,751 tonnes) of paper ash.

Method

Although the initial plan was to use a batch-wise pug-mill mixing plant, a less complex and cheaper method was applied. Standard soil liming equipment (dosing and rotary mixing) was used on layers (or lifts) of 0.5 m (20 in), and then scraped and stockpiled by a bulldozer.

Testing

Validation of the solidification treatment was carried out on both laboratory- and field-scale. Freshly treated sediment was sampled and compacted in a proctor molds, tested immediately, at 14 days and after 28 days by means of CBR (Californian Bearing Ratio). The results can be correlated to field plate tests, which were also carried out to confirm the target values were achieved.

References

- Bujulu P., Sorta A., Priol G., Emdal A. Potential of wastepaper sludge ash to replace cement in deep stabilisation of quick clay
- Characterization and Improvement of Soils and Materials, Session of the 2007 Annual Conference of the Transportation Association of Canada, Saskatoon, Saskatchewan

CASE STUDY: 31
Perth Amboy

Location:	Perth Amboy, NJ
History:	Landfill
Contaminants:	Petroleum hydrocarbons
Scale:	52,000 tons (47,173 tonnes)
Reagents:	Cement
Method:	In-situ
End-Use:	N/A

Site Description

The Perth Amboy refinery operated from the 1920s to 1983, producing gasoline and heating oil, before changing to asphalt refining. In the late 1990s, cleaning of the storage tanks at the refinery was undertaken, and the oil sludge was placed into 2 impoundments – the 8 acre (32,374 m^2) Northfield Basin and the 2 acre (8,093 m^2) Surge Pond.

Characterisation

The sludges were an emulsion of lubrication oil range (C28–C60) petroleum hydrocarbons, asphaltic tars, and water.

Objectives

Chevron, the site owners, wanted to utilize the 8 acre Northfield basin for storm water collection, and to enable this, in-situ solidification of the oily sludge in both impoundments was proposed. This was followed by removal of material from the Northfield Basin to the Surge Pond, capping and closure.

Method

The oily sludge material was encapsulated in a cementitious matrix using treatment cells of 500 yd³ (382 m³). Solidification treatment started along the edge of the impoundments and then continued inward. Portland cement was pneumatically conveyed to each cell and was then mixed into the oily sludge using a long-reach excavator. The solidified material could support the long-reach excavator within 2 days, enabling the solidification process to continue to completion. Upon successful treatment of both sites, excavation and removal of solidified sludge to the Surge Pond for final disposal, was followed by covering with an impermeable clay cap.

CASE STUDY: 32
Portsmouth

Location:	Portsmouth, Virginia
History:	Wood Treating Facility
Contaminants:	Creosote, Pentachlorophenol (PCP)
Scale:	47,000 yd^3 (35,934 m^3)
Reagents:	Portland Cement, Ground Granulated Blast Furnace Slag Cement, Organophillic Clay
Method:	In-situ, Excavator-based
End-Use:	Not Disclosed

Site Description

The Atlantic Wood Industries (AWI) site consists of approximately 48 acres (194,249 m^2) of land on the industrialized waterfront of Portsmouth, Virginia, and 30 to 35 acres (121,406 m^2 to 141640 m^2) of contaminated sediments in the Southern Branch of the Elizabeth River. From 1926 to 1992, a wood-treating facility operated at the site using both creosote and pentachlorophenol (PCP).

At one time, the US Navy leased part of the property from AWI and disposed of waste on-site, including used abrasive blast media from the sand blasting of naval equipment and sludge from the production of acetylene.

Polynuclear aromatic hydrocarbons (PAHs), PCP, dioxins and metals contamination (mainly arsenic, chromium, copper, lead and zinc) have been detected in soils, groundwater, and sediments. The EPA issued a Record of Decision in 1995 for the clean-up of the site, which was amended in 1997. The AWI Site was added to the National Priorities List of most hazardous waste sites in 1990.

Geo-Con was contracted by the Norfolk District of the US Army Corps of Engineers to perform Phase 1B of the cleanup effort which included the stabilization and solidification of over 47,000 cy (35,934 m^3) of DNAPL-impacted soils.

The Treatment Area, or Historic Disposal Area (HDA), consists of 29 separate sections ranging in depth from 8 ft to 27 ft (2.4 to 8.2 m) below ground surface. Due to the amount of debris expected throughout the HDA, Geo-Con selected the excavator/rotary blending SSM system to perform the stabilization of the DNAPL-impacted soils. This system combines the use standard hydraulic excavators with an excavator-mounted rotary-blending unit to homogenously blend the subsurface soils with the chosen reagents.

The impacted soil was blended with a cement-based liquid grout consisting of:

- Portland Cement
- Ground Granulated Blast Furnace Slag Cement
- CETCO Organophillic clay

Two separate mix designs were selected for the project depending on the section being mixed. Each mix design was to attain a minimum UCS of 50 psi (0.3 MPa) and a maximum permeability of 4×10^{-9} m/s.

Geo-Con served as the prime contractor for the project. The scope of work for the project also included:

- Relocation of electrical utilities for the neighbouring Portsmouth Public School District Property to accommodate soil mixing
- Perimeter air monitoring
- Pre-excavation, removal, handling and stockpiling of over 3,000 cy (2,293 m³) of miscellaneous debris
- Final grading
- Installation of an impermeable clay cap
- Restoration

CASE STUDY: 33
Rieme

Location:	**Rieme, Belgium**
History:	**Chemical industry**
Contaminants:	**Acid tar**
Scale:	**170,000 m³ (222,351 yd³)**
Reagents:	**Cement/pozzolan/filler blend**
Method:	**Ex-situ**
End-Use:	**Not Disclosed**

Site Description

In the first half of the 20th century Fina (now Total) has produced white medicinal oil from crude oil by means of concentrated sulphuric acid treatment. This process typically generated a viscous residue, called acid tars, and a clayey residue, called Fuller's Earth.

At the production site in Rieme, located in the port of Ghent (Belgium), these residues have been stored in large open unlined lagoons, containing about 80,000 m³ (104,636 yd³) acid tars, varying from liquid to solid, 20,000 m³ (26,159 yd³) of Fuller's Earth, and 70,000 m³ (91,556 yd³) of contaminated soils.

Characterisation

The liquid acid tars found in the smallest lagoon were actually emulsions, with up to 50 % water content, a pH of around 2, and a sulphur content of about 1.7 %. The more viscous to solid acid tars also contained high water contents (even up to 60 %), were far more acidic (pH 0 to 1) and contained very high sulphur levels between 5 and 10 %. Due to the high water content the calorific value of these tars was limited. An attempt to reveal the complex hydrocarbon composition of the tars was not carried out.

Objectives

Due to the large amount of contaminated materials, and due to the absence of external treatment facilities, it was decided to treat the materials by solidification/stabilisation, and store them in a confined disposal facility on the site.

The Flemish Environment Agency (OVAM) demanded that the treated materials meet the European Waste Acceptance Criteria for hazardous waste landfills (except for the parameter TOC). The disposal facility was constructed following the European rules of a hazardous waste landfill (double bottom liner system, leak detection, double cover system). On top of the chemical criteria Total set out geotechnical and durability criteria. The most relevant criteria are summarized in Tables 1 and 2, below.

Table 1: Main geotechnical criteria of S/S

Parameter	Unit	Objective
Volume increase	%	< 30
CBR-value (ASTM D1883-99)	%	> 11
Compressibility modulus – field static plate loading test	MPa	> 11
Hydraulic permeability	m/s	< 10^{-7}
Strength loss by ageing (wet-dry according to ASTM 558 and freeze-thaw according to ASTM 559)	%	< 15

Table 2: Main chemical criteria

Parameter	Unit	Initial value range	Maximum
pH		0 – 3	4 – 13
Water soluble part	% on dry matter	5 – 20	10
DOC (leached according to EN 12457-3)	mg/l	3,000 – 28,000	90% reduction

Drivers

The groundwater in the vicinity of the site was heavily contaminated by hydrocarbons and sulphuric acid, extending into residential gardens near the site. Total decided to stop the source of this pollution first by treating the acid tar lagoons. The next phase of work, (yet to be done), involved treating the groundwater.

Method

The main challenge for the S/S of the acid tars was achieving the high mechanical strengths while staying within the volume increase boundary condition of 30 %. In order to establish the right mix formulations, various commercially and locally available binders and additives were chosen:

- Cements: OPC (CEM I, ordinary Portland cement), blast furnace slag cement (CEM III)
- Alternative binders or pozzolans: CKD (cement kiln dust), GGBS (ground granulated blast furnace slags), paper ash
- Alkaline minerals: quicklime, steel slag, flue gas desulphurization gypsum
- Filler materials: PFA (pulverized fly ash)

About 90 mix formulations were laboratory tested on the 6 types of lagoon-derived materials, cured for 28 days and tested for CBR, the main trigger criterium. Mixes that passed were then further tested for the chemical criteria, and a specific mix formulation was chosen, based on the same 4 additives (but applied at different dosages dependent on the requirements of the specific lagoon material).

Generally, the more liquid the acid tars were, the more filler/binder material was required, relative to the neutralizing additive. Conversely, the solid acid tars required less solidification agent, but as they tended to be more acid and leached greater they needed greater amounts of neutralizing additive.

A bespoke mixing plant, with a 60 tons (54 tonnes) an hour throughput was designed and constructed, and set up in a closed hall in order to be able to extract the emissions from the acid tars (see picture overleaf). The mixing unit itself was a batch concrete mixing plant (double axis pug-mill system), and is shown schematically (see overleaf).

Challenges

The acid tars emitted extremely high amounts of sulphur dioxide. For safety and nuisance reasons, the emissions had to be controlled and monitored during all phases of the work and the process, as were various preventive measures to protect the equipment (excavators, mixing plant) against corrosion.

Another specific issue was the presence of unexploded ordnance in the lagoons from WW II, which required attention during excavation of the lagoons.

Validation

Internal validation of the treated materials was achieved using the CBR (Californian Bearing Ratio ASTM D1883-99) after 1 month of curing.

Contractually, a more extended list of geotechnical tests was employed. The cured materials were also WAC tested based on the EN 12457–4 leaching test, and checked for several parameters (see Table 2) that were the most relevant.

The same mechanical tests were carried out after durability testing on the treated materials (freeze-thaw and wet-dry ageing tests). A maximal strength loss of 15 % of the required values was allowed.

References

Pensaert, S., De Puydt, S., Janssens, T., Vanpée, N., Vander Velpen, B., De Cock, C., Goorden, G., (2008). The remediation of the acid tar lagoons at Rieme, Belgium. Consoil 2008, Milano, Italy.

Pensaert, S., De Puydt, S., Janssens, T., (2010). The remediation of the acid tar lagoons in the port of Ghent. 2010 International Solidification/Stabilisation technology forum
Sydney, Nova Scotia, Canada, June 2010.

Batch concrete mixing plant (double axis pug-mill system)

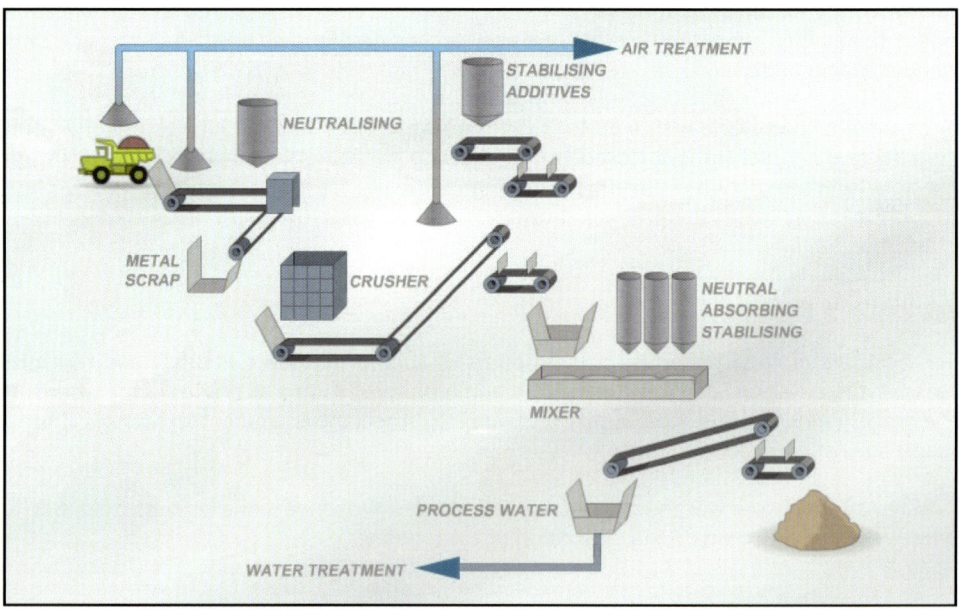

Bespoke mixing plant

CASE STUDY: 34

Roma Street Station

(photo credit wikipedia)

Location: Brisbane, Queensland, AUS
History: Railway station
Contaminants: PAHs
Scale: 7,000 m³ (9,155 yd³)
Reagents: Fly ash & waste ash/coke
Method: Ex-situ
End-Use: Parkland

Site Description

Roma Street Station is a major railway station in Queensland and has been in operation since 1874. Railway operations have affected the environmental quality of the land underlying and adjacent to the station. Part of the area adjacent to the stations was slated to be redeveloped into parkland. Contaminated soil on this parcel needed to be addressed before the redevelopment could be initiated.

Characterisation

Contamination at the site comprises both in-situ and stockpiled soils containing heavy metals, polycyclic aromatic hydrocarbons (PAHs) and total petroleum hydrocarbons (TPH) in concentrations above regulatory threshold values. The existing stockpiled soils originated from work associated with the adjacent Roma Street Station upgrade.

The majority of in-situ contamination at the site is associated with a shallow ash/coke soil layer to approximately 300 mm (12 in) depth, on average across the site. This layer comprises heavy metals and PAHs above regulatory threshold values.

Risk Drivers

The Roma Street Parkland redevelopment was to create an attractive garden/parkland where community and visitor patronage are the final success factors. The key objective in proactive management of PAH-impacted material at the site was to satisfactorily reduce the mobility of the PAH contamination and eliminate exposure pathways so that the associated hazard is reduced to an acceptable level.

Objectives

The remedial approach for the site was to separate clean from contaminated material. Contaminated material was then treated on-site to reduce leachable contaminant concentrations prior to containment within an on-site cell. The stabilisation process incorporated the beneficial reuse of waste ash and ash/coke. The approach of maximising reuse of materials on-site is consistent with the EPA remediation hierarchy which gives preference to on-site solutions. The reuse of on-site waste also reduced treatment costs, limited the volume of reagents brought on-site, and minimised the volume of the on-site cell.

The treated material was placed back into the excavation inside a minimum design cap and containment system. This placement of the treated material on-site was consistent with, and did not interfere with, the following redevelopment of the Roma Street Parklands. The resulting gardens and parkland provide the citizens of Brisbane with a world-class botanical experience in a former industrial setting. This remediation was the first organic stabilisation project completed in Australia.

Method

The stabilisation treatment involved the use of fly ash and waste ash/coke and was designed to reduce the PAH leachability. The target post-remediation PAH leachabilities were:

Compound	Target
Naphthalene	<50 ug/L
2-Methylnaphthalene	<50 ug/L
Acenaphthylene	<10 ug/L
Acenapthene	<10 ug/L
Fluorene	<10 ug/L
Phenanthrene	<10 ug/L
Anthracene	<5 ug/L
Fluoranthene	<5 ug/L
Pyrene	<5 ug/L
Benz(a)anthracene	<2 ug/L
Chrysene	<2 ug/L
Benzo(b) & (k)fluorene	<2 ug/L
Benzo(a)pyrene	<2 ug/L
Indeno(1,2,3-cd)pyrene	<2 ug/L
Dibenz(a,h)anthracene	<2 ug/L
Benzo(g,h,i)perylene	<2 ug/L

The target leachable concentration for total PAHs in treated materials was 0.1 mg/L

Validation

Samples of the treated soil were obtained at a frequency of every 100 tonnes (110 tons) treated and subject to SPLP testing for PAHs. Over 98 % of the treated samples met the performance criteria on the first pass. Those 100 tonne (110 ton) stockpiles not meeting the performance criteria were retreated.

Equipment

A pug-mill system, consisting of a feed hopper with conveyor, pug-mill, reagent silo, and stacking conveyor were used to treat 150-200 tonnes (165-220 tons) of contaminated soil per hour. The treated soil was segregated into 200 tonne (220 tons) stockpiles for confirmation testing.

CASE STUDY: 35
Sag Harbor

Location:	Sag Harbor, New York
History:	Manufactured gas plant
Contaminants:	BTEX, PAHs
Scale:	7,200 yd^3 (5,505 m^3)
Reagents:	Potassium permanganate & Portland cement
Method:	In-situ
End-Use:	Not Disclosed

Site Description

The Sag Harbor Manufactured Gas Plant operated from 1859 to 1930 and originally produced gas from coal or wood resin. It was switched to a water-gas process in 1892. The by-products of gas production that may have either spilled, leaked, or were intentionally disposed of on this site are responsible for contamination of the environment.

The New York State Department of Environmental Conservation (NYSDEC) and the New York State Department of Health (NYSDOH) announced their decision for a remedy, which was signed on March 31, 2006.

The chemicals of concern at this site are residues of the former MGP process and include volatile organic compounds, semi-volatile compounds, and cyanide. The volatile organic compounds of concern are benzene, toluene, ethylbenzene, and xylene (BTEX). The semi-volatile organics of concern are polycyclic aromatic hydrocarbons (PAHs), and the principle waste material at this site is coal tar, a thick, black, oily liquid which was a by-product of the gas production process. Coal tar has been found under most of the site and mostly located in the upper 12 ft (3.7 m). The groundwater, contaminated by contact with the coal tar, moves to the north, almost as far as Sag Harbor Cove.

Geo-Con teamed with Sevenson Environmental Services and was contracted by National Grid to provide a total remediation solution for this MGP site. The complete clean-up effort included:

- Installation of 7,200 yd^3 (5,505 m^3) of In-Situ Soil Mix Wall (SMW)
- Water Table Drawdown (Dewatering)
- Water Treatment and Discharging
- Impacted Soil Excavation and Disposal
- Backfilling

Geo-Con installed over 7,200 yd^3 (5,505 m^3) of vertical barrier wall around the perimeter of the site using the Large Diameter Auger Shallow Soil Mixing (SSM) method. The purpose of the wall was to provide structural support and a groundwater barrier during excavation of the impacted soil. Soil was mixed in place using Geo-Con's custom made Calweld Drill Platform mounted on a Manitowoc 3900 crawler crane. Soil was mixed with Type 1 Portland Cement/water slurry, and was produced in Geo-Con's batch plant. A combination of 7, 8, and 10-ft (2.1, 2.4, and 3 m) augers were used to complete the SMW to depths up to 23 ft (7 m) deep.

Due to the nature of MGP waste and this particular site's location (in the heart of the Village of Sag Harbor), Sevenson/Geo-Con faced a unique set of challenges in the completion of the project. The site was immediately adjacent to many active businesses and occupied residences. Health and Safety, odour, and noise control were paramount throughout the project. Odor control foaming was used continuously to control MGP odours during pre-trenching and SMW installation, and a series of structures equipped with vapour collection systems were used during excavation. Perimeter real-time air monitoring was performed throughout the duration of the project.

The small size and geographical location of the site provided several logistical challenges requiring careful coordination of day-to-day construction activities. Specific trucking routes were established and strictly enforced for all deliveries and trucks hauling waste for disposal.

The project was completed within budget and ahead of schedule and with no health and safety related incidents.

CASE STUDY: 36
Sanford MGP Superfund Site

Location:	Sanford, FL
History:	Manufactured gas plant
Contaminants:	PAHs, DNAPL
Scale:	125,000 yd^3 (95,569 m^3)
Reagents:	Portland cement & blastfurnace slag
Method:	In-situ
End-Use:	Restoration

Site Description

The former Sanford Gasification Plant, a manufactured gas plant (MGP), was located between 830 and 901 West Sixth Street and between Holly Avenue and Cedar Avenue in Sanford, Florida. The remediation area included the former Sanford Gasification Plant and a number of properties downstream from the site and immediately adjacent to Cloud Branch Creek. Cloud Branch Creek which traverses the remediation work areas discharges to Lake Monroe past the confluence of Mill Creek.

Soil remedial activities planned for the Sanford site by Natural Resources Technology, Inc. and WRScompass included the demolition of 3 abandoned structures, excavation of the first 2 ft or 0.6 m (20,000 yd^3 or 15,291 m^3) of unsaturated soils, in-situ solidification (ISS) of 125,000 yd^3 or 95,569 m^3 saturated soils, extensive utility relocates as well as major improvements to Cloud Branch Creek in the form of the installation of nearly 1,000 ft (305 m) of 7 ft (2.1 m) x 7 ft (2.1 m) and 11 ft (3.4 m) x 7 ft (2.1 m) culverts, realignment of the creek as well as 450 ft (137 m) of open channel improvement of the creek.

Risk Drivers

The contaminated soils required removal/treatment to prevent contact with the contaminated soil by the residents and to prevent migration of PAHs to Cloud Branch Creek and Lake Monroe.

Objectives

- Backfill the landfill area; removal and disposal of the unsaturated soil
- Treatment of the saturated soil down to the clay aquitard
- Realignment and improvement of Cloud Branch Creek
- Restore the site

Method

Portland cement and ground granulated blast furnace slag were used to improve the unconfined compressive strength to >50 psi (0.3 MPa) and reduce the hydraulic conductivity to <1×10^{-9} m/sec.

Validation

Samples of the treated soil were obtained at a frequency of every 250 tons (227 tonnes) treated and subject to UCS and permeability testing. Over 98 % of the treated samples met the performance criteria on the first pass. Those columns not meeting the performance criteria were retreated.

Equipment

The ISS operations within the site were conducted using a 4000-series Manitowoc crane equipped with an attached Hain Platform. The crane/platform assembly is supplemented with a swivel-mounted, top-feeding Kelly Bar capable of reaching a depth of 75 ft (23 m) bgs. Augers were attached to the bottom of the Kelly Bar. A 10 ft (3 m) auger and a 12 ft auger (3.6 m) were utilized on this project.

The appropriate amount of water was metered into an initial 5 yd³ (3.8 m³) batch tank equipped with a high-speed, high-shear mixer. The reagents (Portland cement and ground granulated blast furnace slag) were transferred from the silos to the batch tank using the internal screw conveyor to deliver the specified volume of reagent.

The water was added to the mix tank first and the volume of water recorded. Each reagent was added separately to the mix tank. The scales on which the mix tank sets were tarred before each reagent was added to verify that the correct amount of reagent had been added. WRScompass periodically tested each batch being prepared using a mud balance to insure the proper mix design was being met. The batch number, volume of water used, and the weight of each reagent added were recorded on a grout log by the batch plant operator.

When the correct grout composition was achieved, the blended grout was transferred to the auger. The batch plant was also equipped with a second storage tank to allow for temporary storage of a blended batch to allow for uninterrupted production. A high speed mixer in this second tank was to ensure the blended batch does not separate. The pre-determined grout volume was pumped to the treatment area based on the soil density, reagent admixture ratio, and the work area dimensions (i.e. column diameter or panel dimensions).

The ISS Treatment was performed in a series of overlapping columns as per the schematic below. Columns along the perimeter of the ISS area had a 1 ft (0.3 m) overlap in the area known as the triple treatment triangle, while the interior columns had a neat-line overlap.

WRScompass used a TOPCON 3000 Series Total Survey Station during ISS operations. This instrument insured the proper overlaps of the columns, the locations of the columns, the vertical extent of the treatment, and the rate of advancement of the tool. Using the predetermined column locations, WRScompass placed stakes at the center point of each column slated for treatment for the day's production.

The crane operator would set the auger tip immediately over the center stake insuring the proper location. Once the auger was set up at said location, ISS personnel verified the proper column designation and location. Prior to the initiation of the drilling/ISS operations, ISS personnel verified the key parameters for the column (i.e. total anticipated depth, grout volume needed, etc.). This information was recorded in the ISS Master QA/QC Log.

The grout mixing plant personnel were in constant radio contact with the crane operator and the QC personnel to insure proper grout volumes are dispensed and incorporated into the column. The QCO verified vertical depth by surveying the elevation of the top of the Kelly bar (known length) when the auger is at the top of the mixing area and at the bottom of the mixing area. When the terminating depth has been reached and the overall grout volume for the column injected, the auger was extracted and reintroduced to the same column to complete at least 3 mixing passes total per column to achieve a homogeneous mixture. The auger was then moved to the next column unless the finished column is slated for sampling.

In-situ treated material sampling was performed utilizing WRScompass' in-situ sampler. Upon the completion of the ISS column, the in-situ sampler (see overleaf) was lifted by the excavator and advanced to the vertical midpoint of the column. Once the in-situ sampler reached the sampling depth, the sampling chamber was opened using a hydraulic actuator. The sample would then enter the sampling chamber. Once the chamber was filled, it was hydraulically closed and the in-situ sampler was retrieved.

WRScompass' ISS Swell Management Plan was to incorporate the ISS swell into the site's final contours and grades. To the extent practical, all ISS swell was managed on-site and within the ISS treatment limits. WRScompass would begin grading the ISS to the site's final contours and grades before the ISS treated material had started to set. This allowed an on-going determination as to whether or not all of the ISS swell could be managed on-site and within the ISS treatment limits.

Specific Issues

One of the most sensitive and complex elements of the work was the stabilization of approximately 250 lineal feet (76 m) of roadway which carried banks of fiber optic lines. These lines could not be moved which required WRScompass to stabilize the soils underneath these lines. WRScompass designed and implemented a method that successfully treated soils while supporting the fiber optics lines with no loss of service.

390 | APPENDIX B: CASE STUDIES EMPLOYING S/S

In-situ sampler lifted by the excavator and advanced to the vertical midpoint of the column

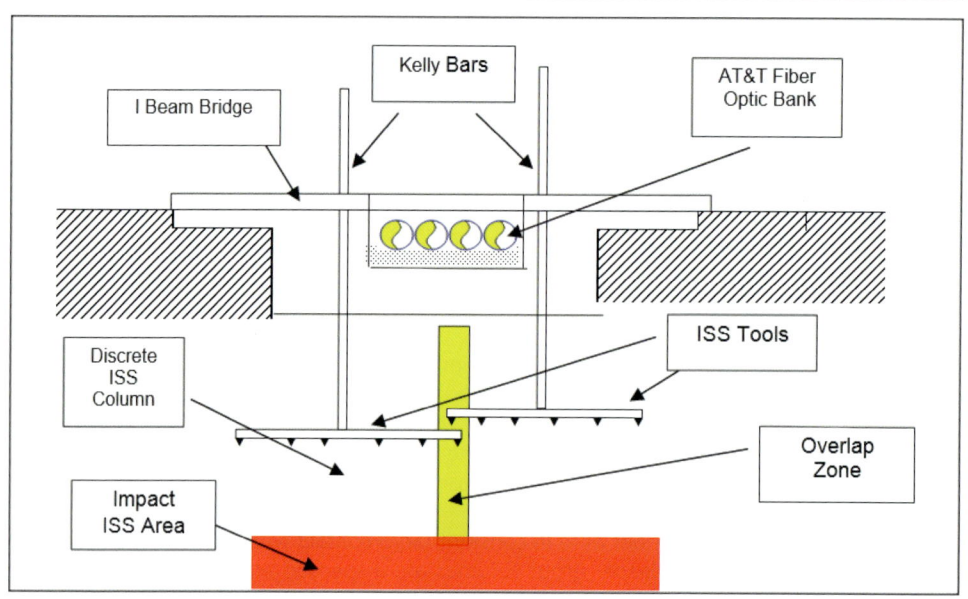

Auger tip immediately over the center stake insuring the proper location

CASE STUDY: 37
Söderhamn

Location:	Söderhamn, Sweden
History:	Wood treatment site
Contaminants:	Metals, PAHs
Scale:	31,000 tonnes (34,171 tons)
Reagents:	Zero-valent iron
Method:	Ex-situ
End-Use:	Public Park

Site Description

The former wood impregnation site at Söderhamn (Sweden) operated between 1937 and 1997, and was found to be contaminated by PAH and heavy metals (As, Cu, Cr). This resulted from the use of creosote and CCA (Chromated Copper Arsenate salts) for the impregnation of the wood. The contamination was found both in the soil and the groundwater, as well as the sediments in the adjacent river.

The presence of various hotspots of heavy metal and creosote contamination was the driver for the remediation.

191,000 tons (173,272 tonnes) of the site were excavated, of which 76,000 tons (68,946 tonnes) were sent off-site as hazardous waste, 31,000 tons (28,123 tonnes) treated by chemical stabilization to be reused as capping material on a local landfill, and the remainder of the soil could be reused on site. The site was turned into a public park space.

The remediation works started in September 2009 and ended in October 2011.

Objectives

The main objective of the remediation of this site was to remove the contaminant sources, and to optimise the beneficial reuse of materials excavated from the site. The soils had to be remediated to below the Swedish standards MKM for metals and PAHs, as indicated in the Table below.

Remediation standards for metals and PAHs at the Söderhamn site

Pollutant	As	Cu	Cr	PAH total
MKM-standard (mg/kg DM)	40	200	250	80

All soils above the remediation standards were excavated and brought to a landfill. The soils not complying to the landfill leaching standard of 2 mg/kg DM for arsenic were chemically stabilised.

Method

31,000 tons (28,122 tonnes) of soil were chemically stabilised in order to reduce the leaching of metals, in particular arsenic. Zero-valent iron powder was added at a dosage of 1.5 %. This dosage was determined after an extensive preliminary lab testing programme, assured a robust process, and avoided reprocessing.

The soil was treated by means of a batch mixing plant (double axis pug-mill system, picture above), with a throughput of about 40 tons (36 tonnes) an hour.

Validation

Every batch of treated soil (representing about 1,000 tons or 907 tonnes) was sampled and tested according to EN 12457-4. The results were compared to the target levels in the Table above.

CASE STUDY: 38
Southeast Wisconsin

Location:	Southeast Wisconsin, USA
History:	Former Manufactured Gas Plant (MGP)
Contaminants:	MGP coal tar residuals
Scale:	33,000 yd³ (25,230 m³)
Reagents:	Portland cement and blast furnace slag
Method:	In-situ auger and bucket mixing
End-Use:	Future redevelopment (not specified)

Site Description

Natural Resource Technology provided remedial alternatives analysis, design, and environmental construction management for a former manufactured gas plant (MGP) facility located near Lake Michigan in a high profile, residential and commercial downtown area. Project planning required extensive permitting and coordination with the City, Wisconsin Department of Natural Resources, state health department and the community.

A solution was designed to maximize remediation of the MGP impacted material, both above and below the water table, by in situ stabilisation/solidification (ISS) and excavation. Both auger and bucket mixing techniques were used. ISS allowed cost-effective treatment of soils to depths greater than what could be achieved with traditional excavation due to site constraints. All ISS swell material was managed on-site, as designed. The site remediation area is approximately 3.5 acres (14,164 m²) in size.

Characterisation

- Remedial Alternatives Analysis
- Remedial Design
- Bench- and Pilot-Scale Studies
- Construction Oversight and Documentation

Risk Drivers/Project Constraints

This project presented multiple spatial and logistical constraints including:

- Multiple condominiums (high-rise and townhome), hotel, restaurants, and marinas are located adjacent to the project site
- A 4-storey vacant 60,000 ft^2 (5,574 m^3) commercial building was located on the site next to where the remediation was performed
- A parking deck structure required demolition before remedial construction could begin.
- Former MGP structures (large foundations, piping, steel and concrete gas holders) required removal prior to ISS
- MGP impacted soil and groundwater extended more than 30 ft (9.1 m) below ground surface near the on-site building and adjacent City streets
- Underground electrical, and City storm and sanitary sewers required rerouting prior to project start

Objectives

- Restore the property for future development
- Minimize potential for direct contact exposure to the community
- Reduce contributions of MGP residuals to groundwater

Method

- Over 33,000 yd^3 (25,230 m^3) were solidified using auger and bucket mixing techniques near the on-site building to 30 ft (9.1 m) below ground surface
- The contractor (ENTACT) utilized a DELMAG RH-32 caisson-type drill rig equipped with a 10 ft (3 m) diameter auger to complete overlapping ISS columns
- A conventional long reach excavator was utilized by ENTACT to complete bucket mixed ISS grids.
- Reagent grout mixing was conducted using an on-site batch plant
- Reagent consisting of a 3:1 mix (slag and Portland cement) mixed in-situ between 8 and 10 % relative to the dry unit weight of soil
- Approximately 21,000 tons (19,051 tonnes) of MGP impacted debris and soil was excavated and landfilled.
- Comprehensive perimeter air monitoring was performed

Validation

Discrete samples were collected from the top, middle or bottom of the selected ISS columns or grids by ENTACT according to the following frequency:

- One discrete sample collected every 1,000 yd^3 (764 m^3) or once per day of ISS production.
- One discrete sample collected every 100 linear feet (30.5 m) around the perimeter of the ISS treatment area.

Samples were tested for unconfined compressive strength and hydraulic conductivity to demonstrate compliance with the performance objectives.

Benefits of Solution

- Cost effective solution when compared with other technologies
- Reduced vapour phase and fugitive particulate emissions during remediation
- Reduced need for long-term management and environmental concerns for future development
- ISS monolith suitable as structural foundation for future construction

CASE STUDY: 39
St Louis

Location:	St Louis, Missouri
History:	Ordnance plant
Contaminants:	Impacted soils
Scale:	1,400 yd^3 (1,070 m^3)
Reagents:	Zero valent iron
Method:	In-situ
End-Use:	Not Specified

Site Description

This project was performed at the former St. Louis Ordnance Plant in St. Louis, MO.

In-situ stabilization (ISS) work was performed on soils that were impacted by contaminated groundwater.

The treatment zone covered approximately 1,400 ft^2 (130 m^2) of surface area with depths from the surface to bedrock ranging from 20 to 30 ft (6.1 to 9.1 m).

The project goal was to mix the contaminated soil with zero-valent iron (ZVI), and involved an innovative application and mixing process that was performed in the dry, i.e. without using drilling fluids.

Treatment involved using a minimum 0.6 % ZVI addition, with CH2M Hill confirming that all soil mixed columns were in compliance with the treatment goals. Dry soil mixing using ZVI at this site is the first now full-scale application of this type carried out.

Geo-Solutions performed all the soil mixing activities utilizing specialty equipment, including a soil mixing rig with a 5 ft (1.5 m) diameter mixing tool, vapour shroud/extraction/treatment system, and ZVI injection tooling.

CASE STUDY: 40
Sunflower Army Ammunition Depot

(photo credit https://en.wikipedia.org/wiki/Sunflower_Army_Ammunition_Plant)

Location:	DeSoto, KS
History:	Ammunition depot
Contaminants:	Explosives, propellants, lead
Scale:	70,000 tons (63,502 tonnes)
Reagents:	Portland cement
Method:	Ex-situ
End-Use:	Restoration

Site Description

The 9065 acre (36.7 km^2) Sunflower Army Ammunition Plant (SFAAP) was a government owned, contractor-operated military installation. Solid Waste Management Units (SWMU) 10 and 11 consisted of the Blender and Roll House Area, F-Line Press Buildings, and the F-Line ditches and settling/blender ponds. The F-Line served as the final production area for N-5 propellant. The F-Line area had 21 ditches, which originate at the production buildings and terminated into 3 pairs of settling ponds. The 21 ditches are generally 10 ft (3 m) deep with relatively steep banks. The 3 pairs of settling ponds were unlined earthen impoundments equipped with stand pipes to permit settling of solids and decanting of water. Two blender ponds were located northwest of the production area. The ditches and ponds received wastewater from the manufacturing process. The wastewater was contaminated with lead salts, propellant waste, nitro-glycerine and nitrocellulose from the manufacturing process.

Objectives

- Remove the contaminated soil
- Treat to reduce the leachability of lead and reduce the reactivity of the explosives and propellants
- Stockpile the treated soil and transport to a local landfill for use as daily cover
- Restore the site

A total of 46,000 tons (41,730 tonnes) of material was excavated from SWMUs 10 and 11. 22,000 tons (19,958 tonnes) of soil was removed from the 62 Roll/Blender House building foundations in Areas A through H. A total of 19,000 tons (17,236 tonnes) of soil was removed from 22 F-Line drainage ditches and ditches A through F. 5,000 tons (4,534 tonnes) of soil was removed from the bottoms and sides of 8 settling/blending ponds. Post-excavation sampling and analysis performed at the designated sample locations verified that the excavation attained the remediation cleanup goals established for this site.

The concentration of propellant per unit volume of excavated material was sufficient to propagate detonation only at some locations along the ditches, so the physical nature and potential extent of energetic hazards associated with propellant warranted special consideration. Therefore, the remedial action was worked at a productive and efficient pace, while maintaining a safe working environment for site personnel. The excavator cabs were outfitted with a specialized lexan blast protective shield, customized to protect the glass windshield and prevent injury to the operator. A 2,000 gallon (7,570 litre) water truck was readily available to provide continuous water mist to the excavation area in an effort to minimize the potential for ignition.

Method

The contaminated soil was treated with Portland cement, producing a treated material that did not exhibit a hazardous characteristic for leachability (primarily lead) or ignitability (due to the presence of explosives and propellant).

Validation

Temporary 500 ton (453 tonne) stockpiles of the treated soil were created. Confirmatory testing to verify that the treated material that did not exhibit a hazardous characteristic for leachability (primarily lead) or ignitability (due to the presence of explosives and propellant) was performed on the sample from each temporary stockpile. All of the confirmatory samples met the performance requirements for leachability.

Equipment

Contaminated soils were blended using a Findlay 393 Hydrascreen with a pre-cutter pulverization unit. Once the material was screened to remove >2 in (>5 cm) debris, the homogeneous material was stabilized in the RapidMix 400 pug-mill using Portland cement. Production rates through the pug-mill of over 1,000 tons (907 tonnes) per day were routinely achieved.

CASE STUDY: 41
Sydney Tar Ponds

Location:	Sydney, Nova Scotia, Canada
History:	Scrap yard
Contaminants:	Heavy metals, PAHs, PCBs, VOCs
Scale:	700,000 tonnes (771,617 tons) (PAH affected)
	50,000 tonnes (55,115 tons) (PCB affected)
Reagents:	Portland cement, fly ash & slag
Method:	In-situ
End-Use:	Restoration

Site Description

The Sydney Tar Ponds are centrally located within Sydney, Nova Scotia, as shown in Figure 1. The Tar Ponds cover an area of approximately 33 Hectares (81 acres or 327,795 m^2). They are what remain of the Muggah Creek tidal estuary after nearly 100 years of steel and coke production activities. The sanitary sewage of the surrounding communities also drained into this estuary prior to the cleanup. Preliminary works by the municipal government now collect and treat the roughly 13 million liters a day that once discharged into the ponds, through a modern treatment facility, which discharges into Sydney Harbour. Two significant tributaries – Wash Brook and Coke Oven Brook – drain the urban watershed and discharge into the Tar Ponds (AECOM, 2008-B).

The Sydney Tar Ponds consists of the North Tar Pond and the South Tar Pond. The dividing line between these 2 ponds is the causeway and bridge at Ferry Street.

Over 100 years of steel making and coking industries in Sydney resulted in deposition of coal tar, fine coal, and sediment within the ponds. The majority of the sediments in the Tar Ponds were transported by Coke Oven Brook from the Coke Ovens site. Contaminants in the sediments include heavy metals, polycyclic aromatic hydrocarbons (PAHs), volatile organic compounds (VOCs), and polychlorinated biphenyls (PCBs).

Figure 1: Location of the Tar Ponds site in Sydney Nova Scotia, Canada (AECOM, 2008-B)

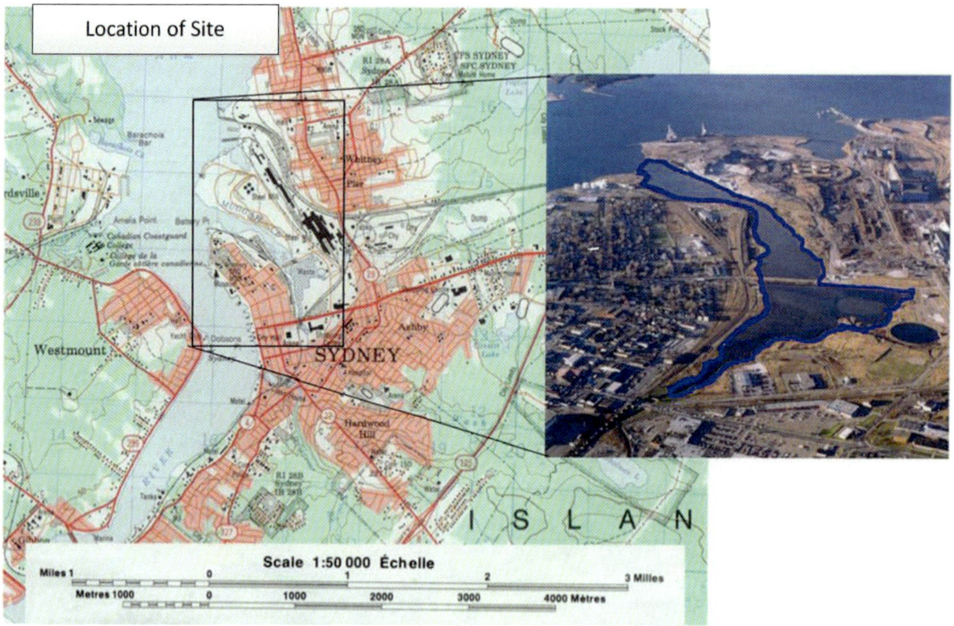

Figure 2: Contaminated Sediment in South Tar Pond

The overall work required to remediate the Sydney Tar Ponds and Coke Ovens Sites was extensive, complex and distributed across a large area. Prior to this remediation project, there had been 2 failed attempts to design and implement a remedial solution. In order to make this project manageable, the project was split into approximately 13 sub-projects or "elements". The Solidification/Stabilization (S/S) of the Tar Ponds was the largest element in both scope and value ($75m).

Project Funding, Stakeholders and Management

On May 12, 2004, after decades of study and 2 failed cleanup attempts the governments of Canada (federal) and of Nova Scotia (provincial) provided $400m in funding to clean up the sites (MOA 2004). The funding for the clean-up was split between the Canadian federal government (70 %) and the Nova Scotia provincial government (30 %). A number of major stakeholders maintained an active interest in the progress of the project and provided the necessary checks and balances to ensure its success.

As project implementer, the Sydney Tar Ponds Agency's (STPA) fundamental role was to direct, procure, and implement the services needed to complete the project, while protecting the interests of the funding partners. Other key project team members included:

- Independent Engineer (CRA) - validate the technical merits of the project, and to report on costs incurred, cost to complete, and contract compliance
- Stakeholder groups such as the Environmental Management Committee (EMC), Community Liaison Committee (CLC), Operations Advisory Committee (OAC), Remediation Monitoring Oversight Board (RMOB)
- Design Engineer (AECOM) - design the project, and oversee the execution of the design
- Independent Quality Assurance Consultant – Stantec Consulting Ltd

Design of Solution

The Design Engineer (AECOM) was engaged in October 2006 and immediately began work on selecting a remedial option for the North and South Tar Ponds, as well as for the other project elements. The remedial option chosen for the Tar Ponds sediments was in-situ Solidification/Stabilization (S/S) with cement. In addition, an engineered channel would be constructed through the Tar Ponds to permit the Coke Ovens Brook and Muggah Creek to flow through the remediated Tar Ponds to Sydney Harbour. After S/S was completed the area would be covered with an engineered cap. To permit S/S in the dry an extensive water control management system would be constructed to control any water intrusion from the ocean and capture/pump incoming streams sequentially around the working areas (AECOM, 2008-B).

Remediation

Pre-S/S Work
Prior to S/S of the North and South Tar Ponds there were a number of tasks that had to be carried out including the construction of access roads along the perimeter of the ponds. The two key tasks that were required prior to S/S activities were as follows:

Bench-Scale Testing and Pilot Study
As part of the Bench-Scale Testing, the Design Engineer (DE) collected numerous sediment samples from both the North and South Tar Ponds. In the laboratory, the DE mixed sediment samples with potential S/S reagents such as cement, slag and/or flyash. The goal of the Bench-Scale Testing was to find reagent(s) and sediment mixtures that could meet the project S/S performance criteria in the laboratory. Utilizing the information gained from the Bench-Scale Testing, a Pilot Program was carried out. The Pilot Project utilized an in-situ S/S approach using interlocking steel sheet pile cells in the North and South Tar Pond. Seven cells were constructed in the South Tar Pond and 6 cells were constructed in the North Tar Pond, each with an approximate surface area of 27 m² (291 ft²). After the cells were homogenized using

an excavator, known quantities of various combinations of cement, slag and/or flyash were mixed with the cell sediments. After mixing, samples were taken and analyzed with respect to meeting the desired performance requirements (see Tables 1 and 2). The Pilot Program indicated that there were a number of reagent variations that had the potential of meeting the performance criteria. A report of the Pilot Program was prepared and made available to the potential bidders on the S/S project.

Dewatering of the S/S Work Areas
The Tar Ponds are a tidal estuary with 2 large brooks (Coke Ovens and Wash Brook) flowing into the South Tar Pond. To carry out S/S it was necessary to dewater the work areas prior to the S/S contractor beginning work in the area. To achieve this pre-S/S dewatering, a contractor (Beaver Marine/MB2 Joint Venture) was engaged by STPA to implement the Design Engineer's dewatering design. To create dry work areas in the South and North Tar Ponds a multi-stage sequencing of pump stations was used. The pumping stations were comprised of inlet works, pumping equipment, 48 in (122 cm) pipelines for conveyance of the flows and discharge facilities with energy dissipation structures. The dewatering design initially involved pumping the 2 brooks around the South Tar Pond with the discharge introduced into the North Tar Pond. Once the flows had been diverted, pumping of the remaining water above the sediments was the responsibility of the S/S contractor. After completing the S/S of the South Tar Pond and construction of the channel, the pumping infrastructure would be moved to drain the upper portion of the North Tar Pond and then again to drain the lower portion of the North Tar Pond. The pumping stations were a combination of electric and diesel pumps and had to be able to handle a combined high peak flow in the range of 14 m^3 per second or 3,698 gallons per second (AECOM, 2008-A).

Figure 3: South Tar Pond Dewatering

Solidification/Stabilization (S/S)
The S/S tender was issued in early 2009 and the successful contractor Nordlys Environmental LP was awarded the contract to carry out the work in June of 2009. Nordlys Environmental LP was a limited partnership between a local contractor J&T Van Zutphen Construction Inc. and ECC of Massachusetts. The contract was a Performance Based Contract (PBC) in that the contractor only received a payment for a specified S/S volume of sediment (or cell) when it successfully passed the 3 performance criteria (see Table 1). The contractor engaged experts

to carry out Quality Control (QC) on all the cells while STPA engaged an Independent Quality Assurance Consultant, Stantec Consulting Ltd, to perform Quality Assurance (QA) verified through random testing of approximately 10 % of the cells. Failure of either the quality control (QC) or the quality assurance (QA) of any of the performance criteria meant the cell failed and had to be remixed prior. Table 2 lists the 46 parameters tested using the Synthetic Precipitation Leaching Procedure. Table 3 compares the frequency of QC and QA tests for each test type for each cell.

Table 1: Contractual Performance Criteria for S/S Work by Contractor (Ingraham, 2011)

Property	Test Method	Criterion
Unconfined Compressive Strength (UCS)	ASTM D1633 Method B (modified)	≥ 0.17 MPA (25 psi)
Hydraulic Conductivity (Perm)	ASTM 5084 (Flex Wall)	$\leq 1 \times 10^{-9}$ m/sec
Leachability (SPLP)	Modified Synthetic Precipitation Leaching Procedure 1312 (monolithic structural integrity procedure) to check 48 site-specific compounds	Site specific leachate criteria (SSLC)

Table 2: Parameters tested using Synthetic Precipitation Leaching Procedure (SPLP) during the remediation

SPLP Parameters			
Modified TPH (Tier 1)	Copper	Strontium	Dibenzo(a,h)anthracene
Aluminum	Iron	Thallium	Fluoranthene
Antimony	Lead	Tin	Fluorene
Arsenic	Lithium	Uranium	Indeno(1,2,3-cd)pyrene
Barium	Manganese	Vanadium	Naphthalene
Beryllium	Mercury	Zinc	Perylene
Boron	Molybdenum	Benzo(b)fluoranthene	Phenanthrene
Cadmium	Nickel	Benzo(g,h,i)perylene	Pyrene
Chromium	Selenium	Benzo(k)fluoranthene	Total PAH
Cobalt	Silver	Chrysene	Total PCB

Table 3: Sampling and Testing Requirements for Quality (initial tests at 28 days; retests at 56 days if needed) (Ingraham, 2011)

Test Type	QC Frequency (Minimum)	QA Frequency (Minimum)	Cell Thickness & Sample Frequency	
			Sediment ≤ 2m	Sediment > 2m
UCS	All (~250m³) cells; 8 cylinders/cell	Every 10th cell; 8 cylinders/cell	2 finite location homogeneous samples	1 homogeneous sample from top & 1 from bottom
Perm	All (~250m³) cells; 2 cylinders/cell	Every 10th cell; 2 cylinders/cell	1 composite sample	1 composite sample
SPLP (Pre- & Post-mixing)	Cells 1-20: every cell; Cells 21-40: every 2nd cell; Cells 41+: every 4th cell	Cells 1-20: every 10th cell; Cells 21-40: every 20th cell; Cells 41+: every 40th cell	1 composite sample	1 composite sample

Mixing the Cells

Each cell was mixed following the method agreed upon by the contractor, the Design Engineer and the STPA. The contractor used hydraulic excavators to mix the solidifying agents (Portland cement) and gravel/slag, if required, according to an approved recipe. The following summarizes the steps utilized for in-situ mixing:

- Contractor provides location of a cell to be mixed
- Design Engineer (DE) confirms predicted bottom of cell using contractor provided center of cell and Digital Terrain Model
- DE visually confirms with the contractor the Field Confirmed Bottom of Sediment Elevation while In-Field
- Cell volume sized to received 1 tanker load of cement
- Contractor's excavator operator mixes the cement into the impacted sediments in the cell. Cells are interlocking
- While each cell is being mixed the DE observes operations to ensure contractor compliance with the contract documents and standard operating procedures (i.e. Daily Diary, Mixing Oversight Checklist)
- Contractor takes bulk samples from which specimens are taken and tested
- Where any non-compliance is found a Request for Action (RFA) or Non-Conformance Report (NCR) is issued to ensure the Contractor corrects the issue
- In a timely manner the DE responds to any Request for Information (RFI) the contractor may have to ensure compliance to design
- DE confirms cell quantity and compliance to contractual requirements
- Additional Oversight Tasks Include: Confirming odour observations reporting, water management observations, dust control, noise, safety and environmental protocols, quality control observations, contract administration, change order observations, noise and dust management observations

Figure 4: In-situ S/S with dust control conveyance boxes and odour control foam (white) and pressure washer application of bio-solve

Special Challenges

As one might expect there were a number of special challenges associated with carrying out a large S/S project in a tidal estuary in the center of a city. Below is a summary of some of the challenges encountered and the techniques utilized to overcome these challenges:

1. **Water Management Issues associated with high groundwater flows and close proximity of ocean.** Although an issue throughout the project, this was a major challenge in the final phase (Phase 3) which located adjacent to the ocean. A combination steel sheet pile wall and armour stone barrier provided a physical separation and fairly impermeable barrier but the tidal water moved relatively freely through the existing slag area bordering the east side of the pond. The contractor strategically placed clay and S/S berms in areas to aid dewatering efforts. This enabled S/S and channel work to begin in specific locations while effective water management was being carried out in other locations.
2. **Constitution of impacted sediments varied throughout the North and South Tar Ponds.** The type, location, and amount of contaminant as well as the water content of the sediment varied throughout the north and south ponds. In order to overcome these issues, the contractor was

constantly observing and analyzing the sediment for these changes. This permitted the contractor to recognize issues prior to mixing and make approved adjustments to mix deign and duration of mixing.

3. **Sediment depths deeper than predicted in some areas.** As previously stated, the contractor's primary mixing methodology was bucket mixing. In deeper areas (> 16 ft or 4.9 m) bucket mixing began to be problematic resulting in some cells not meeting the criteria on the first S/S attempt. After discussions with stakeholders, the contractor began utilizing an ALLU type mixer in deeper areas. The Allu mixer proved to be very effective in deeper areas.

4. **Odours.** Odour generation during S/S became a significant issue due to a combination of sediment disturbance/drying, the exothermic cement/sludge reaction and general warm weather mixing. Given that the work was being carried out in the middle of Sydney, significant odour generation was not acceptable. Excavation of S/S material for channel construction was especially odorous. A comprehensive odour monitoring and management program was developed which involved utilizing odour suppression materials such as foams, liquid sprays and surface cover materials. In addition, work practices were altered to minimize odour generation. For example, sediments were permanently relocated out of the channel footprint prior to S/S.

5. **Dust.** As with odour, significant dust generation within the city was not acceptable. Cement dust can pose a potential health and safety risk on-site and off-site. As part of the dust control methodology, the contractor had a "dust budget" with pre-defined dust generation limits. As the dust budget was approached or exceeded, the contractor would stop work and alter work procedures to lessen dust generating practices. In a conscious effort, the contractor utilized a cement conveyance box which was fitted on top of the sediment to be mixed and then powder cement was blown into the box prior to mixing. The contractor then utilized significant wetting techniques during the mixing process.

6. **Cell Sediment Volume and Cement.** Initially each cell was of the same spatial size, however in the field it was soon realized that mixing a fraction of a cement tanker load into a cell was not practical. The procedure was modified to a volume-based approach so that the volume of the cell required 1 tanker load of cement per cell.

Documenting Results

The contract required a "cradle to grave" document trail. At completion of the project it was estimated that over 70,000 pages of quality control documentation alone was generated and reviewed. Each of the 3,486 cells has a unique set of data associated with it (See Table 4).

Summary and Future Site Use

Stabilization of the final phase of the S/S project (Phase 3) was completed on January 16, 2013, with a total of 253,671 m^3 (331,789 yd^3) mixed. Overall for all 3 phases of the project 679,016 m^3 (888,119 yd^3) of sediment were treated and over 100,000 laboratory quality control/assurance tests were carried out. The S/S component of the project was completed ahead of schedule and under budget at a cost of $73.5m.

After completion of the S/S component of the project, the S/S material was capped with a low permeability clay cap and then a Future Site Project was carried out which created what

Table 4: Documentation captured for each cell (Ingraham, MacNeil, McNeil, MacDonald, MacCormack, & Francisco, 2011)

Documentation	Data Contained	
Mixing Oversight Checklist	DE verification of Key Cell Data and conformance to Contract and SOPs	
Cell Profile	Center Co-ordinate	Wet Density
	Cell Co-ordinates	Sediment Thickness
	Bottom Elevation	Cell Surface Area
	Pre-Top Elevation	Wet Volume
	Post-Top Elevation	Additional Soil Volume
In-situ S/S (ISS) Report	Mix Time	Hours of Equipment Used for Mixing
	Reagent Quantities	
Cement Receipt	Cement quantity	
Test Reports	UCS (x4), PERM and SPLP (x46 if applicable) test results	

Figure 5: Open Hearth Park located on the former Sydney Tar Ponds site

is now called Open Hearth Park. As seen in Figure 5, Open Hearth Park includes a common area, an outdoor concert venue and skating area, an all weather sports field, natural turf field, a playground, a bike training facility, off-leash dog park, public art displays and a trail network complete with bridges that once again provides connectivity between downtown Sydney and the communities of Whitney Pier, Ashby and Sydney`s North End.

Below is a quote from the **Honourable Diane Finley, Minister of Public Works and Government Services** in 2013 on the $400m Sydney Tar Ponds Cleanup:

"This has been the most successful contaminated site remediation project in Canada's history and we're proud that it was completed on time and on budget."

Bibliography

1. AECOM, (2008-A). Remediation of the Tar Ponds and Coke Ovens Sites Design and Construction Oversight Services 100 Percent Design Report TP6A. Sydney, NS: AECOM available through Sydney Tar Ponds Agency.
2. AECOM, (2008-B). Remediation of the Tar Ponds and Coke Ovens Sites Design and Construction Oversight Services 100 Percent Design Report TP6B. Sydney, NS: AECOM available through Sydney Tar Ponds Agency.
3. Ingraham, D., (2011). Managing Quality on the $400M Sydney Tar Ponds and Coke Ovens Cleanup Project. 3rd International/9th Construction Specialty Conference (pp. CN-061-1 to CN-061-10). Ottawa, ON: Canadian Society of Civil Engineers.
4. Ingraham, D., (2011). Sydney Tar Ponds and Coke Ovens Remediation Update - Phases I & II and Moving Forward. Presentation to Randle Reef Contaminated Sediment Remediation Project Members and Environment Canada. Sydney, NS.
5. Ingraham, D., & McNeil, W., (2010). Quality on a Multi-Year Remediation Project. 2010 S/S-Tech: International Solidification/Stabilization Technology Forum. Sydney, NS.
6. Ingraham, D., & McNeil, W., (2013). A Strong Quality Program - Imperative to Environmental Remediation Success. Remediation Technologies Symposium 2013 - RemTech(TM) 2013. Banff, AB.
7. Ingraham, D., MacNeil, J., McNeil, W., MacDonald, J., MacCormack, S., & Francisco, R., (2011). 100 % Performance Demands Collaborative Quality Management. RemEast 2011 Managing Challenging Environment Remediation Projects – Case Studies. Halifax, NS.
8. MacNeil, J., Ingraham, D., Gangopadhyay, S., & Van Zutphen, V., (2012). Managing Program and Project Requirements During Sediment Stabilization of the Sydney Tar Ponds. Real Property Institute of Canada - Federal Contaminated Sites National Workshop. Halifax, NS.
9. Parker, M., Mcneil, W., Ingraham, D., & White, D., (2012). Sydney Tar Ponds and Coke Ovens: Integrating Broad Socio-Economic Considerations into a Contentious Remediation Project. 37th CLRA / ACRSD Annual General Meeting "Seeding Change – Cooperative Reclamation". Sydney, NS.

CASE STUDY: 42
Umatilla Ammunition Depot

Location:	Hermiston, OR
History:	Explosive disposal area
Contaminants:	Explosives, metals
Scale:	32,800 tons (29,755 tonnes)
Reagents:	Portland cement & activated carbon
Method:	Ex-situ
End-Use:	Restoration

Site Description

Five sites at the Umatilla Army ammunition depot were selected by the U.S. Army and the U.S. EPA for remedial action. Three of the sites were contaminated with explosives. One of the sites was the TNT Sludge Burial and Burn Area where TNT-containing sludges from the Explosive Washout Plant may have been dumped and burned. The other site consisted of open burning trenches and pads where a variety of debris, ordnance waste, and explosives sludges were burned.

The remediation contaminants of concern for the sites were antimony, arsenic, barium, beryllium, cadmium, chromium, lead, cobalt, thallium, copper, nickel, silver, zinc, 1,3,5-TNB, 2,4-DNT, RDX, 2,4,6-TNT, HMX, and Tetryl. The remedial action selected for the contaminated soil at these sites was stabilization, after bioremediation of the explosives was demonstrated to be cost-prohibitive.

Objectives

- Remove the contaminated soil
- Treat to reduce the leachability of metals and explosives
- Place and compact the treated soil back into the excavation area
- Cap the site with a low permeability clay
- Restore the site

Method

The contaminated soil was treated with Portland cement and activated carbon to produce a treated material that met the performance criteria shown below.

Compound	Leaching Requirement (µg/L)
Barium	100,000
Chromium	5,000
Copper	140,000
Lead	5,000
Zinc	1,100,000
1,3,5-TNB	180
2,4-DNT	130
RDX	200
2,4,6-TNT	200
HMX	40,000

Validation

Temporary 75 ton (68 tonne) stockpiles of the treated soil were created. Confirmatory testing to verify that the treated material met the leachability performance criteria was carried out. 8 (1.8 %) of the 437 production lots failed for TCLP-leachable explosives. This material was reprocessed to meet the TCLP-leachable explosives criteria.

Equipment

A pug-mill system, consisting of a feed hopper with conveyor, pug-mill, reagent silo, and stacking conveyor were used to treat 150-200 tons (136-181 tonnes) of contaminated soil per hour. The treated soil was segregated into 75 ton (68 tonnes) stockpiles for confirmation testing.

CASE STUDY: 43
Valero Paulsboro Refinery

Location: Paulsboro, NJ
History: Refinery
Contaminants: Pb, As, Cd, V, Zn, TPH
Scale: 18,000 yd³ (13,762 m³)
Reagents: Proprietary binder
Method: In-situ
End-Use: Restoration

Site Description

Valero Energy Corporation closed the North Recycle Pond (NRP) area located on the Paulsboro Refining Company LLC's refinery in Paulsboro, New Jersey under an Administrative Consent Order (ACO) with the New Jersey Department of Environmental Protection (NJDEP). The overall project scope not only involved the closure of the NRP area through in-situ soil stabilization with chemical reagents, but also excavation of sediment in the West Recycle Pond A (WRP-A) and stabilization of this material in the NRP.

Characterisation

The total volume of NRP and WRP-A sediments requiring stabilization were 10,000 and 8,000 yd³ (cy) (7,645 and 6,116 m³), respectively. The sediments primarily had elevated levels of volatile organics and metals above the NJDEP Soil Cleanup Criteria (SCS) and/or Groundwater Quality Standards (GWQS).

Risk Drivers

Public housing was constructed on top of soil contaminated from the past foundry activities. The contaminated soils required removal to prevent contact with the contaminated soil by the residents.

Objectives

The first phase of work involved excavation of the WRP-A sediment and relocating the material to the NRP. To accomplish this, it was necessary to pre-condition the WRP-A sediment with reagent (Calciment). The WRP-A also required dewatering using a 1,200 gallon (7,570 litres) per minute well point system to lower the water table to target depths. Since the WRP-A received the majority of the refinery runoff, a storm water bypass system was installed consisting of 300 linear feet (91 m) of 34 in (86 cm)x 54 in (137 cm) elliptical RCP piping to intercept storm water flow from the A Sump and divert it to adjacent pond WRP-B.

The conditioned WRP sediment was transported over 1 mile (1.6 km) on refinery roads and deposited into the NRP. Excavators then blended this relocated sediment with existing NRP sediment. Once the sediments were homogenized, the surface was surveyed. Also, prior to stabilization, the NRP was dewatered, which was accomplished using sumps and dewatering trenches.

Stabilization was performed typically in 700 cy (535 m^3) cells using the reagent mix design that was established via bench-scale testing performed by WRScompass prior to mobilization. The stabilization reagents used included moist ferrous sulfate, Portland cement, and granular blast furnace slag, which were placed directly over a treatment cell then mixed with an excavator to homogenize the reagent with the sediment throughout the cell. Post-treatment samples from each cell were analyzed to verify achieving the treatment criteria for unconfined compressive strength and permeability as well as SPLP testing for select metals of concern to satisfy stringent NJDEP impact to groundwater cleanup criteria. Post-treatment activities included importing and placing 23,000 cy (17,584 m^3) of common fill over the treated sediment in the NRP and installing a stone cover. Finally, a storm water inlet was installed within the graded pond with a 15 in (38 cm) RCP gravity pipe, which discharged to a nearby reconstructed lift station.

Method

The sediment material in the WRP-A was solidified with Calciment™ (a proprietary cement kiln dust source) to allow excavation and transport of the solidified sediment to the NRP. The combined solidified WPA sediment and the NRP sediment was stabilized with a combination of Portland cement, ground granulated blast furnace slag, and ferrous sulphate to chemically immobilize the metals so that the SPLP leachate of the final treated material met the New Jersey Class II Groundwater Quality Standards.

Validation

Samples of the treated soil were obtained at a frequency of every 700 cy, (535 m^3) treated and subject to SPLP testing for metals along with UCS and permeability testing. Over 90 % of the treated samples met the performance criteria on the first pass. Those treatment cells not meeting the performance criteria were retreated.

Equipment

An excavator was used to mix the WRP-A sediment with the Calciment. The same excavator was used to mix the combined solidified WRP-A sediment and NRP sediment with the Portland cement, ground granulated blast furnace slag, and ferrous sulphate to produce the final treated material.

CASE STUDY: 44
Waukegan

Location:	Waukegan, Illinois
History:	Not Specified
Contaminants:	TCE
Scale:	8,900 yd³ (6,804 m³)
Reagents:	Peat moss/bentonite slurry
Method:	In-situ
End-Use:	Not Specified

Site Description

The purpose of this project was to use single auger soil mixing to treat TCE impacted soils in-situ. Geo-Solutions (GSI) was subcontracted by CH2M Hill to be the prime contractor for this phase of the remediation.

GSI completed all portions of the work relating to soil mixing and subcontracted some of the general site preparation and earthwork items.

The soil mixing required the installation of 224 9 ft (2.74 m) diameter soil mixed columns to depths of 18–25 ft (5.5–7.6 m) below workpad elevation. Each column was mixed with zero valent iron (ZVI) and bentonite slurry. The ZVI is meant to reduce the TCE to less harmful constituents, whereas the bentonite slurry lowered permeability, thereby reducing groundwater flow through the contaminated zone. A representation of the column layout for the soil mixing work.

All of the soil mixing work was completed using GSI's Delmag RH-18 drill rig, batch plant, and silos (see photo overleaf).

The mix area post-mixing (shown above) was almost completely inaccessible to equipment. The material was allowed to consolidate over a few months before the final geotextile cap was placed.

APPENDIX B: CASE STUDIES EMPLOYING S/S

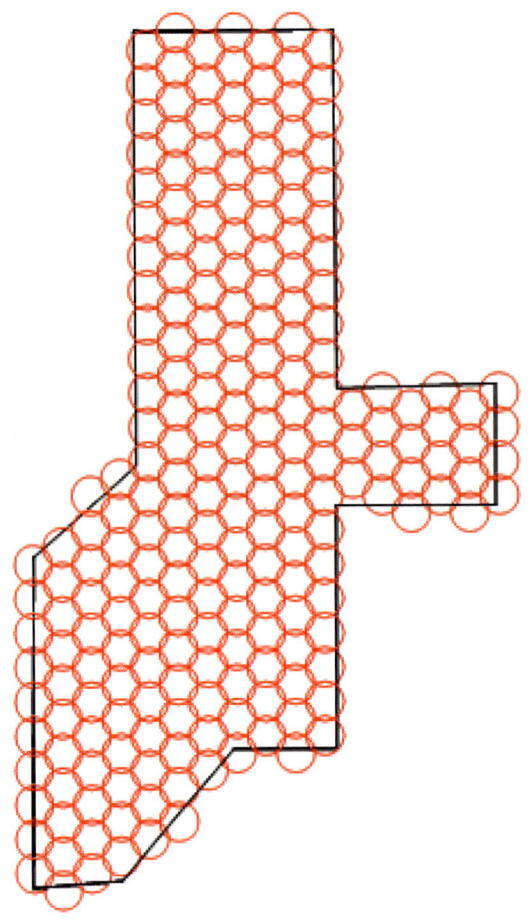

CASE STUDY: 45
West Doane Lake Site

Location:	Portland, OR
History:	Chemical works discharge pond
Contaminants:	PAHs, DNAPL
Scale:	22,000 yd^3 (16,820 m^3)
Reagents:	Portland cement & additives
Method:	In-situ
End-Use:	Restoration

Site Description

West Doane Lake is a long slender lake, approximately 1,000 ft (305 m) long, that is oriented north-south, adjacent and parallel to the BNSF embankment and adjacent to a former pesticide manufacturing facility. West Doane Lake has received soil, sediment, surface water, and groundwater from the former pesticide manufacturing facility. The southern portion of the lake is approximately 60 ft (18.3 m) wide, and the northern portion of the lake is approximately 40 ft wide. The southern portion of the lake is deeper than the northern portion, with typical water depths of 1 to 2 ft (0.3 to 0.6 m). The northern portion of the lake is often dry during the summer months.

Characterisation

The West Doane Lake sediments primarily consist of very soft to soft, black to gray, overbank silt deposits. Permeability test results indicate a range of 10-8 to 10-11 m/sec for particle sizes ranging from clayey sands to high plasticity silts. Coarse materials, believed to have sloughed off the railroad embankment, were often encountered on the northern edge of West Doane Lake, along with black and gray sands, believed to be foundry sands from a nearby property. The impacted sediment depth was approximately 11 ft (3.3 m). Debris (brick, gravel, wire, concrete, and battery casings) was observed on the eastern edge of West Doane Lake from historic filling activities. Discontinuous non-aqueous phase liquid (NAPL) blebs were observed in multiple borings completed within West Doane Lake.

Contaminants of interest (COIs) include chlorinated pesticide, chlorinated herbicides, PAHs, VOCs, and metals.

Risk Drivers

West Doane Lake may be a potential continuing source of constituents of interest to potential human and ecological receptors at the site. Both Oregon DEQ and DFW consider West Doane Lake to be an important resource for migratory birds and water fowl and consider the "maintenance of a healthy aquatic ecosystem to support migratory birds and waterfowl" as an endpoint for evaluating West Doane Lake. ISS at West Doane Lake will significantly reduce the potential contribution of COIs via groundwater because the treatment will effectively bind the COIs within the stabilized monolith to minimize leaching.

Objectives

- Removal of the surface water
- Treatment of the sediment down to 14 ft (4.3 m) bgs
- Restore the water level

The work performed by WRScompass consisted of the following activities:

- Site preparation activities were performed that included pre-work topographic surveying and existing conditions surveys of adjacent facilities, locating existing utilities using a combination of utility locator services, installing erosion controls, setting up a support zone and contamination reduction zone, performing clearing and grubbing, and constructing a material staging area and a batch plant area
- A dam was placed on the inlet side of West Doane Lake and the water level lowered until the contaminated sediment was exposed
- ISS was performed to the required depth of 14 ft (4.3 m) using an excavator-mixing method. A grout plant was set up and the required reagent grout was produced in an on-site batch plant then conveyed to the ISS treatment cell, where it was added on a per weight basis using a pre-determined mix design of 18 % by weight for Portland cement and 3 % bentonite. Activated carbon and organoclay were added dry to the surface of each treatment cell at 2 % and 3 % by weight, respectively. ISS was performed on over 22,000 cy (16,820 m³) of soil
- A multilayer capping system of imported earthen and geosynthetic material with storm water collection was installed on the treated sediment. A structural fill layer was placed over the stabilized sediment and WDL banks to seal the surface and provide a working subgrade. The structural fill was overlain by a geosynthetic clay liner (GCL) and a 60-mil (60/1000 inch) HDPE geomembrane, a 1 ft (0.3 m) thick layer of sand, and a layer of filter geotextile. Structural fill was placed over the geotextile in compacted layers to the desired final cap grades
- All equipment and temporary facilities were decontaminated and removed from the site and support areas were restored

Method

A combination of Portland cement, bentonite, activated carbon, and organoclay were used to chemically immobilize the chlorinated pesticides, chlorinated herbicides, PAHs, VOCs, and metals contamination and to improve the unconfined compressive strength to >50 psi (0.3 MPa) and reduce the hydraulic conductivity to <1×10^{-10} m/sec.

Validation

Samples of the treated soil were obtained at a frequency of every 500 yd³ (382 m³) treated and subject to UCS and permeability testing. The ISS performance criteria were >50 psi (0.3 MPa) UCS and <1×10^{-10} m/sec permeability. All of the treated samples met the performance criteria on the first pass.

Equipment

The ISS operations were conducted using an excavator and a batch plant. The batch plant was used to prepare the reagent grout. The appropriate amount of water was metered into an initial 5 yd³ (3.8 m³) batch tank (equipped with a high-speed, high-shear mixer) and recorded. The bentonite and organoclay were added to the batch tank from supersacks. Each reagent was added separately to the mix tank and the scales on which the mix tank sets were tarred before each reagent was added to verify that the correct amount of reagent had been added. When the correct grout composition was achieved, the blended grout was transferred to the excavator. The pre-determined grout volume was pumped to the treatment area based on the soil density, reagent admixture ratio, and the treatment cell dimensions to add 3 % by weight each of bentonite and organoclay.

Portland cement was pneumatically transferred onto the surface of each treatment cell, along with supersacks of activated carbon to achieve 18 % and 2 % by weight, respectively.

When both the grout and the dry reagents had been added to a treatment cell, the excavator mixed the sediment and reagents until visibly homogenous.

CASE STUDY: 46
X-231B Pilot Study

Location:	Portsmouth, OH
History:	Oil biodegradation unit
Contaminants:	TCE, 1,1,1-TCA
Scale:	210 yd^3 (160 m^3)
Reagents:	Portland cement & activated carbon
Method:	In-situ
End-Use:	Pilot-scale trial

Site Description

The X-231B Oil Biodegradation Unit is located in the Portsmouth Gaseous Diffusion Plant (PORTS), a U.S. Department of Energy production facility in Piketon, Ohio. The X-231B Unit encompasses 0.8 acres (3237 m^2) and was reportedly used for the treatment and disposal of waste oils and degreasing solvents from 1976 to 1983. From 1989 to 1990, efforts were made to close the X-231B Unit in compliance with RCRA requirements. Site characterization activities revealed the presence of several VOCs [e.g. trichloroethylene (TCE) and 1,1,1-trichloroethane (TCA)] in fine-textured soils from the ground surface to a depth of 25 ft (7.6 m).

Risk Drivers

TCE at levels higher than the Federal drinking water standard was found in the shallow groundwater directly beneath and 750 ft (229 m) down gradient from the X-231B Unit. Concerned over the continuous release of contaminant VOCs into the groundwater, the Ohio Environmental Protection Agency (Ohio EPA) required that soil remediation be included in the closure of the X-231B Unit.

Objectives

The objective of the pilot test was to demonstrate that in-situ S/S could immobilize the leachable VOCs while improving the physical strength and permeability of the treated material.

Method

A combination of Portland cement and activated carbon were utilized to chemically immobilize the VOCs, while producing >50 psi (0.3 MPa) unconfined compressive strength and <1×10^{-9} m/s hydraulic conductivity within 28 days of curing.

Validation

During treatment of each column, the temperature, pressure, and VOC headspace was determined within the column. Post-treatment samples were obtained from the top, middle, and bottom of each column. All samples met the TCLP leachate (<0.005 mg/L TCE), UCS (>50 psi or 0.3 MPa) and hydraulic conductivity criteria (<1×10^{-9} m/s) criteria.

Equipment

A mechanical system was employed to mix unsaturated or saturated contaminated soils while simultaneously injecting treatment or stabilization agents. The main system components include the following:

- A crane-mounted soil mixing auger
- A treatment agent delivery system
- A treatment agent supply
- An off-gas collection and treatment system

The mixing system used in the demonstration was manufactured and operated by Millgard Environmental Corporation, Livonia, MI. It comprised a track-mounted crane with a hollow kelly bar attached to a drilling tool, known as the MecTool™, consisting of one 5 ft (1.5 m) long horizontal blade attached to a hollow vertical shaft, yielding an effective mixing diameter of 10 ft (3 m). Depths of 40 ft (12.2 m) can be achieved with this equipment.

A grout of the treatment reagents were injected through a vertical, hollow shaft and out into the soil through 0.50 in (1.3 cm) diameter orifices in the rear of the soil mixing blades. Treatment is achieved in butted or overlapped soil columns.

The ground surface above the mixed region was covered by a 14 ft (4.3 m) diameter shroud under a low vacuum to contain any air emissions and direct them to an off-gas treatment process. The off-gas treatment system consisted of activated carbon filters followed by a HEPA filter.

APPENDIX B: CASE STUDIES EMPLOYING S/S

CASE STUDY: 47
Zwevegem

Location:	Zwevegem, Belgium
History:	Various industrial activities
Contaminants:	Metals
Scale:	20,000 tons (18,143 tonnes)
Reagents:	Alkaline fly ash
Method:	Ex-situ
End-Use:	N/A

Site Description

Contamination occurred on a site in the South-West of Belgium as a result of various industrial activities such as electrochemical plating. Two separated areas could be distinguished, each underneath former production halls. In the so-called 'area B' the main problem was Zn, Pb, Cu and Ni found in the shallow groundwater (up to 4 m BGL), together with some cyanide. 'Area C' on the other hand, showed mainly Zn, Pb and Ni. Traces of Cd, Cr and Ni were also present. At both areas the groundwater was moderately acidic.

The spreading of the contaminated groundwater plume was defined as the driver for remediation.

Characterisation

Table 1 below shows the composition of one of the groundwater hotspots at area C. Groundwater contamination resulted from leaching of contaminants from soils as illustrated in Table 2.

The texture of the soil is loamy sand, which makes a classic pump-and-treat remedial approach difficult, and ex-situ treatment of the soil by washing economically not feasible (due to the high fines content).

Table 1: Groundwater concentrations at area C.

Pollutant	pH	Cd	Cr	Cu	Pb	Ni	Zn
Concentration (mg/l)	3.73	0.011	0.173	0.952	87.7	1.14	282

Table 2: Leaching of soils from areas B and C (expressed in mg/l leachate, EN 12457-4).

	AREA B	AREA C
pH	4.04	4.95
As	< 0.01	< 0.01
Cd	< 0.001	< 0.001
Cr	< 0.01	< 0.01
Cu	0.11	0.021
Hg	< 0.0005	< 0.0005
Pb	0.57	8.14
Ni	0.097	0.06
Zn	1.49	14.3

Objectives

The remediation target for the site was derived via a risk assessment, with groundwater pollution in the vicinity as the risk driver. This resulted in a groundwater quality target in and around the remediated site x10 the 'Bodemsaneringsnorm' (Intervention Value) for each of the heavy metals.

These targets are listed in Table 3 below. The same target values were applied to evaluate the leachability of the stabilised soils.

Table 3: Target groundwater concentrations for heavy metals.

Pollutant	As	Cd	Cr	Cu	Hg	Pb	Ni	Zn
Target concentration (mg/l)	0.2	0.05	0.5	1	0.01	0.2	0.4	5

Method

Initially, the environmental consultant for the project proposed stabilisation of the heavy metals by means of adding lime milk via in-situ soil mixing, to precipitate heavy metals by increasing pH. However, this in-situ technique has disadvantages, such as uncertainty over product durability over time resulting in re-acidification), sensitivity to obstructions (stones, foundations, debris), negative impact on geotechnical soil stability, and the difficulty in accurate dosing of the lime.

For all these reasons an alternative method based on ex-situ mixing was selected and approved in joint agreement with OVAM (Environment Agency of Flanders), the consultant and the problem owner. The additive used was an alkaline fly ash applied at a dosage of 2.5 %. The soil was treated by means of a continuous mixing plant, with a throughput of 200 tonnes (220 tons) an hour.

Validation

Every daily batch of treated soil (representing about 1000 tons) was sampled and tested according to EN 12457-4. The results were compared to the target levels in Table 3 above.

As can be seen, the stabilised soils complied with the site-specific reuse criteria. In addition the VLAREA-criteria were met, which are the criteria for free reuse of the soil as secondary building material in Flanders. The treated soil was compacted during backfilling, and achieved a compressibility modulus of 11 MPa, as tested by the field plate test.

Table 4: Overview of the full-scale stabilisation results (20 validation samples for area B, 21 validation samples for area C). Values are expressed in mg/l (leachate or groundwater)

	AREA B			AREA C			Reuse criteria	
	Initial	Average stabilised	Max stabilised	Initial	Average stabilised	Max stabilised	10XBSN	Vlarea
pH	4.0	11.9	12.4	5.0	11.9	12.4		
As	< 0.01	< 0.01	0.012	< 0.01	< 0.01	0.012	0.2	0.08
Cd	< 0.001	< 0.001	0.001	< 0.001	< 0.001	0.001	0.05	0.003
Cr	< 0.01	< 0.03	0.047	< 0.01	< 0.03	0.04	0.5	0.05
Cu	0.11	< 0.023	0.046	0.021	< 0.024	0.116	1	0.05
Hg	< 0.0005	< 0.0005	0.0005	< 0.0005	< 0.0006	0.0028	0.01	0.002
Pb	0.57	< 0.005	0.008	8.14	< 0.005	0.005	0.2	0.13
Ni	0.097	< 0.005	0.033	0.06	< 0.002	0.002	0.4	0.075
Zn	1.49	< 0.02	0.031	14.3	< 0.021	0.031	5	0.28

INDEX

A

acids, 39, 42, 177, 178
Acid Sludge Tar. See AST
activated carbon, 3, 31, 39, 43, 179, 203
additives, 9, 151, 158, 159
adsorption, 39, 179
agents, reducing, 42
alkaline reagents, 145, 203
ancillary equipment, 62, 82, 97, 108, 111, 117
antimony, 25, 35
arsenic, 35, 36, 38
asphalt, 74, 75, 216, 226, 227
AST (Acid Sludge Tar), 39, 41
auger, 31, 51, 53, 54, 55, 56, 61, 93–94, 96, 98–99, 101, 106, 108, 116, 118, 122, 125, 135, 196, 204

B

Backhoe, 108, 110
barium, 35, 37
batches, 69, 132
batch mixing plant, 53–54, 56, 57, 93, 97, 108, 109, 111, 113, 117, 124, 125, 196
Belgium, 78
Bench-Scale Testing, 86, 118, 121, 122, 155, 164, 182, 183, 184–85, 191, 193, 194, 195, 197, 203, 206, 207, 209, 211
bentonite, 55, 131, 181, 190, 221
benzene, 3, 41, 42
beryllium, 35, 36
binder system, 3, 5, 125
bioremediation, 3, 30, 154
boreholes, 44, 46–48
BOSS systems, 109, 111, 113, 115, 117
bottom ash, 171, 176, 181
boulders, 44, 54, 158
British Standard (BS), 45–46, 161
BTEX, 7
bucket mixing, 80, 108, 116–18, 120, 122, 196, 199
bulk samples, 47, 134, 137, 191

C

cadmium, 25, 35, 36, 177, 178
calcium aluminosilicates, 173, 175, 176
calcium hydroxide, 175, 176, 178
calcium oxide, 171, 173, 174, 176
Canada, 1, 25–26, 78, 180, 181
capillary break, 216, 225, 227, 228, 230, 231, 232
capping materials, 227, 229
caps, 12, 139, 154, 167, 207, 208, 209, 210–11, 215, 217–19, 221, 225–32, 235, 236
carbon, 174, 176, 178, 179, 182

CASSST (Codes and Standards for Stabilisation and Solidification Technology), 24
cement, 3, 7, 11, 12, 23, 38, 42, 145, 159, 181, 182, 185, 189
Cement kiln dust. See CKD
CERCLA, 17, 18
chemical fixation, 35, 177, 178
chromium, 35, 37, 42
CKD (Cement kiln dust), 171, 173–74, 180
clay, 9, 39, 41, 42, 106, 117, 123, 169, 171, 211, 216, 217, 218–19, 223, 227–28, 230, 232
    compacted, 219, 221, 223, 225, 228, 230, 232
coal tar, 117, 118, 133, 149, 153
cobbles, 44, 104, 158, 195, 216
COCs, 19, 143, 153, 155, 166, 167, 204, 205, 235, 236
Columns (S/S), 102, 106, 122, 126, 131, 132, 134, 135, 149, 196, 205, 212
Common reagents, 171
compaction, 50, 85, 86, 87, 90, 146, 183–84, 218, 219, 228
compliance, 19, 86, 121, 123, 153, 154, 155, 163
compressive strength, 10, 51, 147
conceptual site model. See CSMs
construction, 121, 123, 124, 126, 153, 155, 159, 161, 166, 207, 209, 216, 218, 219, 220, 227, 229, 230, 231, 232
construction quality assurance. See CQA
construction quality control. See CQC
contact, 119, 159, 160, 161, 171, 172, 173–74, 176, 177, 178, 179, 180, 181, 233
    eye, 172, 174, 176, 177, 178, 179
    skin, 172, 176, 177, 178
containment, 18, 151, 154, 158, 160
    secondary, 61, 79, 80
contaminants, 3, 5, 6–7, 17, 19, 21, 28, 29, 30, 48, 154, 156, 158, 159, 160–61, 163, 183, 184, 233, 235
contaminated land, 1, 10, 11, 18, 23, 25, 27, 48, 143, 156–57
contaminated soil and waste, 1, 2, 23, 24, 27, 49, 79–80, 117, 154, 155, 167, 169, 184, 189, 235
copper, 25, 35, 37, 39
costs, 24, 31, 32, 47, 48, 49, 56, 57, 101, 108, 154, 155, 182, 184–85, 189, 194, 195, 204, 205, 206
CQA (construction quality assurance), 121, 124, 127
CQA process, 123, 124, 128
CQA program, 122, 124, 125
CQC (construction quality control), 121, 124, 145
CQC documentation, 125, 126
creosote, 3, 7, 144, 167
CSMs (conceptual site model), 27, 28, 31, 153, 156, 157, 158
curing, 105, 114, 120, 121, 128, 132, 133, 134, 137, 141, 143, 144, 161, 182, 189, 191, 193, 197
cyanide, 36, 38

D

data quality objectives (DQOs), 156
debris, 44, 47, 49, 54, 55, 63, 64, 66, 70, 74, 75, 79, 82, 83–84, 137, 138, 141, 158, 195
density, dry, 86–87, 89, 123, 161, 218
desiccation, 216, 223, 227, 229, 231
design, 28, 29, 44, 48, 122, 123, 124, 127, 128, 129, 159, 160, 207, 209, 211, 215, 218, 227, 229, 231
dioxins, 3, 4, 39, 144, 167

disadvantages, 47, 50, 51, 62–66, 70, 72, 74, 75, 76, 77, 80, 81, 82, 88, 106, 108, 115, 116, 205, 206
disposal, 17, 18, 24, 25, 39, 50, 106, 115, 119, 135, 147, 159, 162, 190, 191
distribution, 48, 63, 64, 66, 156, 159
DNAPL, 156
Documentation, 1, 2, 5, 10, 18, 49, 121, 126, 133, 163, 233, 235, 236
dozers, 86, 88, 97, 108, 111, 117, 137, 193, 212, 213, 214
DQOs (data quality objectives), 156
drainage, 209, 216-218, 227, 228, 229, 230
Dredged Bay Mud. See DBM
drum mixers, injection tillers and rotary, 108
dry matter. See DM
dry powder, 61, 109, 113, 171, 172, 173, 174, 175, 176, 179
durability, 8, 30, 31, 86, 87, 160, 163, 203
dust, 29–30, 80, 169, 172, 174, 175, 176, 179, 196

E

EA, 24, 31, 32, 160, 161, 162, 163
earthen materials, 209, 210, 215, 216, 217, 229
earthen pits, 79, 80
emissions, 4, 23, 51, 196
enclosures, 79, 82
Environmental Protection Agency. See EPA
EPA (Environmental Protection Agency), 2, 11, 12, 17, 19, 24, 31, 166, 174, 184
equipment, 49, 51, 54, 55, 56, 57, 72, 74, 84, 93, 94, 101, 102, 108, 113, 115, 116, 117, 193–94, 196
  screening, 55, 82, 83, 84
  tracked, 228
erosion, 216, 218, 229
Europe, 11, 15, 22, 24, 26, 32, 162, 174, 180, 181
evaluation, 27, 123, 126, 127, 128, 131, 155, 158, 182, 197, 208, 209
evapotranspiration, 216, 217, 225, 228, 230, 231, 232
excavation, 49, 50, 55, 115, 167, 169, 205, 209, 211, 236
excavator, 2, 55, 56, 59, 69, 70, 72, 79–82, 94, 97, 108, 110–11, 115, 116–17, 119, 135, 136, 212, 213, 214
  excavator bucket, 70, 81, 115, 117, 135
  bucket mixing, 69, 70, 72, 81
  mounted rigs, 94, 96, 101
explosives, 39
ex-situ, 2, 19, 31, 32, 49, 50, 55, 56–57, 72, 73, 74, 75, 76, 77, 85, 86, 87, 89, 151, 153
  mixing, 55, 56, 59, 72, 76, 82, 83, 85, 108, 125–26, 135, 193, 196, 197, 211
  S/S equipment and application, 56, 59, 61, 63, 65, 67, 69, 71, 73, 75, 77, 79, 81, 83, 85, 87, 89, 91
treatment, 2, 49, 50, 55, 79, 158, 167, 195, 196, 205, 212

F

failure, 87, 122, 127, 132, 134, 143, 144, 146, 147, 149, 151, 154, 165, 190, 194, 205–6, 235
FD (field demonstration), 156, 191
Feasibility Study, 159, 160
ferrous metal, 85

field-scale, 164, 191, 194, 197
  demonstration (FD), 156, 191
  samples, 191
fiscal years. See FY
flexible membrane liners. See FMLs
fly ash, 3, 7, 43, 167, 171, 174–75, 180, 181, 182, 184, 185
FMLs (flexible membrane liners), 165, 166, 167, 169, 170, 219
formulations, 3, 24, 31, 47, 155, 183, 185, 187, 189, 190, 204
foundations, 159, 209, 219
France, 5, 22, 24, 236
freeze-thaw, 5, 30
full-scale treatment, 134, 154, 182, 184, 191, 193, 204, 205
FY (fiscal years), 17, 19

G

GCL (Geosynthetic clay liner), 147, 165, 166, 167, 216, 221, 222–25, 227–30, 232
generators, 17
geo-membrane, 210, 216, 219, 221, 222, 223–25, 228, 230, 232
Geosynthetic clay liner. See GCL
grading, final, 212, 213, 214, 220
gravel, 210, 216
  surface, 223
ground, 2, 23, 26, 47, 48, 49, 90, 93, 106, 125, 159, 172, 176, 203, 219, 231
  -conditions, 27, 47, 48
  -water, 27, 28, 29, 46, 47, 48, 50, 106, 153, 154, 155, 156, 158, 159, 160, 161, 163, 165, 169, 197, 235
  -monitoring wells, 234
Groundwater Quality Standards (GWQS), 153
grout, 2, 97, 102, 105, 106, 109, 113, 114, 115, 116, 120, 197
guidance, 1, 11, 24, 31, 151, 157, 191, 231, 233
GWQS. See Groundwater Quality Standards

H

halides, 36, 37, 38
Handling failure, 145, 149
hazardous constituents, 39, 42, 178–79
hazardous wastes, 1, 12, 17, 19, 22, 23, 31, 210
hexavalent chromium, 172, 174
hopper, 61
hydrated lime, 80, 171, 172–74, 181
hydraulic conductivity, 153, 156, 159, 160, 161, 162, 169
hydrogeology, 27, 156, 157, 159, 163
hydroxide, 36–38, 178

I

identification and description of soils, 45–46
impacted soils, 153, 159
infiltration, 30, 217, 218, 225, 228

information sources, 10, 11
infrastructure, 27, 159, 209
in-situ, 2, 19, 22, 47, 49, 55, 69, 72, 73, 74, 75, 76, 77, 126, 135, 151, 158, 159, 207, 211
   applications, 31, 49, 123, 125
   auger, 44, 93, 101, 102, 104, 117, 133, 135, 149, 198
   bucket mixing, 115, 117, 196, 199
   excavator, 72
   mixing, 23, 53, 56, 57, 76, 93–94, 97, 102, 106, 107, 108, 122, 135, 195, 197
   S/S, 3–4, 19, 50, 51, 54, 55, 56, 57, 73, 74, 75, 76, 77, 113, 115, 119, 122, 125, 153, 211
   S/S equipment, 51, 54, 93, 95, 97, 99, 101, 103, 105, 107, 109, 111, 113, 115, 117, 119
In-Situ Stabilization. See ISS
   treatment, 2, 49, 51, 54, 107, 113, 132, 137, 205, 208
Integrated Pollution, Prevention and Control (IPPC), 23
IPPC (Integrated Pollution, Prevention and Control), 23
iron, 38, 41, 172, 174, 176, 177, 178
ISS (In-Situ Stabilization), 159

L

laboratory testing, 134, 143
lagoons, 80, 167
landfill, 23, 24, 85, 162, 173, 190, 210
layers, 10, 37, 47, 54, 216, 217, 218, 223, 229
   erosion control/topsoil, 210
leachability/leachate, 2, 26, 43, 144, 161, 162, 164, 171, 173, 175, 176, 208
leaching tests, 154, 155, 162, 164, 189
lime, 8, 9, 11, 23, 43, 175, 176, 177, 181
lime kiln dust, 174, 181, 182
Lime kiln dust. See LKD
limestone, 171, 172, 176
liquid reagents, 69, 72, 73, 74, 75, 76, 79
LKD (Lime kiln dust), 174, 181, 182
loaders, 56, 80, 81, 82, 97, 111, 117
   front-end, 72, 73, 76, 108
Long-Term Response Action (LTRA), 231
LTRA (Long-Term Response Action), 231

M

management, 17, 25, 50, 102, 208, 209, 213, 215
manufactured gas plant. See MGP
material handling equipment, 55, 56
material performance goals, 3, 151, 153, 154, 155, 160, 164
mercury, 35, 37, 178
metals, 3, 7, 18, 21, 35, 36, 37, 38, 42, 70, 79, 81, 84, 85, 171, 173, 175, 176, 177, 178
   immobilise, 171, 172, 173, 174, 175, 176
MGP (manufactured gas plant), 32, 149, 153
mixer, 55-57, 59, 61, 64-67, 69, 79, 113, 137
   ribbon, 65–66
mixing, 49, 55, 56, 59, 69, 73, 74, 75, 76, 79, 81, 94, 106, 109, 115, 125, 134, 135, 193, 197
   continuous, 59, 61, 64, 65

binders/reagents, 49
energy, 55, 105, 114, 120
head, 56, 69, 72, 94, 96, 109, 111, 118
pits, 59, 79, 80
reagents, 183
tanks, 79, 81, 96, 109
moisture content, 49, 57, 63, 64, 65, 87, 122, 123, 134, 155, 158, 161, 183, 196, 211
molds, 105, 114, 120, 137, 138, 141, 143, 144
monitoring, 121, 122, 156, 207, 209, 211, 213, 215, 217, 219, 221, 223, 225, 227, 229, 231, 233, 235, 236
monolith, 5, 30, 147, 162, 169, 208-209, 212, 213, 215-216, 218-219, 221, 225, 228-231, 232, 235

N

Naphthalene, 41
NAPL (non-aqueous phase liquids), 153, 154, 156, 159
nickel, 35, 37, 38
nitrate, 38
non-aqueous phase liquids. See NAPL
non-hazardous waste, 23

O

obstructions, 101, 106, 108, 115, 116, 119, 158
odour control, 61, 73, 79, 80
off-site disposal, 72, 73, 74, 76, 80, 81, 82
oily sludge, 52
Operational Issues, 102, 115, 119, 129
organics, 19, 31-32, 35, 39, 41-43, 145, 171, 173, 175, 176, 179, 203, 216
  free-phase, 48, 171, 173, 175, 176
organic stabilisation, 39, 43, 175, 176
organoclay, 179
overlapping columns, 102, 103
oxidation, 39, 41, 42

P

paddle aerators, 59, 76, 77
PAHs (polycyclic aromatic hydrocarbons), 7, 25, 39, 153
particle size, 44, 63, 64, 66, 82, 134, 221
pathways, 19, 27, 28, 30, 41, 160, 161
PC. See Portland Cement
PCBs, 7, 39, 153
PCP, 3, 4, 167
performance,
  criteria, 1, 2, 49, 50, 86, 87, 121, 144, 145, 151, 152–55, 157, 159, 161, 163, 164, 165, 169, 189, 194
  long-term, 207, 208, 229, 236
  samples, 113, 119, 132, 133, 135, 137, 143, 144, 145, 147, 149, 165

standards, 123, 126–27
tests, 133, 153, 154, 161, 164
permeability, 7, 86, 87, 133, 134, 137, 143, 145, 146, 153, 155, 158, 160, 162, 164, 185, 188, 189, 190, 191
phosphates, 37, 38, 171, 177, 178, 180
pilot-scale, 122, 123, 125-126, 128, 182, 193-197, 203, 207-208
pits, 47, 79, 80, 213
placement, 74, 75, 76, 79, 80, 81, 82, 85, 86, 87, 88, 160, 161, 183, 184, 212, 213, 215, 216, 228
plastic barrier, 172, 174, 176, 177, 178, 179
plasticity, 57, 63, 64, 66, 161
pocket penetrometer, 140, 141, 164, 192, 193
polycyclic aromatic hydrocarbons. See PAHs
Portland Cement (PC), 8, 12, 14, 43, 79, 131, 160, 167, 171–74, 180, 181, 208
practitioners, 1, 5, 101, 102, 111, 113, 182, 184, 189
pre-excavation, 55, 195
preparation, 55, 80, 81, 82, 121, 123, 128, 132, 141, 191, 221
production rates, 1, 57, 90, 101, 113, 126, 204
project phase, 158, 159, 160, 194
pug-mill, 3, 4, 44, 51, 55, 59, 61–63, 65, 67–69, 78, 125, 133, 135, 137, 183–84, 196
pumps, 56, 93, 96, 108, 109, 111, 177, 178, 196

Q

QA. See Quality Assurance
QC. See Quality Control
Quality Assurance (QA), 105, 114, 120–21, 123, 125, 127, 129, 131, 132, 133, 135, 137, 139, 141, 143, 145, 147, 149, 193, 196
Quality Control (QC), 67, 72, 76, 101, 102, 105, 113, 114, 116, 117, 118, 119, 120–21, 143, 145, 149, 193, 194, 195, 197
quicklime, 8, 172, 174, 181

R

raw materials, 9, 179
RCRA (Resource Conservation and Recovery Act), 17, 18, 169, 174
reaction,
  cementitious, 28, 86
  chemical, 2, 160, 161, 171, 173, 175, 176
  reagent-waste, 194
reagent
  addition, 56–57, 61, 101–2, 105, 113, 114, 118, 120, 122, 126, 132, 143, 191, 213
  costs, 57, 108, 115, 185
  delivery, 57, 76, 96, 109, 113, 125
  densities, 122, 125, 211, 212
  dosage, 101, 113
  formulas, 182, 212
  grout, 55, 56–57, 213
  injection, 116, 211
  mix designs, 207
  mixing/activation, 211
reagents, 2, 57, 67, 69–70, 72, 73–77, 79, 80, 94, 96–97, 109, 117, 121, 123, 125, 184, 185, 193,

    195, 211
    cementitious, 164
    inorganic, 42, 43
    organic, 43
    selection, 184, 195
    silos, 3, 53, 57, 61, 196, 202
    slurry, 54, 93, 108, 117, 211
    storage, 56, 57
    tank, 51
receptors, 19, 27, 28, 29, 30, 61, 73, 79
reduction, 30, 39, 41, 42, 43, 160, 163, 178
regulations, 10, 25, 174
regulatory agencies, 2, 207, 209, 211, 231, 236
remediation, 1, 11, 12, 15, 17, 18–19, 27, 29, 31, 33, 47, 94, 109, 123, 144, 145, 147, 149, 196, 203
removal, 18, 85, 106, 116, 119, 147, 149, 158, 159, 209, 217, 233
Resource Conservation and Recovery Act. See RCRA
Restoration, 12, 158
retreatment, 127, 145, 147, 149, 165, 205–6
ribbon blenders, 59, 61, 65–66, 69
Risk Framework, 28
risks, 1, 3, 5, 23, 25, 26–30, 39, 48, 156, 160, 194
    assessment, 27, 29, 157, 160
    management, 17, 19, 21, 23, 24, 25, 26–27, 29, 31, 33
    potential, 27, 28, 29, 122
    unacceptable, 26, 27, 28, 157
rotary drum mixers, 108, 109, 111, 115
rototillers, 73, 74

S

sample collection, 102, 105, 114, 120, 121, 128, 132, 141, 205
sample preparation, 121, 132, 134, 137, 182
samples, 113, 119, 123, 124, 126, 127, 132, 133, 134, 135–36, 141, 143, 144, 145, 146, 147, 181, 191, 193, 204–6
    fresh, 134, 135
    replicate, 127, 143, 144, 145, 164
    representative, 182, 204, 235
    re-test, 128
sampling, 47, 54, 121, 125, 128, 129, 132, 135, 137, 156, 189, 204, 205, 235
sand, 8, 9, 10, 167, 171, 210, 216, 219
screen, 44, 82, 83, 137, 138
screw auger, 64, 67–68, 72, 76
screw mixers, 59, 61, 64, 65, 67, 69, 78
sediments, 19, 25, 26, 49, 121, 156
selenium, 35, 38
Shredding equipment, 55, 84
silica, 8, 10, 174, 176
silos, 51, 57, 96, 109, 125, 172, 174, 175, 176, 177, 178, 179
site conditions, 43, 46, 49, 156, 158, 196
site development, 46
site investigation, 43, 44, 47, 48, 158, 159, 160, 204
slag, 131, 171, 185, 189

slopes, 221, 223, 229, 230
sludge, 3, 11, 39, 49, 66, 101, 109, 113, 116, 167, 206
slurry, 51, 54, 115, 117, 118, 183, 193
soil, 3, 24, 25, 29, 41–49, 74, 87–88, 102, 106, 113, 115, 117, 118, 123, 145, 158, 159, 206, 211, 222–23
  classification, 43, 44
  clayey, 228, 230, 232
  cohesive, 61, 91, 161
  compacted, 87, 218
  dense, 106, 158
  high moisture content, 102, 113
  layers, 217, 225
  plastic, 70, 74, 75, 219
  storage, 225
  types, 21, 27, 43, 93, 108, 117, 122, 126, 158, 209, 212, 228
soil vapour extraction, 19, 30
specifications, 3, 127, 146, 147, 151, 165, 167, 169, 209–10, 219
specimens, 132, 137, 162, 164, 185
SPLP (Synthetic Precipitation Leaching Procedure), 26, 144, 162, 164
spoil, 106, 115, 118, 119
S/S, 1, 3, 5, 10–15, 17, 18–33, 35, 39, 49, 121, 122, 124, 153–61, 165, 166–67, 182, 193, 233, 235, 236
  columns, 102, 106, 125
  contractor, 194
  equipment, 48, 55, 194, 195
  full-scale, 164, 194, 205, 206
  materials, 5, 8, 86, 90, 123, 127, 134, 135, 147, 148, 151, 158, 160–63, 165, 207, 209, 211, 213, 235, 236
  mixing equipment, 55, 125
  monolith, 207, 208, 209, 211, 212, 213, 215, 216, 221, 225–27, 234, 235
  operations, 121, 123, 126, 131, 213
  projects, 43, 48, 121, 132, 154, 181, 182, 207, 226
  reagents, 51, 82
  selection, 17, 19, 20, 22, 31, 158
stockpiles, 56, 57, 175, 205
storage, 17, 113, 119, 132, 134, 141, 172, 174, 175, 177, 178, 179, 217, 225, 228
strength, 7, 8, 133, 134, 143, 145, 146, 147, 149, 153, 155, 160, 161, 164, 165, 169, 171, 185, 189, 191
strength-gain, 155, 156
sulfides, 35–38, 171, 178, 180
surface waters, 26, 28-29, 156, 160, 165, 208, 215, 217, 228-229, 231
swell, 56, 106, 125, 194, 208, 211, 212

T

tanks, 9, 56, 79, 81, 82, 159, 177, 178
target depth, 108, 135
TCLP (Toxicity Characteristic Leaching Procedure), 151, 162
technical specifications, 121, 123, 124, 126, 209
test methods, 121, 132, 137, 143, 151, 162, 163
test pits, 46–48, 197, 203
Thallium, 35, 38
tillers, 59, 72, 76, 79
  rotary, 115, 116

TOC (total organic carbon), 43
topsoil, 216
total organic carbon (TOC), 43
total petroleum hydrocarbons. See TPH
TPH (total petroleum hydrocarbons), 7
tractors, 72, 73, 76
treatability, 3, 153, 154–55, 159, 181, 182–84, 189, 191, 193, 204, 206
treatability study, 67, 145, 155, 158, 190, 204, 205, 206
treatment cell, 69, 72, 76, 118
treatment pad, 72, 76

U

UCS (Unconfined Compressive Strength), 26, 145, 149, 160, 161, 185, 187
UK, 5, 7, 22, 23, 24, 32, 78, 236
Unconfined Compressive Strength. See UCS
Unified Soil Classification System. See USCS
United States, 17, 31, 78, 174, 180, 181, 225
USCS (Unified Soil Classification System), 44, 217
utilities, 30, 156, 159

V

valence states, 35–38
Validation, 27, 30
vapour extraction, in-situ soil, 19
vegetative growth, 216, 217, 229
vibration, 83, 88, 90, 113, 119
VOCs (volatile organic compounds), 7, 153, 203
volatile organic compounds. See VOCs

W

waste, 1, 5, 12, 22, 24, 32, 39, 42, 61, 63-64, 66-67, 69–70, 72, 73–77, 79-82, 84-85, 164, 169, 172–73, 182, 183-184, 193–97
  characteristics, 194, 195
  particles, 171, 173, 175, 176
  products, 181
  reagents, 180, 181
  treatment, 1, 22, 24
water, 23, 25, 27, 29, 39, 43, 47, 49, 55, 86, 125, 131, 141, 160, 171, 172, 173–79, 193, 211, 225
  infiltration, 219, 221, 225
  table, 19, 33, 50, 51, 55, 57, 106, 108, 147, 159, 163, 233
water vapour, 171, 172
weathering, 7, 9

Z

zero-valent iron. See ZVI
zinc, 25, 35, 38, 42
ZVI (zero-valent iron), 54

Hygge *['hyga] /HUE-gah/* is something we all seek – but seldom find. It is a Danish word meaning a "complete absence of anything annoying, irritating or emotionally overwhelming, and the presence of and pleasure from comforting, gentle and soothing things". It is especially associated with Christmas time, barbeques on long summer evenings and sitting around a cosy fire surrounded by lit candles on a cold, rainy night. There's nothing more hygge than gathering round a table with loved ones and a glass or two of wine, discussing the big and small things in life…

Hygge Media is inspired by the essence of hygge, creating a warm friendly atmosphere and enjoying the good things in life with good people. Hygge Media is a boutique publishing and design consortium, specialising in producing high quality specialist and technical magazines and books.

The philosophy of Hygge Media is to only work with people we want to work with; people whom we trust; highly skilled people who are invested in the projects we are working on.

Our team is comprised of professional, passionate people drawn from our professional networks. People that we have either worked with in former roles or who have been recommended to us by trusted colleagues and friends.

Working in this way means that we will only ever offer the best service possible.

<u>Acknowledgements</u>

Publisher	Alex Stacey
Graphic Design & Typesetting	Ruth Shedwick and Adam Thomas
Cover	Shay O'Donnell
Proof Reading	Dr Michael Mellors

<u>Enquiries</u>
info@hyggemedia.com
www.hyggemedia.com

NOTES

NOTES

NOTES

NOTES

NOTES

NOTES

NOTES

NOTES

NOTES

NOTES

NOTES

NOTES